About the Author

GRAEME HUNTER studied b...
at the University of Glasgow......
with the degree of Ph.D in 1980. He
carried out post-doctoral research at
Stanford University and the University of
Toronto. In 1988, Dr Hunter became an
Assistant Professor in the Department of
Oral Biology at the University of Alberta.
Since 1991, he has held the position of
Associate Professor in the Faculty of
Medicine & Dentistry at the University of
Western Ontario.

Dr Hunter's current research interests
are in the areas of biomineralization and
the history and philosophy of biology.

OTHER TITLES OF INTEREST

VITAL FORCES

*The Discovery of the
Molecular Basis of Life*

VITAL FORCES

The Discovery of the
Molecular Basis of Life

Graeme K. Hunter

ACADEMIC PRESS

A Harcourt Science and Technology Company

San Diego   San Francisco   New York   Boston
London   Sydney   Tokyo

Academic Press
A Harcourt Science and Technology Company
Harcourt Place, 32 Jamestown Road, London NW1 7BY, UK
http://www.academicpress.com

Academic Press
A Harcourt Science and Technology Company
525 B Street, Suite 1900, San Diego, California 92101-4495, USA
http://www.academicpress.com

ISBN 0-12-361810-X   Hardback
0-12-361811-8   Paperback

Library of Congress Catalog Card Number: 99–67772

A catalogue for this book is available from the British Library

Designed and typeset by Kenneth Burnley, Wirral, Cheshire
Printed in Great Britain by MPG, Bodmin, Cornwall
00 01 02 03 04 05 MP 9 8 7 6 5 4 3 2 1

To Miranda, Paola and Amelia

'The meanest living cell becomes a magic puzzle box full of elaborate and changing molecules, and far outstrips all chemical laboratories of man in the skill of organic synthesis performed with ease, expedition, and good judgement of balance.'

Max Delbrück, *A Physicist Looks at Biology* (1949)

List of Plates

Plate section appears between pages 204 and 205.

Contents

Plate section appears between pages 204 and 205.

Preface

This is the story of a scientific revolution. It began around 1770, when the French chemist Antoine Lavoisier commenced the work that would demonstrate the common nature of living processes and chemical reactions, and ended around 1970, when the solving of the genetic code made it possible, in general terms, to describe the molecular interactions that underlie all forms of life. From the origin of chemistry to the advent of genetic engineering took only two centuries: Justus Liebig performed some of the first chemical analyses of organic molecules; his great-grandson, Max Delbrück, helped to determine the mechanism of gene replication. This 200-year period defines what I will refer to as the 'biochemical revolution'.

A revolution must be against something. The scientific revolution of the sixteenth and seventeenth centuries overthrew Ptolemaic astronomy and Aristotelean mechanics. The chemical revolution of the late eighteenth century overthrew the phlogiston theory of combustion and the four-element theory of matter. The biochemical revolution of 1770–1970 overthrew the vitalistic belief that the characteristic features of living organisms were manifestations of a special force operating only in living organisms and known variously as *pneuma*, *archeus*, *Lebenskraft*, *élan vital*, entelechy, 'biotonic laws', etc. By discrediting vitalism, the biochemical revolution achieved for biology what the scientific revolution had achieved for physics.

It is not difficult to see why belief in a vital force was so common throughout most of human history. After all, only living things exhibit such quintessentially vital properties as growth, reproduction, assimilation, sensibility and consciousness. Surely the laws that explained an overflowing bathtub or a falling apple could not also explain the nest-building behavior of a bird, far less the cathedral-building behavior of humankind. The 'natural' idea that separate laws governed the animate and inanimate worlds appears first to have been threatened by the chemical revolution: Lavoisier's rejection of the phlogiston theory was based on an analogy between respiration and combustion; his law of the conservation of matter was derived from the study of alcoholic fermentation; his chemical analyses revealed the common elemental compositions of plant and animal tissues.

In the 1830s, further cracks appeared in the foundations of vitalism when it became clear that the 'ferments' (enzymes) of living organisms had the same effects on reactions as inorganic catalysts. Like the findings of Lavoisier, this indicated analogy, not identity, between the living and non-living. Around the same time, however, the German chemist Friedrich Wöhler threw a bridgehead across the philosophical chasm by achieving a feat long thought to be impossible – the chemical synthesis of an organic molecule. By mid-century, the hope was openly stated that all the vital phenomena could be explained by physics and chemistry.

As more and more of the animate world fell to physical and chemical explanation, vitalism began its long retreat. It was never a rout, however; the adherents of vitalism periodically regrouped around a new 'special' form of matter – protoplasm, 'living protein' and, most importantly, colloids. Nor should it be supposed that the debate about the nature of life was one in which the vitalists were always in the wrong: Theodor Schwann's cell theory, Liebig's theory of fermentation and John Northrop's view of bacteriophage as autocatalytic enzymes are instances in which materialism was taken to excess.

Vitalism's last stand came in the quixotical quest by the physicist Max Delbrück, inspired by Niels Bohr, to find the 'paradox' at the heart of genetics. The discovery of the structure of the gene, its mode of replication and control over protein synthesis, completed by 1970, finally made untenable the belief that biological chemistry was fundamentally different from the ordinary kind. Delbrück himself had abandoned the quest in 1953, coincidentally at almost exactly the time the structure of DNA was solved, but another decade and a half would be required to work out the mechanism of gene expression.

The 200-year period of the biochemical revolution can be conveniently broken up into four periods in which different names were given to the chemical analysis of life. The first of these periods, from about 1780 to 1850, was that of 'animal chemistry', characterized by the elemental analysis of organic compounds. Progress in animal chemistry was made possible by the identification of chemical elements, the development and refinement of techniques for determining the elemental compositions of compounds, and the application of these techniques to a wide range of animal and vegetable tissues. The second period, that of 'physiological chemistry' (roughly 1850 to 1900), was characterized by the development of theories, particularly valence theories, describing the ways in which atoms were arranged in compounds, and the application of these theories to small organic molecules. By the end of the century, structural formulas had been determined for many classes of biological molecules, including the amino acids, simple carbohydrates and nucleic acid bases. The third period, that of 'biochemistry', about 1900 to 1940, was characterized by the analysis of the interconversions of simple organic molecules within living cells. However, the rediscovery of Gregor Mendel's laws of inheritance at the very beginning of the century, and the identification of chromosomes as the bearers of the hereditary elements shortly thereafter, brought the phenomena of inheritance and embryonic development within the ambit of biochemistry. This led to such important developments as the recognition that genes are composed of deoxyribonucleic acid (DNA) and that they function by directing the production of enzymes. The final stage of the biochemical revolution, the period of 'molecular biology' (about 1940 to 1970), was characterized by the structural analysis of complex organic molecules, in particular by X-ray diffraction techniques. The realization that proteins and nucleic acids were gigantic polymers suggested that many important properties of these molecules would be defined by their three-dimensional structures, and were therefore not amenable to 'biochemical' analysis. By 1970, the structures of at least some nucleic acids and proteins had been determined.

Even this brief summary of its main stages makes clear that the biochemical revolution of 1770–1970 encompasses elements of organic chemistry, physiology, genetics and physics. It would be a gargantuan task to write a definitive history of all the interwoven strands that resulted in our contemporary view of the molecular nature of life, and it is not attempted here. Rather, what I have tried to do is derive the historical origins of our understanding of the central mechanisms of transmission and expression of hereditary information. As it turns out, these mechanisms are essentially represented by the structure and function of proteins and nucleic acids: the nucleic acids acting as the bearers of hereditary properties; the proteins, generally speaking, by acting as catalysts for specific biological reactions.

Although it was not possible to talk about the chemical nature of the hereditary material or the mechanism of action of enzymes until, say, 1870, it was only because of earlier breakthroughs that such speculations were then possible. The discovery of the molecular basis of life therefore includes nineteenth-century attempts to determine the structures of organic molecules, which proved that life has a chemical basis, and to define the nature of fermentation, which showed that life chemistry was directed along certain channels by the action of specific organic catalysts.

In the course of the research for this book, there emerged two major themes that underlie the history of the biochemical revolution. The first is the distinction between genetic information and the structures formed when that information is expressed. In various forms, this concept arose from several scientific disciplines in a number of different countries. In the late nineteenth century, the French physiologist Claude Bernard attempted to explain the vital phenomena by distinguishing between the 'legislative' and 'executive' forces of living systems; the British 'student of heredity' Francis Galton concluded from his observations on familial resemblances that humans have both 'latent' and 'patent' characteristics; and the German zoologist August Weismann proposed that multicellular organisms arose by a division of labour between reproductive 'germ-plasm' and structural 'somatoplasm'. By the time the Danish botanist Wilhelm Johannsen formalized these distinction by introducing into genetics the terms 'genotype' and 'phenotype', there was already strong evidence to suggest that genes function by producing enzymes. Only in 1945, however, was this formally proposed, in the 'one gene–one enzyme' hypothesis of George Beadle. Enzymes therefore represented Johannsen's 'phenotype'. Around the same time that the one gene–one enzyme hypothesis clarified the nature of the phenotype came the first (modern) suggestion that the 'active ingredient' of genes could be DNA rather than protein. By 1953, when the double-helical structure of DNA was proposed by James Watson and Francis Crick, it was generally accepted that hereditary information was carried by nucleic acids. However, the final twist in this theme came in 1965, when John Kendrew, by restating the distinction between legislative and executive as 'information and conformation', incorporated the recognition that, whereas nucleic acids are linear information strands, proteins are three-dimensional, stereospecifically interacting molecules.

The second major theme of the biochemical revolution is the concept of the aperiodic polymer, or macromolecule composed of non-repeating subunits. This was first explicitly stated by Albrecht Kossel around the time that Johannsen proposed the terms 'genotype' and 'phenotype'. Like the concept of phenotype and genotype, Kossel's idea that complex biological molecules are composed of different arrangements of *Bausteine* (building blocks) also had its roots in the nineteenth century. The earliest glimmer of this idea can perhaps be discerned in Justus Liebig's 1846 disproof of the theory that all proteins contained an identical 'radical' to which various amounts of phosphorus and/or sulfur were attached. By the 1870s it had become clear that proteins consisted largely or entirely of subunits called amino acids, and that proteins from different sources contained different amounts of the various amino acids. When Franz Hofmeister and Emil Fischer proposed the polypeptide structure of proteins in 1902, the way was clear for Kossel to suggest that proteins were 'mosaics' or 'railroad trains' of amino acids. The importance of aperiodic polymers in genetics was only clearly stated in 1944, when the physicist Erwin Schrödinger proposed in his book *What Is Life?* that the gene was an 'aperiodic crystal'. Schrödinger's book attracted to the study of the gene both Erwin Chargaff, who demonstrated in 1950 that DNA, like protein, was an aperiodic polymer, and Francis Crick, who saw more clearly than anyone else the importance of aperiodic polymers in encoding hereditary information.

The year 1970 represents a watershed in the development of biological science because the breaking of the genetic code made possible a comprehensive description of the molecular mechanisms of life. The solving of the genetic code also represented a fusion of the two concepts discussed above. By 1953, it was clear that the genotype corresponded approximately to DNA, and the phenotype approximately to protein, and that both of these molecules were aperiodic polymers. The expression of phenotype from genotype therefore represents a translation of nucleotide sequence in DNA into amino acid sequence in protein. It took another dozen years to decipher the correspondence between what Crick called 'the two great polymer languages'. At that point, the aperiodic polymer and the distinction between genotype and phenotype became textbook information rather than research-guiding concepts.

In writing any historical work, certain stylistic choices have to be made. One of these concerns the amount of space to be allocated to the lives of the historical figures involved rather than the events in which they participated. In contemporary history of science, the biographical aspect is often minimized. No doubt this is a reaction against a regrettable earlier tradition of personality cults and hagiography. Taken to an extreme, however, the current historiographic fashion may create the impression that scientists are the helpless agents of social forces, and therefore their motivations and biases are irrelevant. In fact, scientific research, like any creative activity, is an intensely personal matter. Its protagonists identify closely with their findings, as exemplified by the *ad hominem* reactions of individuals such as Justus Liebig, Louis Pasteur and Linus Pauling to criticisms of their work. Such visceral reactions have, no doubt, affected the progression of scientific thought throughout history. The antagonism between the American chemists Phoebus Levene and Walter Jones may have been in part responsible for the slow development of nucleic acid chemistry in the early part of the twentieth century. Similarly, Max Delbrück's antipathy for biochemist John Northrop may well have contributed to his distaste for the reductionist approach of biochemistry. The poor relationship between the British crystallographers Lawrence Bragg and William Astbury may have resulted in John Randall, rather than Astbury, being awarded a biophysics unit and thereby hastened (or perhaps delayed!) the discovery of the double helix. Similar examples of personal factors impinging upon scientific 'progress' will be encountered throughout this book. In science, one might say, the personal is epistemological.

For this reason, I believe it important that history of science should recognize the human side of scientific research, and I have therefore attempted to describe the elucidation of the molecular basis of life in large part through the life stories of the scientists involved. In that sense, the present work owes more to Suetonius than to Tacitus. Clearly, two hundred years of science involve a large number of individuals. In order to keep the *dramatis personae* within manageable limits, I have concentrated upon the major figures involved. For the sake of narrative coherence, therefore, many peripheral events have been omitted.

In most cases, I have found myself in agreement with previous writers concerning the importance of individual scientific contributors to the history and prehistory of biochemistry and molecular biology. In some instances, however, I have been forced to conclude that the significance of a particular scientist or group of scientists has been overrated or underrated. In the former category are, for example, Delbrück, whose ability to inspire his fellow scientists seems to have been matched only by his unfortunate tendency to endorse erroneous theories, and Liebig, whose views on vitalism, fermentation and animal chemistry surely do not justify his traditional status as a 'father of biochemistry'. On the other hand, many workers emerge from this account with their reputations significantly enhanced. Among these are Torbjörn Caspersson, a crucial figure in the recognition of the genetic roles of the nucleic acids; Maurice Huggins, whose contributions to protein chemistry have been unfairly attributed to Linus Pauling; and Phoebus Levene, the traditional scapegoat for the failure to recognize DNA as the genetic material, but in fact a giant of nucleic acid chemistry.

Historians have long been aware of the fallacy of depicting the present as an inevitable consequence of the past. In the history of science, this Whiggish tendency may be influenced by the fact that the development of science, unlike the development of human civilization, is, in one sense of the word, progressive. One could make a good argument that Periclean Athens represented a higher form of civilization than the present-day United States of America, but it would be far harder to justify the view that the scientific world-view of a century ago was closer to physical reality than that of today. Even those philosophers of science who adopt the professional position that science is a purely cultural artifact with no basis in external reality do not scorn the use of synthetic drugs or electronic devices based upon the findings of that science.

In the history of science, therefore, the risk is not so much the glorification of the present as the oversimplification of the past. Even a cursory reading of the older scientific literature reveals a plethora of failed theories, forgotten controversies and obsolete terminology. Consider the following quote, from a 1902 paper by Leopold Spiegel: 'According to the side chain theory these enzymes act on added substances after having become attached to the haptophoric groups directly or by means of an amboceptor.' I imagine that no scientist working today would recognize the terms 'side chain theory', 'haptophoric groups' and 'amboceptors'. Similarly, the biochemist Malcolm Dixon wrote in 1974 of his early days in science: 'Some of us worked on things like gamma-glucose, pnein, physin and thio-X whose very names are now forgotten.' Indeed, the older scientific literature echoes with many such names, as alien and yet strangely familiar as the minor principalities of medieval Europe: inogen and biogen, chyme and enchyme, *Abwehrferment* and *Atmungsferment*, chromonucleic acid and plasmonucleic acid . . .

The paradox here is that any attempt to make sense of the past represents an oversimplification; but any attempt to present the past in all its complexity would not be history, not even mere 'chronicle writing'. I have attempted to steer a middle course by presenting, wherever possible, the major competing theories and the evidence upon which these were based. In the early twentieth century, for example, there coexisted at least three major concepts of protein structure: the polypeptide, colloid and cyclol theories. As described below, the polypeptide theory endured and its alternatives were discarded. However, by no means all scientific disputes are resolved by knock-outs. The nineteenth-century debate between Liebig and Pasteur over the nature of fermentation is a good example of the Kantian idea of thesis and antithesis resulting in a synthesis, as the eventual solution lay somewhere between the two positions. In other instances, biochemical thought advanced by a series of incremental stages. A good example of this is the series of 'factor theories' of inheritance in the late nineteenth century. The ideas of Charles Darwin, August Weismann and Hugo de Vries represent a progressive refinement of theory to conform more closely to experimental observation.

From consideration of controversies like these, two points relating to scientific method become clear: scientific theories are always based on incomplete information, and are therefore more inductive than deductive; and, as a corollary of this, most incorrect scientific theories are based on excellent reasoning. For example, compare Max Bergmann's periodicity hypothesis of protein structure (see Chapter 8) with Francis Crick's 'central dogma' (see Chapter 13). Based on the incomplete evidence available at the time, both theories were perfectly feasible. When it became possible to determine the amino acid sequences of proteins, Bergmann's hypothesis was discredited; when the mechanism of protein synthesis was worked out, Crick's hypothesis was supported. The essence of science is to make generalizations (hypotheses) from particulars (observations); in the case of all non-trivial hypotheses, however, it is more likely than not that a counter-example will subsequently invalidate the generalization.

Clearly, then, success in scientific research requires an element of luck. It would be absurd to conclude, however, that *only* luck is required. Some breakthroughs in the biochemical revolution involved a clear, even obvious, research plan, and a massive commitment of effort. Such instances include Levene's thirty-year odyssey in search of the chemical structure of DNA and Max Perutz's equally lengthy X-ray diffraction analysis of hemoglobin. In a similar conceptual vein, although on a smaller logistical scale, are James Sumner's crystallization of urease and George Beadle's *Neurospora* program.

In many cases, however, the defining feature of the successful scientist appears to be an intuitive sense of which alternative is likely to be correct. A good example of this is Lavoisier, who abandoned the phlogiston theory of combustion as soon as it came into conflict with experimental observations, while contemporaries such as Joseph Priestley and Henry Cavendish clung stubbornly to it. In some cases, scientific intuition can be caught in the act. Schrödinger's 'aperiodic crystal' and Pauling's 'conditions under which complementariness and identity might coincide' must have seemed oxymorons at the time. How can a crystal be aperiodic? How can identical molecules be complementary? In retrospect, however, it is clear that both Schrödinger and Pauling had partial insights into new forms of chemistry. Perhaps the best example from the biochemical revolution of a scientist operating on the intuitive level is Francis Crick. His studies on the genetic code reveal an uncanny ability to predict what experimentation would later demonstrate – almost as if he were in tune with nature to the extent that he could sense which mechanisms were 'biological' and which were not.

If a common factor among scientific revolutionaries can be discerned from the history of the biochemical revolution, it is that they are often newcomers to the field that they revolutionize. Lavoisier, Amedeo Avogadro and Mathias

Schleiden were lawyers who turned to science; Pasteur, Jacobus van't Hoff and Joseph Le Bel were in their twenties when they made their great contributions to stereochemistry; Delbrück, Schrödinger and Crick were imported into biology from physics.

Due acknowledgment of the achievements of scientific genius should not obscure the existence of scientific incompetence. The history of the biochemical revolution is littered with sloppy experimentation and lazy or self-serving reasoning. These include the preposterous metabolic schemes of Liebig's 'animal chemistry', the failure of professional botanists to match the rigorous experimental design of the amateur Gregor Mendel, the poorly controlled osmotic experiments that suggested a low molecular weight for proteins, and the speculative excesses of Astbury and Dorothy Wrinch.

However, the scientific method proved robust enough to eventually reject these experimental and conceptual failures. Nothing in science can be proved; nor do I believe, with all due respect to the philosopher of science Karl Popper, that anything can be disproved. The strength of science is that it is based on the reproducibility of observation. To be generally accepted, therefore, a scientific theory must be based on observations that can be replicated by anyone who has the appropriate equipment to repeat the experiment. It is this self-correcting mechanism, this constant reference to the external physical reality, that separates science from the humanities – and thereby distinguishes the history of science from other historical disciplines.

The identification of the pneumococcal transforming principle demonstrates in miniature many of the characteristics of scientific enquiry. It involved the chance observation that bacteria could be converted from one serotype to another; many failures in trying to produce transformation *in vitro* and in attempting to purify the transforming 'principle'; an eventual success involving careful experimentation and the use of newly developed techniques; a mean-spirited vendetta by a former collaborator; and a long struggle to convince the many skeptics.

The elucidation of the chemical basis of life, to the point that individual genes can now be altered at will, represents one of the most significant scientific achievements of all time. To the historian of science and the historically minded scientist alike, the origins of the biochemical revolution are therefore of great interest. This book is the story of the men and women, the theories and the experiments, the successes and the failures, that produced the modern conception of life as a molecular process.

Acknowledgments

I would like to thank all the colleagues, friends and family who encouraged me to attempt this project and whose interest helped sustain me throughout its execution. Thanks, also, to the following undergraduate students at the University of Western Ontario who provided invaluable assitance: Mufaddal Pirbhai, Jeremy Harb and Eniola Idowu helped with the research, Adrienne Pedrech with the figures and Vincent Yeung with copyright issues. I am very grateful to Dr Mel Usselman for his critical reading of parts of this work and to my editor at Academic Press, Dr Tessa Picknett, for her advice and support. I thank my wife, Francine, for proof-reading the manuscript, and not sparing my feelings.

Vital Forces is a work of historical synthesis, not one of primary scholarship. The source materials consulted in the preparation of this book were all published ones: biographical writings, historical studies and the scientific literature itself. For this reason, I am indebted to those historians and scientists who have previously written on the history of biological chemistry. Historical and biographical works of particular interest or relevance are listed at the end of the text.

Finally, I wish to thank in advance any readers who do me the service of drawing my attention to any errors.

GRAEME K. HUNTER
University of Western Ontario, London, Canada
August, 1999

Chapter 1
The Revolution in Chemistry Has Come to Pass

Lavoisier

In February 1773, Antoine Lavoisier was twenty-nine years old. Trained as a lawyer, he had only five years earlier become a member of the Royal Academy of Sciences in Paris, and, as an associate member, occupied only the middle rank of that institution. His scientific work comprised a geological survey and some mundane chemical studies on minerals. However, the young Lavoisier possessed unbounded ambition and, perhaps, the first inklings of an idea that was to overthrow the entire theoretical system of eighteenth-century chemistry. To his private laboratory notebook, Lavoisier confided that he intended no less than 'an immense series of experiments destined to bring about a revolution in physics and chemistry'.

Antoine Laurent Lavoisier was born in Paris in 1743, the son of a lawyer and of prosperous bourgeois stock on both sides. He attended the Collège des Quatres Nations from 1754 to 1761, then transferred to the faculty of law, being admitted to the prestigious Order of Barristers in 1764. During his legal training, Lavoisier also studied science, attending classes on geology, physics, astronomy, mathematics, botany and anatomy. Most importantly, he attended the outstanding chemistry lectures of Guillaume-François Rouelle at the Jardin du Roi.

Lavoisier had acceded to his family's wishes in becoming a lawyer, but he was determined to make his mark as a scientist. To do this, he realized that he must become a member of the Royal Academy of Sciences, which had been founded by Louis XIV in 1666 to promote the advancement of scientific knowledge. The Academy had approximately fifty members, divided into six sections: geometry, astronomy, mechanics, anatomy, botany and chemistry. Each section had three pensioners (full members who were paid stipends), two associates and two adjunct members. The Academy was the principal forum for 'natural philosophy' in eighteenth-century France; its *Memoirs* provided a means of disseminating research findings, while its building provided meeting rooms and laboratory space.

Lavoisier began his campaign for membership in the Academy in 1764 by entering a competition to light the streets of Paris, and was awarded a special medal. He also read several papers at the Academy. More important were the lobbying efforts of his influential family. A vacancy in the chemistry section occurred in 1766, but Lavoisier failed to be elected. Two years later another vacancy occurred; this time he was elected as a supernumerary adjunct. Lavoisier became an associate member of the Academy in 1772, and a pensioner in 1778.

The same year he was elected to the Royal Academy of Sciences, Lavoisier bought a share in the *Ferme Générale* (Company of General Farmers), an agency that contracted with the Crown to collect indirect taxes and to sell monopoly items such as tobacco and salt. He was assigned by his fellow 'tax farmers' the task of overseeing the tobacco trade in areas east and north-west of Paris. Lavoisier proved to be an energetic and efficient administrator, and in 1775 King Louis XVI made him a director of a new royal office, the *Régie des Poudres* (Gunpowder Administration). This position brought with it an apartment and a laboratory at the Paris Arsenal. Although his income from these activities allowed Lavoisier the financial security to pursue his scientific studies, they also took up the majority of his time. He did his chemical work between eight and ten in the morning and between seven and ten in the evening, and on the one full day a week he could spare from his other duties.

When Lavoisier began his studies in the early 1770s, chemistry was still dominated by ideas that dated from the philosophy of antiquity and the alchemy of the Middle Ages. In the chemistry lectures that Lavoisier attended, Rouelle taught that matter consisted of four elements, characterized by different properties: earth, which was cold and dry; water, which was cold and wet; fire, which was hot and dry; and air, which was hot and wet. Transformation of substances was possible, based on altering the proportions of the four elements. Alchemy, a kind of empirical chemistry with a supernatural rather than a rational mode of explanation, had used the four-element theory to explain various chemical transformations.

To this mixture of abstract theory and empirical practice, the eighteenth century had added an important new concept – phlogiston. In 1718, the Bavarian chemist Georg Ernst Stahl had proposed that combustible bodies contained this 'principle of inflammability' which was released when these substances were burned. Stahl believed that highly flammable substances such as charcoal and 'spirit of wine' (alcohol) were composed almost entirely of phlogiston, as their combustion left little residue. The limited amount of combustion that occurred in an enclosed volume was due to the limited capacity of air to take up the 'principle of inflammability'. Phlogiston may now appear a somewhat ludicrous interpretation of combustion phenomena, but at the time represented an extremely important scientific concept; Douglas McKie has described the phlogiston theory as 'the first great and systematizing generalization in chemistry'.

By the late eighteenth century, it was becoming difficult to accommodate chemical observations within the four-element theory. In his 1727 book, *Vegetable Staticks*, the English chemist Stephen Hales reported experiments showing that heating various substances released 'air', suggesting that air was 'fixed' in these substances. The idea of fixed and free elements could be incorporated into the four-element theory; for example, certain crystals were known to contain 'fixed' water. More problematic, however, were findings that suggested the existence of airs with different properties. In 1756, the Scottish chemist Joseph Black showed that 'fixed air' was released not only by combustion but also by respiration, fermentation and heating mild alkalis. Fixed air did not support life, leading Black to the conclusion that it was different from 'common air'. Ten years later, the English chemist Henry Cavendish isolated and characterized an 'inflammable air' that was produced by the action of dilute acids on metals. In 1772, Daniel Rutherford showed that removal of fixed air from air that had been depleted by respiration or combustion left another new air, 'noxious air'. Between 1772 and 1777, the English chemist Joseph Priestley identified no fewer than seven new airs. It was against this background of exciting new findings in 'pneumatic chemistry' that Lavoisier began to study combustion phenomena.

Combustion and Calcination

In the summer of 1773, the 'revolution in physics and chemistry' began when Lavoisier read to the Academy a series of four memoirs on the fixation of air in combustion and calcination. These experiments described were performed in a 'pneumatic chamber' – a bell jar inverted over water, which allowed changes in air volume to be measured. In this chamber, Lavoisier burned phosphorus and sulfur, which converted them to 'acid of phosphorus' and 'acid of sulfur', respectively. He also heated metals, which converted them to a powder or 'calx'.[a] To characterize the airs produced in these transformations, Lavoisier used the standard tests of the day: the ability to support the flame of a candle or the respiration of a small animal or bird;

---

a   Lavoiser wrote that metals undergoing this process 'lose their metallic splendour, and are reduced into an earthy pulverulent matter'.

and the ability to form a precipitate with limewater or be absorbed by caustic alkali (sodium hydroxide), both characteristics of fixed air.

Using these methods, Lavoisier showed that the weight of phosphorus or sulfur consumed during combustion was less than the weight of acid produced. This observation was an apparent violation of the phlogiston theory, according to which substances should lose, not gain, weight during combustion. Where did the extra matter come from? Lavoisier showed that the combustion of sulfur or phosphorus was accompanied by a decrease in the volume of atmospheric air contained in the pneumatic chamber, such that the loss of weight of the air corresponded to the gain in weight of the acid. However, not all the air could be removed by combustion – the maximum volume of air that could be lost in this way was one-fifth to one-sixth of the total. In agreement with the previous findings of Rutherford and Black, the air remaining after combustion of phosphorus and the calcination of lead – which Lavoisier called 'mephitic air' or '*mofette*' – did not support respiration or further combustion.

Heating of metals – 'calcination' – appeared to be accompanied by changes similar to those observed in the burning of phosphorus and sulfur. Lavoisier demonstrated that converting lead to its calx increased its weight and caused a corresponding reduction in air volume. The calx of lead could be converted back to metallic lead by heating it with charcoal. This process produced the fixed air of Black. Like *mofette*, fixed air did not support combustion, but it killed animals more quickly.

From these studies, Lavoisier concluded that atmospheric air, in the classical theory an element, actually consisted of two different airs: *mofette*, which did not support combustion but was less toxic than fixed air; and an 'elastic fluid', comprising approximately one-fifth of ordinary air by volume, which was incorporated into the acids of phosphorus and sulfur and the calx of lead. The nature of this second air was mysterious. However, in October of 1774, Priestley visited Paris and met Lavoisier and other eminent chemists. He described his recent finding that heating 'red mercury' (mercuric oxide) produced an 'air' which supported the flame of a candle better than atmospheric air. This property was also exhibited by the 'nitrous air' (nitric oxide) that Priestley had previously characterized, so the air produced from red mercury did not appear to be particularly interesting. In March 1775, however, Priestley thought to test the ability of this air to support respiration and found, to his great surprise, that a mouse survived longer in it than in 'common air'. This was a property he had never observed in any of his new airs. He concluded that the air produced by heating red mercury was 'five or six times better than common air, for the purpose of respiration, inflammation, and, I believe, every other use of common atmospherical air. As I think I have sufficiently proved, that the fitness of air for respiration depends upon its capacity to receive the *phlogiston* exhaled from the lungs, this species may not improperly be called *dephlogisticated air'* [emphases in original].

As a convinced phlogistonist, Priestley interpreted his findings in terms of Stahl's theory. If his new air was more respirable than 'common air', then it must be a better receptacle for phlogiston – dephlogisticated air lacked the phlogiston that common air received from fires, breathing and fermentations.

After meeting Priestley, Lavoisier began his own studies on 'dephlogisticated air'. To produce this substance, he heated red mercury in a retort with a long, curved neck that was attached to a jar inverted over water. The properties of the air produced were later described by Lavoisier in the vivid language of eighteenth-century chemistry: 'A part of this air being put into a glass tube of about an inch diameter, showed the following properties: A taper [candle] burned in it with a dazzling splendour, and charcoal, instead of consuming quietly as it does in common air, burnt with a flame attended with a decrepitating noise, like phosphorus, and threw out such a brilliant light that the eyes could hardly endure it.'

Lavoisier showed that converting a metal to its calx involved the uptake of Priestley's new air. Significantly, he did not adopt the term 'dephlogisticated air', preferring to call it 'the purest part of the air', 'eminently respirable air' or, later, 'vital air'. Lavoisier also showed that heating red mercury in the presence of charcoal produced not 'vital air' but fixed air. This simple observation allowed him to deduce the composition of fixed air. His logic was simple: red mercury consists of the metal and vital air; heating of red mercury in the presence of charcoal consumes the charcoal, producing the metal and fixed air; therefore, fixed air must consist of charcoal and vital air.

Priestley and Lavoisier therefore put quite different interpretations on the same observations. Priestley believed that 'dephlogisticated air', 'common air' and 'mephitic air' differed in the amount of phlogiston. Dephlogisticated air supported combustion better than common air because the latter contained some phlogiston. Mephitic air was completely saturated with phlogiston. As Priestley wrote in 1775:

In the course of my inquiries, I was, however, soon satisfied that atmospherical air is not an unalterable thing; for that the phlogiston with which it becomes loaded from bodies burning in it, and animals breathing it, and various other chemical processes, so far alters and depraves it, as to render it altogether unfit for inflammation, respiration, and other purposes to which it is subservient; and I had discovered that agitation in water, the process of vegetation, and probably other natural processes, by taking out the superfluous phlogiston, restore it to its original purity.[1]

Lavoisier, however, concluded that common air was composed of two quite distinct substances: vital air, which supported respiration, combustion and calcination; and *mofette*, which remained when the vital air was removed by these processes. Vital air could be combined with a number of other substances – with metals, to produce calces; with sulfur or phosphorus, to produce acids; and with charcoal, to produce fixed air.

If he had ever embraced it at all, Lavoisier had already abandoned the phlogiston theory. As early as sometime in 1773–4, he wrote in his private notes that: 'the heat of animals is sustained by nothing else than the matter of fire which is disengaged by the fixation of air in the lungs.' Lavoisier's 'matter of fire', unlike phlogiston, was released not by combustion, but by the fixation of air.

Respiration and Fermentation

Joseph Black had shown that respiration and the combustion of charcoal both produce fixed air; thus did chemistry and biochemistry have a common origin. Lavoisier now decided to investigate how air was changed by respiration. As fixed air is absorbed by water, these experiments were done in a bell jar inverted over mercury. An unfortunate bird was introduced into the bell jar by passage through the mercury and left until it died of asphyxiation. To demonstrate the presence of fixed air, the air vitiated by respiration was exposed to 'caustic alkali' (sodium hydroxide). This reduced the volume of air by approximately one-sixth and converted the caustic alkali to 'mild alkali' (sodium carbonate). The air remaining after the removal of fixed air was *mofette*. The results of these studies were presented to the Academy in a 1777 memoir entitled 'Experiments on the Respiration of Animals and on the Changes which Air Undergoes while Passing Through the Lungs'. This memoir reported that air vitiated by respiration contains fixed air, but air vitiated by the calcination of mercury does not.

Later that year, Lavoisier read a 'Memoir on the General Nature of Combustion' that summarized the progress made to date and attempted a new theoretical synthesis that 'may explain in a very satisfactory manner all the phenomena of combustion and of calcination, and in part even the phenomena which accompany the respiration of animals'. The four general features of combustion were: it involved the release of fire or light; it occurred only in the presence of 'pure air'; pure air was decomposed and the weight of the burning substance increased by exactly the same amount; and the addition of air changed the burned substance into an acid.

Lavoisier then made an open attack on the phlogiston theory. Like Stahl and the phlogistonists, Lavoisier accepted the existence of an elemental 'matter of fire', describing it in terms that would also be applicable to phlogiston: 'the matter of fire or of light is a very subtle, very elastic fluid which surrounds all parts of the planet which we inhabit, which penetrates bodies composed of it with greater or less ease, and which tends when free to be in equilibrium in everything.' However, he did not believe that this matter of fire was released from burning bodies: 'if we demand of the partisans of the doctrine of Stahl that they prove the existence of the matter of fire [phlogiston] in combustible bodies, they necessarily fall into a vicious circle and are obliged to reply that combustible bodies contain the matter of fire because they burn and that they burn because they contain the matter of fire. Now it is easy to see that in the last analysis this is explaining combustion by combustion.'[2]

Instead, Lavoisier proposed that the matter of fire – which he called 'caloric' – was released by vital air when it became fixed. Burning of phosphorus or sulfur to produce the corresponding acids, conversion of a metal to its calx, the production of fixed air in respiration – all of these processes involved the fixation of vital air, and also generated heat. According to Lavoisier, the fixation of air produced heat because the vital air gave up the caloric that previously maintained it in the state of an 'elastic fluid'. This new entity, caloric, was therefore not only involved in the change of chemical form that occurred during combustion and calcination, but also in changes of physical state: airs contained more caloric than liquids, which in turn contained more than solids.

According to Lavoisier's new theory of combustion, vital air was composed of two elements: caloric, and a base which he referred to as the 'acidifying principle' or (later) the 'oxygen principle'. When charcoal was burned in vital air, the air was decomposed into the oxygen principle, which combined with charcoal to form fixed air, and caloric, which was released. When phosphorus was burned, the oxygen principle of vital air became fixed with it to form acid of phosphorus. Caloric was therefore related not only to the ability of substances to exist in different physical states (solids, liquids and gases), but also to their ability to exist in different chemical states (free and fixed). Because it put a large number of observations into a new theoretical system, Lavoisier's oxygen theory can be regarded as the high-water mark of his chemical revolution.

In his experimentation, Lavoisier was assisted by his wife, Marie, who, unlike Lavoisier, read English. Madame Lavoisier, who studied drawing with the great painter Jacques Louis David, also made the fine engravings that accompanied her husband's manuscripts. Lavoisier's collaborators included a number of his fellow academicians. It was one of these collaborators, the brilliant young mathematician Pierre-Simon de Laplace, who devised an apparatus to measure caloric. The ice calorimeter, invented by Laplace in 1782 and first used during the winter of 1782–3, consisted of a central chamber surrounded by two layers of ice. The inner ice layer was connected to a funnel so that the amount of ice melted could be measured. An animal or a burning substance was placed in the central chamber, and the amount of heat produced was measured by the amount of meltwater collected.

When Lavoisier's theory of combustion was put to a quantitative test, a problem emerged. In 1783, Lavoisier and Laplace compared the amounts of charcoal and vital air consumed in combustion with the amount of fixed air produced. These experiments were performed using a bell jar of vital air inverted over mercury. Following combustion, the fixed air was absorbed with caustic alkali. The amount of vital air consumed could be measured by the loss of volume, the amount of fixed air produced by the increase in weight of the caustic alkali, and the amount of charcoal consumed by weighing. The result was disappointing – the weight of fixed air produced was approximately one-third less than the weights of charcoal and vital air consumed. Another anomaly appeared in the calorimetry experiments. If respiration and combustion represented the same process, as Lavoisier believed, these processes should produce similar amounts of heat. However, the combustion of charcoal and the respiration of a guinea pig that produced the same amount of fixed air did not melt the same amount of ice. The guinea pig produced significantly more heat. Nonetheless, Lavoisier and Laplace were convinced that the fixation of air was the source of animal heat. In their 1783 'Memoir on Heat', they wrote: 'Respiration is therefore a combustion, very slow to be sure, but perfectly similar to that of charcoal . . . the air which we breathe serves two purposes equally necessary for our preservation; it removes from the blood the base of fixed air [carbon], an excess of which would be most injurious; and the heat which this combination releases in the lungs replaces the constant loss into the atmosphere and surrounding bodies to which we are subject'.

Something was wrong with Lavoisier's system. Respiration and combustion clearly had much in common, but why did they produce different amounts of heat? And why did the combustion of charcoal in vital air not produce an equivalent amount of fixed air? As in 1774, the British chemists were one step ahead of Lavoisier experimentally, although well behind him theoretically. Once again Lavoisier could take a new finding from Britain and use it to advance his own theoretical system.

In June 1783, Lavoisier learned that Cavendish had shown that water is produced by burning a mixture of inflammable air and either dephlogisticated air or common air. Cavendish had used a number of ingenious experimental designs to effect the combustion of these airs and collect the 'dew' that resulted. In the account of this work presented to the Royal Society the following year, he wrote: 'By the experiments with the globe it appeared, that when inflammable and common air are exploded in a proper proportion, almost all the inflammable air, and nearly one-fifth of the common air, lose their elasticity, and are condensed into dew. And by this experiment it appears, that this dew is plain water, and consequently that almost all the inflammable air, and about one-fifth of the common air, are turned into pure water.'[3]

The conclusion that water is a compound of inflammable air and vital air seems inescapable. However, to a dedicted phlogistonist this was literally unthinkable, and Cavendish concluded instead that 'dephlogisticated air' was in fact dephlogisticated water, and inflammable air was phlogisticated water! 'From what has been said there seems the utmost reason to think, that dephlogisticated air is only water deprived of its phlogiston, and that inflammable air, as was before said, is either phlogisticated water, or else pure phlogiston; but in all probability the former.'

Lavoisier, of course, had no such attachments to the phlogiston and four-element theories. According to Lavoisier's theory of combustion, the interpretation of Cavendish's experiment was obvious – water was a compound in which vital air was the acid and inflammable air the base. Later that same month, assisted by a number of other academicians, Lavoisier staged a demonstration of the synthesis of water at the Academy. The amount of water produced was approximately equal to the amount of the gases consumed, vindicating Lavoisier's view of water as a compound.

If water could be made from vital and inflammable air, it should be capable of being broken down into these airs. It was Laplace who realized that this must be occurring when inflammable air is produced by the action of dilute acids on metals. To test his colleague's theory, Lavoisier performed the simple experiment of placing iron filings in water. The iron became calcined, and inflammable air was produced. The latter substance was of great interest, owing to its possible use in balloon flight. In December 1783, Jacques Charles ascended to an altitude of 1700 fathoms (3000 metres) in a balloon filled with inflammable air.

To perform a large-scale decomposition of water, Lavoisier recruited another mathematician, Jean-Baptiste Meusnier, who was serving as an army engineer. Lavoisier and Meusnier took a cannon barrel with a four-inch (10 cm) internal diameter and heated it in the middle while water was trickling through it. In March 1784, they produced inflammable air from the lower end of the gun barrel and observed calcination of the metal within. Using a copper-lined barrel containing water and charcoal, they produced a mixture of inflammable and fixed air, demonstrating combustion without air.

Lavoisier's 1784 'Memoir on the Acid Called Fixed Air or Chalky Acid, and Which I Shall Henceforth Designate by the Name Acid of *Charbon*' summarized a large number of experiments on the combustion of charcoal and of wax, the reduction of metal calces by charcoal, and the gun-barrel studies. The common factor was the production of the 'acid of *charbon*' and the main conclusion was its composition – 28 parts *charbon* and 72 parts vital air.

These studies on the synthesis and decomposition of water resolved the anomalies that Lavoisier and Laplace had earlier observed in their quantitative measurements of combustion. If charcoal contained some inflammable air, then the production of water during its combustion could explain why the amount of fixed air produced was less than the amounts of vital air and charcoal consumed. Similarly, if whatever was being consumed in the body during respiration contained more inflammable air than charcoal did, this could explain the different amounts of heat produced during combustion and respiration. Lavoisier's 1785 'Memoir on the Alterations which Take Place in the Air in the Ordinary Circumstances of Society', this time presented to the Royal Society of Medicine, described respiration not only as the fixation of vital air by *charbon*, but also as the fixation of vital air by inflammable air. This correction for the production of water in respiration and certain forms of combustion removed the last hurdle to the general acceptance of Lavoisier's system.

Apart from respiration, the biological process most amenable to experimental analysis in the eighteenth century was vinous fermentation. This process converted the sugar of grapes into spirit of wine, and also produced fixed air. Lavoisier believed that fermentation involved a combination of the *charbon* from sugar with the vital air of water. The inflammable air remaining in some way gave rise to spirit of wine, also a highly inflammable substance.

To test this theory, Lavoisier attempted in 1785 to analyze the composition of spirit of wine. For this analysis, he devised a highly ingenious apparatus consisting of a pneumatic chamber containing a spirit lamp and into which vital air could be periodically introduced. When the flame was extinguished owing to the accumulation of fixed air, the operator absorbed the fixed air with caustic alkali. The amounts of vital air and spirit of wine consumed and the amount of fixed air produced could therefore be determined. The amount of water produced could not be measured, but this could be calculated, as by this time Lavoisier knew that water consisted of 85% vital air and 15% inflammable air. This experiment showed that spirit of wine contained 28% charbon, 8% inflammable air and 64% vital air – the first elemental analysis of any 'vegetable' substance and the beginning of quantitative organic chemistry and biochemistry.

If Lavoisier were correct in thinking that fermentation involved the production of water, he should have found that the weights of fixed air and spirit of wine produced were greater than the weight of cane sugar consumed. However, he found an exact correspondence between these values – sugar was apparently converted in fermentation to carbonic acid and spirit of wine without the incorporation or elaboration of water. His new interpretation of this process was that sugar was separated into two parts: one, carbonic acid, containing most of the oxygen; the other, 'alkohol', being relatively oxygen-poor and therefore combustible.

Lavoisier was a greater theoretician than he was experimenter, but he was always prepared to drop or revise his theories if they were not in qualitative or quantitative agreement with his experimentation. Thus it was in the case of the vinous fermentation:

> I had formally [*sic*] advanced, in my first memoirs upon the formation of water, that it was decomposed in a great number of chemical experiments, and particularly during the vinous fermentation. I then supposed that water exists ready formed in sugar, though I am now convinced that sugar only contains the elements proper for composing it. It may be readily conceived, that it must have

cost me a good deal to abandon my first notions, but by several years reflection, and after a great number of experiments and observations upon vegetable substances, I have fixed my ideas as above.[4]

His theory of combustion and respiration had to be revised to incorporate the production of water; now his theory of fermentation had to be revised to eliminate the decomposition of water! Typically, though, Lavoisier turned this set-back into a new breakthrough. No doubt inspired by his mathematical collaborators, such as Laplace and Meusnier, he expressed the process of vinous fermentation as a chemical equation:

$$\text{must [juice] of grape} = \text{carbonic acid} + \text{alcohol}$$

The use of balanced equations held out the prospect of a rigorous test of chemical theories: 'We may consider the substances submitted to fermentation, and the products resulting from that operation, as forming an algebraic equation; and by successively supposing each of the elements in this equation unknown, we can calculate their values in succession, and thus verify our experiments by calculation, and our calculations by experiment reciprocally.'[4]

The Elements of Chemistry

In 1787, Lavoisier presented to the Academy a 'Memoir on the Necessity for Reforming and Perfecting the Chemical Nomenclature'. This memoir, written at the suggestion of Louis Guyton de Morveau of Dijon and co-authored by Lavoisier's Paris colleagues Antoine François Fourcroy and Claude Louis Berthollet, was subsequently reworked into the book *Méthode de nomenclature chimique* (*Method of Chemical Nomenclature*). The impetus behind the new nomenclature that Lavoisier and his colleagues were proposing was to legitimize the new compositional chemistry by replacing the existing archaic terminology of compounds. The alchemical names included 'powder of algaroth', 'salt of alembroth', 'pompholix', 'phagadenic water', 'colcothar', 'martial ethiops', 'ruft of iron', 'butter of antimony', 'flowers of zinc', 'Pharaoh's serpent' and 'vitriol of Venus'. Guyton's aim was to replace these colorful but confusing terms with a system in which compounds were named according to their constituent elements. Among the most important terms suggested in the nomenclature of 1787 were 'oxygen gas' for vital air, 'hydrogen gas' for inflammable air, 'azote' for *mofette* (the British chemists adopted the name 'nitrogen') and 'carbon' for the base of carbonic acid. These were all elements, which were defined, following the century-old example of Robert Boyle, as 'substances which we cannot decompose'.

Other elements included light and caloric, the sixteen known metals, and a number of organic 'radicals'. Gases were considered to be combinations of bases and caloric; thus 'oxygen gas' was composed of the element oxygen combined with the element caloric. Calces were renamed 'oxides' – compounds of the metallic elements with oxygen. Other oxygen compounds were named according to the element with which oxygen combined – sulfates, phosphates, carbonates, nitrates, etc.

The new chemistry was described more fully in Lavoisier's 1789 book, *Traité élémentaire de chimie, presenté dans un ordre nouveau et d'après les découvertes modernes* (published in English the following year as *Elements of Chemistry, in a New Systematic Order, Containing All the Modern Discoveries*). This began with a statement of Lavoisier's philosophy that would sound pretentious were it to come from the pen of almost any other scientist: 'I have imposed upon myself, as a law, never to advance but from what is known to what is unknown; never to form any conclusion which is not an immediate consequence necessarily flowing from observation and experiment; and always to arrange the facts, and the conclusions which are drawn from them, in such an order as shall render it most easy for beginners in the study of chemistry thoroughly to understand them.'[4]

The chemical nomenclature presented in *Elements of Chemistry* was updated from the 1787 version by dropping the alkalis and organic radicals from the list of elements. It was already clear that the organic radicals contained several elements. Those from the plant kingdom were oxides of carbon, hydrogen and sometimes phosphorus; those from the animal kingdom were oxides of carbon, hydrogen, phosphorus and 'azote'.

The *Elements of Chemistry* also contained a statement of the law of conservation of matter, long implicitly accepted by chemists but formally stated here for the first time:

> We may lay it down as an incontestible axiom, that, in all the operations of art and nature, nothing is created; an equal quantity of matter exists both before and after the experiment; the quality and quantity of the elements remain precisely the same; and nothing takes place beyond changes and modifications in the combination of these elements. Upon this principle the whole art of performing chemical experiments depends: We must always suppose an exact equality between the elements of the body examined and those of the products of its analysis.[4]

'The Authors of All the Evils That Have Afflicted France'

In 1791, Lavoisier wrote: 'All young chemists adopt the theory and from that I conclude that the revolution in chemistry has come to pass.' However, another revolution, the political one that broke out the same year that Lavoisier's book was published, constrained and eventually cut short Lavoisier's scientific work. At first Lavoisier was free to continue his research, and, in association with a new student, Armand Séguin, he tried to analyze the chemical changes that take place during respiration. While Séguin lacked the intellectual brilliance of Laplace, he attempted to compensate for this by a willingness to act as experimental subject for a number of strenuous and unpleasant procedures. In November 1790, Séguin and Lavoisier presented to the Academy their 'First Memoir on the Respiration of Animals'. This included a statement of their general view of respiration:

> Thus, to confine myself to simple ideas that everyone can grasp, I would say that animal respiration is nothing else but a slow combustion of carbon and hydrogen, which takes place in the lungs, and that from this point of view animals that respire are veritable lamps that burn and consume themselves.

> In respiration, as in combustion, it is the air of the atmosphere which furnishes the oxygen and the caloric. In respiration it is the blood which furnishes the combustible; and if animals do not regularly replenish through nourishment what they lose by respiration, the lamp will soon lack its oil; and the animal will perish, as a lamp is extinguished when it lacks its combustible.[5]

In this memoir, Séguin and Lavoisier also noted that doing physical work (Séguin on a treadmill) caused a greater change in oxygen consumption than lowering the temperature (a guinea pig in a calorimeter). Therefore, respiration is required for muscular work as well as for the production of animal heat.

Lavoisier was a moderate revolutionary who was elected to the Paris Commune and served with the National Guard. In the early years of the revolution, he retained and even enhanced his status. In 1791 he was appointed as a commissioner of the new National Treasury and asked to prepare a report to the National Assembly on the state of the nation's finances. That same year, he became treasurer of the Royal Academy of Sciences. However, when the monarchy fell and the Reign of Terror began, not only Lavoisier's positions but his life were put in jeopardy. He had incurred the enmity of the radical journalist, Jean-Paul Marat, over a

'Treatise on Fire' that the latter had written, and Marat now vented his spleen: 'I denounce this Corypheus of the Charlatans, Sieur Lavoisier, son of a land-grabber, chemical apprentice, pupil of the Genevese stock-jobber, tax-farmer, regent of powder and saltpetre, administrator of the Discount Bank, secretary of the king, member of the Academy of Sciences.'

In August 1792 Lavoisier was forced to resign from the Gunpowder Administration and vacate his laboratory at the Arsenal. The Royal Academy of Sciences, now despised as a creation of the old feudal regime, was abolished a year later. Lavoisier and his fellow 'tax farmers' were arrested in November 1793. At their trial on 8 May 1794, the prosecutor ranted that 'the measure of the crimes of these vampires is filled to the brim . . . the immorality of these creatures is stamped on public opinion; they are the authors of all the evils that have afflicted France for some time past.' Lavoisier and twenty-seven other tax farmers were found guilty of conspiring against the people of France. That same day, they were taken to the Place de la Révolution and guillotined; their bodies were placed in unmarked graves in the Parc Monceaux. Antoine Lavoisier was fifty years old. The mathematician Joseph Lagrange lamented: 'It took them only an instant to cut off that head, and a hundred years may not produce another like it.'

'The Founder of Physiological Chemistry'

Mikulas Teich wrote that the replacement of the phlogiston theory with a 'new theoretical system based on oxygen . . . was an event in the history of chemistry comparable to that of the establishment of the theory of gravitation in the history of astronomy.' Douglas McKie wrote that in the *Elements of Chemistry*, 'Lavoisier had done for chemistry what Newton had done for mechanics a century earlier in the immortal *Principia*.' However, Lavoisier was wrong in thinking that his work would effect a revolution in physics and chemistry; instead it wrought a revolution in chemistry and biology. As Frederic Lawrence Holmes noted: 'If he had not become as firmly identified as the leader of the general "Chemical Revolution", Lavoisier would probably be well known as the founder of physiological chemistry, the parent field for the huge domain now encompassed by biochemistry.' What was termed animal chemistry, then became physiological chemistry and is now known as biochemistry and molecular biology, can be traced back to the brilliant and tragically shortened scientific career of Antoine Lavoisier.

Lavoisier did have some luck on his side. First, he was a wealthy man, able to afford the sophisticated, custom-made pieces of equipment that his novel experimental designs required. Second, he lived in Paris, arguably the leading center for chemical research at the time, where he had the resources of the Royal Academy of Sciences at his disposal, and the services of such able collaborators as Berthollet, Fourcroy and Laplace. Third, he started his chemical studies at a time when experimental findings, particularly those of Black and Priestley, virtually necessitated a new theoretical interpretation.

Lavoisier started with one big idea – that the fixation of air converted flammable substances into acids. His early researches showed that the combustion of phosphorus, sulfur and carbon absorbed some part of the air and produced phosphoric, sulfuric and carbonic acid, respectively. From this it was a small step to realize that the calcination of metals also involved the fixation of air. To take this further, however, Lavoisier had to discard the two major chemical theories of the time. His demonstration that atmospheric air was composed of two quite different substances, oxygen and nitrogen, contradicted the four-element theory. By this time, the four-element theory was more a convenience than an intrinsic belief, so it probably cost Lavoisier little to discard it. However, it should be remembered that Priestley continued to believe that air was an element, even when his experiments were clearly inconsistent with that interpretation, and that Cavendish refused to admit that water was a compound, even after he had both decomposed and synthesized it!

His observation that combustion increased the weight of the burned substance brought Lavoisier into conflict with a much more significant chemical doctrine, the phlogiston theory. Within a few years, however, he had not only abandoned the phlogiston theory for the purposes of interpretation of his own work, but was prepared to publicly attack it. It is important to realize that the new entity Lavoisier proposed, caloric, was no closer to the modern idea of heat energy than was phlogiston. It is now accepted that combustion produces heat because the new chemical bonds formed release more energy than it took to break the old bonds. Therefore, a burning candle produces heat because the formation of the bonds in the products of the combustion, carbon dioxide and water, produces more energy than is required to break the bonds in the materials consumed in the combustion, wax and oxygen. According to the modern view, heat is neither contained in the candle, as envisaged in the phlogiston theory, nor in the oxygen, as envisaged in the caloric theory. The importance of the caloric theory to the development of chemical thought was therefore not that it was formally correct. Rather, its importance was that it concentrated attention on the common factor in combustion, oxygen, and that it related chemical change – combustion and calcination – to physical changes of state – conversion of a gas to a liquid or a solid.

Lavoisier's ability to reverse his theoretical field was related to the profound faith he had in his experimental findings. This faith was sometimes excessive, as when he stated that his elemental composition of water was accurate to 0.5%, when in fact it was off by 4%. His theoretical flexibility allowed Lavoisier to discard not only the theories of others but also his own. When the weights of combustion products did not add up to the weights of the materials consumed, Lavoisier revised his theoretical system to include water with fixed air as products. Conversely, when his measurements on fermentation did not allow for the expected consumption of water, he redefined fermentation as the conversion of sugar to alcohol and carbon dioxide.

In short, Lavoisier was able to effect a revolution in chemistry because he was able to adopt new experimental findings by others, while rejecting their interpretations, and because his unparalleled research skills gave him an unshakeable faith in experiment over theory. Above all, he was able to create a new chemical world-view because he was prepared to destroy the old one.

Chapter 2
The Maze of Organic Chemistry

Elemental Analysis

The chemical revolution of Antoine Lavoisier and his contemporaries can be regarded as the foundation of all subsequent chemical analyses of living systems. Lavoisier's analogy between respiration and combustion initiated the scientific study of the phenomenon of 'animal heat', leading, as the nineteenth century progressed, to important findings in bioenergetics, metabolism and biophysics. The process of alcoholic fermentation, shown by Lavoisier to be a chemical transformation of sugar to alcohol and carbon dioxide, became the intellectual terrain for a furious battle between vitalists and materialists (see Chapter 4). Perhaps most importantly, the demonstration that the tissues of living organisms could be analyzed by the same chemical techniques used for inorganic matter made possible the science of 'animal chemistry' which was the forerunner of biochemistry and molecular biology.

If vital processes such as combustion and fermentation were to be described by chemical equations, and if organic materials such as egg white, blood clots and kidney stones were to be subjected to chemical analysis, techniques for the elemental analysis of organic compounds were required. For this reason, Lavoisier's ingenious analytical methods were as important as his revolutionary theories in putting biology on a chemical basis.

Prior to the chemical revolution, the major technique for the analysis of organic materials was dry distillation. This was the process of gradually heating substances to produce a series of liquids that were characterized as 'volatile', 'mucous', 'salty', 'acidic', 'alkaline' or 'oily'. Acids varied from 'strong' to 'weak' and oils were characterized as 'thick', 'gross', 'attenuated' or 'empyreumatic'. For example, Pierre Joseph Macquer, whose promotion had created the vacancy at the Royal Academy of Sciences for which Lavoisier unsuccessfully competed in 1766, described sugar as 'an acid united with a large quantity of a very attenuated and mucilaginous earth, and with a certain quantity of sweet and not volatile oil, which is in a state perfectly saponaceous'. Lavoisier's later studies on cane sugar (sucrose) led him to conclude that 'the proportion in which these ingredients exist in sugar, are nearly eight parts of hydrogen, 64 parts of oxygen, and 28 parts of charcoal, all by weight, forming 100 parts of sugar'. One needs only compare these two definitions to realize how far chemistry had progressed in a single generation.

The difference between the old and new methods of organic analysis was not just that the latter were more 'exact', but also that they embodied different views of the nature of matter and chemical reactions. As noted by Frederic Holmes: 'The inadequacy of conceptions about chemical change may have been largely responsible for the tendency in this [mid-eighteenth century] period to regard the process of analysis as merely the separation of constituents previously present in the organic substances.' Lavoisier, in contrast, did not believe that sugar contained water, although the elements of water were present in sugar in the correct proportion, but only 'the elements proper for composing it'. To Lavoisier, therefore, compounds such as sucrose were *combinations* of elements, not *mixtures* of elements.

The analytical techniques developed by Lavoisier were largely based on the decomposition of substances to gases. As such, these techniques were ideally suited to organic substances, because these contained oxygen, hydrogen and nitrogen, all of which are gaseous in the elemental state. The other major element of organic materials, carbon, could easily be converted to carbonic acid gas (carbon dioxide). However, there was a lack of organic compounds available in pure form for analysis. In the plant kingdom, the best characterized materials were the 'vegetable acids'. Ten of these substances, including tartaric, lactic, oxalic, citric and malic acids, had been isolated by the Swedish chemist Carl Wilhelm Scheele between 1770 and 1785.

Except for the purposes of answering specific questions, such as the nature of vinous fermentation, Lavoisier was not much interested in animal or vegetable chemistry – his strength was in quantitative analysis, not the qualitative techniques required for the isolation of novel substances. However, many of his colleagues specialized in this new branch of chemistry. Antoine François Fourcroy had by 1785 identified four common components of animal tissues: fibrin, casein, albumin and gelatin. Claude Louis Berthollet, according to Holmes 'next to Lavoisier himself the most capable investigative chemist in France', showed that same year that animal substances contained '*mofette*' (nitrogen), which was transformed into 'volatile alkali' (ammonia) during acid distillation. Berthollet also showed that volatile alkali is a compound of nitrogen and hydrogen. Within groups of compounds such as the vegetable acids, the differences in composition were subtle; Fourcroy wrote in 1789 that the 'intermediate principles differ from each other only in the number and the proportions in which the primary substances [elements] are combined in them'.

As the groups of substances studied by Fourcroy and Berthollet differed among themselves not in the presence or absence of different elements, but rather in the relative amounts of the same elements, the ability to distinguish between them chemically depended upon exact methods of elemental analysis. Following the death of Lavoisier, the first major breakthrough in analytical technique was the 1810 method of Joseph Gay-Lussac and Louis Jacques Thenard. Their idea was to substitute a solid oxidizing agent for oxygen gas. The substance to be analyzed was made into pellets with potassium chlorate and dropped into a combustion chamber. The oxygen and carbon dioxide produced were measured directly, and the water produced calculated by the loss of oxygen from the chlorate. Gay-Lussac and Thenard used this method in 1811 to analyze the elemental composition of sugar, obtaining values that are reasonably close to those used now (Table 2.1). Clearly the new method represented a considerable improvement over that of Lavoisier.

Table 2.1: Elemental analyses of sucrose

	Weight %		
	C	H	O
Lavoisier, 1790	28	8	64
Gay-Lussac and Thenard, 1811	42.5	6.9	50.6
Modern	42.1	6.4	51.5

In 1812, the Swedish chemist Jöns Jacob Berzelius improved upon the analytical technique of Gay-Lussac and Thenard by introducing a condenser to collect the water. Gay-Lussac made another improvement two years later by using copper oxide instead of potassium chlorate, allowing nitrogen to be measured as the elemental gas.

Using these new methods, Gay-Lussac and Thenard found that, whereas sugars and starches contained hydrogen and oxygen in the same proportions as in water, oils and resins contained hydrogen in excess of this proportion. Vegetable acids, in contrast, contained an excess of oxygen. Fibrin, casein, albumin and gelatin, the four animal substances isolated by Fourcroy, yielded hydrogen, oxygen and nitrogen in the proportions necessary to form water and ammonia, suggesting that these 'albuminous' substances were related.

By around 1820, elemental analysis had become a standard technique of chemistry. Possibly the most skilled practitioner of this art was the English physician William Prout. In 1817, Prout reported that the composition of urea was 46.65% nitrogen, 26.65% oxygen, 19.975% carbon and 6.67% hydrogen. The modern values are 46.67% nitrogen, 26.67% oxygen, 20.0% carbon and 6.67% hydrogen. One would be tempted to label this analysis a fluke were it not for the existence of similar examples of Prout's analytical genius (see below).

Prout's elemental analyses led him to distinguish three classes of foods: saccharine (carbohydrates), oleaginous (fats) and albuminous (proteins). It was well known by then that the sugars and starches contain hydrogen and oxygen in the proportion of water; Prout proposed that the members of this class differed by the relative amounts of water and carbon. To explain the high proportion of hydrogen in the oleaginous substances, Prout proposed that these were combinations of water and 'olefiant gas' (ethylene). Among the albuminoids, he distinguished gelatin from albumin, casein and fibrin by its relatively low content of carbon. For Prout, the surprising thing was that such a variety of substances could be produced from 'a material so unpromising and so refractory as charcoal'.

The Atomic Theory

The work of Lavoisier and his colleagues had shown that compounds consisted of combinations of elements, and the analytical techniques they developed allowed the relative weights of the elements in a compound to be calculated. To relate this kind of information to the structures of compounds, however, required another concept, scarcely less revolutionary than those of Lavoisier – the atomic theory of John Dalton.

Dalton was born in 1766 in Eaglesfield, in the north-west of England. He never attended university or any equivalent course of instruction, but in 1793 became professor of mathematics and natural philosophy at New College in Manchester. In 1800, Dalton resigned from New College to open a private 'mathematical academy'. In 1808, he published the first volume of *A New System of Chemical Philosophy*; further volumes followed in 1811 and 1827.

The 'New System' rested on a number of premises. The first, that matter was composed of tiny indivisible particles, or atoms, was taken from the Greeks, particularly Lucretius. Dalton's atoms were spherical and uniform for any particular element: 'We may conclude that *the ultimate particles of all homogeneous bodies are perfectly alike in weight, figure, etc.* In other words, every particle of water is like every other particle of water; every particle of hydrogen is like every other particle of hydrogen, etc.' [emphasis in original]. This allowed atoms to be packed together 'like a pile of shot'. As gases vary widely in density, however, the sizes of the 'ultimate particles' must be different for each element. Thus Dalton's second premise: 'the following may be adopted as a maxim, till some reason appears to the contrary; namely, – *That every species of pure elastic fluid has its particles globular and all of a size; but that no two species agree in the size of their particles*, the pressure and temperature being the same' [emphasis in original].[6] The third premise was that elements combine in integral proportions to form elements. For example, carbonic acid (carbon dioxide) contained exactly twice as much oxygen as did carbonic oxide (carbon monoxide). Dalton had originally noted the phenomenon of multiple proportions in 1805 for the oxides of nitrogen; other examples were subsequently discovered by British and French chemists.

The aim of the New System was to use these principles to understand the compositions of compounds. The key to reformulating chemistry on an atomic basis was the ability to determine the relative weights of atoms; knowing such atomic weights, one could easily calculate how many atoms of different elements made up a compound:

> Now it is one great object of this work, to shew the importance and advantage of ascertaining *the relative weights of the elementary particles, both of simple and compound bodies, the number of simple elementary particles which constitute one compound particle, and the number of less compound particles which enter into the formation of one more compound particle* [emphasis in original].[6]

Dalton described compounds containing one atom of each of two elements (AB) as 'binary', those of the form AB_2 or A_2B as 'ternary' and those of the form A_3B or AB_3 as 'quaternary'. He proposed that two rules governed the combination of elements into compounds. The first was: 'When only one combination of two bodies can be obtained, it must be presumed to be a *binary* one, unless some cause appear to the contrary.' Binary compounds were, he thought, more stable than ternary or quaternary ones because atoms of the same type repelled one another. A corollary of this rule was that if two compounds form from the same elements, one will be binary and the other ternary, as in the case of carbon monoxide and carbon dioxide. Dalton's second rule of combination was that a binary compound should always be denser than a mixture of its constituent elements. The rationale for this was that if elements A and B combine into compound AB, the same weight of matter will occur in half the number of particles.

Dalton used these principles to calculate the atomic weights of a number of elements. For example, the relative weights of oxygen and hydrogen could be calculated from the elemental composition of water. Gay-Lussac had determined that water consisted of 87.4% oxygen and 12.6% hydrogen, a ratio of 7:1. According to Dalton's first law of combination, water must therefore be a binary compound, HO, as no other compounds of hydrogen and oxygen were (then) known. Therefore, the atoms of oxygen must be seven times as heavy as those of hydrogen. Similarly, only one compound of nitrogen and hydrogen, ammonia, was known. Assuming that ammonia was a binary compound, NH, the atomic weight of nitrogen, relative to hydrogen as 1, could be calculated as being 5. By similar considerations, Dalton was able to calculate the weights of other elements. The first (1808) part of the New System listed atomic weights relative to hydrogen for twenty elements, including carbon (5), phosphorus (9), sulfur (13) and a number of metals. The second (1811) part extended this list to thirty-six elements, and the third (1827) part to thirty-seven. The atomic weight values of the organic elements (H, C, N, O, P and S) were not revised, except that by 1827 carbon was listed as 5.4 instead of 5, and nitrogen was listed as '5±, or 10?'.

Dalton pointed the way towards a new chemistry, one that spoke not of the weights of elements in compounds, but of the number of atoms in a molecule. One could therefore visualize compounds as aggregates of different-sized spherical particles. The central concept of the New System was atomic weight, as this was the means by which elemental compositions could be converted to numbers of atoms.

However, Dalton was mistaken in two assumptions. First, he rejected – after having initially believed it – the idea that equal volumes of gases contained equal numbers of particles. He did so because he thought that this belief was inconsistent with atoms being of different sizes, and because the changes in volumes observed when some gases form compounds were not consistent with these gases having the same numbers of atoms. Dalton also turned out to be wrong in thinking that compounds were ordinarily binary. His assumption that water and ammonia were the binary compounds HO and NH, respectively, led him to underestimate the atomic weight of oxygen by a factor of two and that of nitrogen by a factor of three. To determine the atomic weight of carbon, Dalton assumed – correctly – that carbon monoxide was binary (CO) and carbon dioxide ternary (CO_2). However, as his weight for oxygen was too low by half, the value he calculated for carbon was similarly in error. Dalton's combining rules also worked well for the oxides of sulfur and phosphorus; he correctly identified 'sulphurous oxide' as SO, sulfurous acid as SO_2 and sulfuric acid as SO_3. As in the case of carbon, however, the initial error in the calculation of the atomic weight of oxygen caused Dalton to underestimate the values for sulfur and phosphorus.

Dalton was aware of the limitations of his theory. He concluded the final part of the New System with these words:

> I have spent much time and labour upon these compounds [of nitrogen and oxygen], and upon others of the primary elements carbone [sic], hydrogen, oxygen, and azote, which appear to me to be of the greatest importance in the atomic system; but it will be seen that I am not satisfied on this head, either by my own labour or that of others, chiefly through the want of an accurate knowledge of combining proportions.[7]

Unlike Lavoisier, Dalton was born into humble circumstances; also unlike Lavoisier, he died as a highly honored scientist. He served as president of the Manchester Literary and Philosophical Society for the last twenty-seven years of his life and was elected as a fellow of the Royal Society in 1822 and a foreign associate of the (no longer Royal) Academy of Sciences in 1830. Civic honors included a government pension and the erection of a statue. When Dalton died in 1844, his body lay in state for four days as 40 000 Mancunians filed past the coffin.

The Law of Combining Volumes

Dalton had correctly identified 'an accurate knowledge of combining proportions' as the key to determining the relative weights of elements. The law of combining volumes, expounded by Gay-Lussac in 1809, provided a simple means of doing so, as well as strong empirical support for Dalton's theory of multiple proportions. Gay-Lussac observed that gases capable of reacting with one another combine in integral units of volume. For example, the formation of ammonium chloride involved the reaction of equal volumes of muriatic gas (hydrogen chloride) and ammonia: '100 parts of muriatic gas saturate precisely 100 parts of ammonia gas, and the salt which is formed from them is perfectly neutral, whether one or other of the gases is in excess.' Similar relationships could be calculated from the data of others. Berthollet had shown that ammonia itself was formed from one volume of nitrogen and three volumes of hydrogen. From Humphry Davy's analyses of the three oxides of nitrogen, Gay-Lussac calculated a nitrogen/oxygen volume ratio of 2:1 for nitrous oxide, 1:1 for 'nitrous gas' and 1:2 for 'nitric acid'.

One puzzling feature of Gay-Lussac's studies – and a major reason for Dalton's skepticism – was the change in volume that occurred when two gases reacted to form another gas. For example, the volume of ammonia formed was half that of the combined volumes of nitrogen and hydrogen. Such a 'condensation' in volume was to be expected, as the atoms of the gas were also 'condensing'. In other reactions, however, there was no change in volume and in some there was even an increase. Gay-Lussac concluded that such changes in volume were independent of the reaction between the atoms. This anomaly did not detract from the main conclusion: 'I have shown in this Memoir that the compounds of gaseous substances with each other are always formed in very simple ratios, so that representing one of the terms by unity, the other is 1, or 2, or at most 3.'

Avogadro's Law

The full significance of Gay-Lussac's law of combining volumes was realized by Lorenzo Romano Amedeo Carlo Avogadro di Quaregna di Cerreto, professor of physics at the Royal College in Vercelli. In 1811, Avogadro wrote with reference to Gay-Lussac's findings: 'The first hypothesis to present itself in this connection, and apparently even the only admissible one, is the supposition that the number of integral molecules in any gases is always the same for equal volumes, or always proportional to the volumes.' This was a logical deduction from Dalton and Gay-Lussac; if compounds contained integral ratios of different atoms (Dalton), and gases combined in integral volumes (Gay-Lussac), then equal volumes of gases must contain the same number of atoms. Thus, because hydrogen combined with oxygen in the volume ratio 2:1, water must contain two atoms of hydrogen for every one of oxygen (H_2O or H_4O_2, etc.). Similarly, the three oxides of nitrogen discussed by Gay-Lussac must have the compositions N_2O, NO and NO_2 (or multiples thereof). The correspondence between volume and number of atoms – Avogadro's law – was merely the starting point for Avogadro's enquiry, however. More to the point was the use to which this law could be put. As Avogadro pointed out, if equal volumes of gases contained equal numbers of gas molecules, differences in densities between gases would be solely due to the atoms of one being heavier than the atoms of another. To determine the relative weights of gas atoms, therefore, it was only necessary to divide their densities. The densities of hydrogen and oxygen had been measured as 0.073 and 1.10 relative to atmospheric air, so Avogadro was able to calculate that the weight of the oxygen atom was approximately 15.1 times that of the hydrogen atom. From the density of nitrogen, 0.969, Avogadro calculated an atomic weight for this element of 13.2 relative to hydrogen.

Avogadro also addressed the changes in volume that accompanied reactions between gases. If the law of combining volumes meant that atoms combined into compounds in integral ratios, as he believed, then one would expect that the volume of the compound gas should be identical to that of the element that was present in the compound in the lower proportion. For example, the reaction of one volume of nitrogen with three volumes of hydrogen should produce one volume of ammonia ($1N + 3H \rightarrow 1NH_3$). In fact, this reaction produced a volume of ammonia equivalent to twice the volume of the nitrogen consumed. Similarly, the reaction of one volume of oxygen with two volumes of hydrogen should produce one volume of water ($1O + 2H \rightarrow H_2O$), but in fact produced two. Avogadro's explanation was that the gas molecules were not composed of single atoms, but rather two or more atoms united into single molecules: 'Thus, for example, the integral molecule of water will be composed of a half-molecule of oxygen with one molecule, or, what is the same thing, two half-molecules of hydrogen'. Avogadro's assumption that molecules of gaseous oxygen and hydrogen consisted of two atoms joined together explained why their reaction produced a volume of water vapor twice that of the oxygen reacted by reformulating the reaction as $2H_2 + O_2 \rightarrow 2H_2O$. The beauty of this hypothesis was that the number of atoms in the gas molecule was not a matter of guesswork, but was represented by the change in volume that occurred during the reaction. The formation of water doubled the volume of oxygen, so oxygen must be diatomic; likewise, as the formation of ammonia doubled the volume of nitrogen, nitrogen must be diatomic also.

Strictly speaking, Avogadro's method of determining atomic weights could only be used for elements that were gases at normal temperature and pressure, and his method of determining whether these gases were monoatomic or molecular could only be used for gases that reacted to make another gaseous product. However, by making additional assumptions, a similar approach could be used to calculate atomic weights of non-gaseous elements. For example, from the volume change that accompanied the conversion of sulfurous acid plus oxygen to sulfuric acid, Avogadro was able to calculate a theoretical density of 2.32 for sulfur gas. Dividing this by the density of hydrogen, he obtained a value of 31.7 for the atomic weight of sulfur. Using similar reasoning, an atomic weight of 38 was obtained for phosphorus. Assuming that carbonic acid gas (carbon dioxide) was analogous to the sulfurous and phosphorous acids – that is, that carbonic acid was CO_2, sulfurous acid SO_2 and phosphorous acid PO_2 – Avogadro was able to calculate a theoretical density for carbon gas, and from that an atomic weight for carbon of 11.4. From the reaction of hydrogen and 'oxymuriatic acid' (chlorine) to form muriatic acid (hydrochloric acid, HCl), an atomic weight of 33.9 was calculated for chlorine. The fact that many metals occurred in two or more oxidation states allowed Avogadro to determine an atomic weight of 362 for mercury, 94 for iron, 206 for lead, 198 for silver and 123 for copper.

Because of uncertainties about the compositions of metal oxides – a problem that also plagued later generations of chemists – Avogadro's values for the metallic elements were generally too high by a factor of approximately two. However, as shown in Table 2.2, his atomic weights for the other elements were remarkably accurate.

Table 2.2: Atomic weight values relative to hydrogen

Element	Avogadro 1811	Dalton 1827	Modern
Carbon	11.4	5.4	12
Nitrogen	13.2	5 or 10?	14
Oxygen	15.1	7	16
Phosphorus	38	9	31
Sulfur	31.7	13 or 14	32.1
Chlorine	33.9	–	35.5

Avogadro modestly concluded that his hypothesis was 'at bottom merely Dalton's system furnished with a new means of precision from the connection we have found between it and the general fact established by M. Gay-Lussac'. It was, in fact, much more than this – the foundation stone of a new chemistry, one based on molecules rather than on weights of elements. Because Avogadro's concept of diatomic gases was heretical to many chemists, however, it would be fifty years before this foundation would be built upon.

Atomic Weight

Making chemistry into an atomic science required a revolution in thinking similar to that achieved by Lavoisier. The key concept – the Rosetta stone of the new chemistry – was atomic weight. With reliable atomic weights, chemists could convert the elemental compositions of compounds into the numbers of atoms of different elements present. Such empirical formulas would in turn illustrate relationships between compounds and lead to an understanding of molecular structure and chemical reactivity.

By 1820, all the concepts necessary to derive reliable atomic weights were available: Dalton's theory of multiple proportions, Gay-Lussac's law of combining volumes, Avogadro's concept of molecular gases. Unfortunately, although the first two propositions were widely and rapidly accepted, the third was not. Without the insight of Avogadro, calculation of atomic weights required some guesswork. Different chemists guessed differently: in Sweden, Jöns Jacob Berzelius used the atomic weight values $H = 1$, $C = 12$, $O = 16$; in Germany, Justus Liebig used $H = 1$, $C = 6$, $O = 8$; in France, Jean-Baptiste Dumas used $H = 1$, $C = 6$, $O = 16$. To make things worse, because Berzelius did not believe that metal oxides could be of the form M_2O, he doubled the true values for the atomic weights of silver and several other metals. As many organic compounds were prepared for analysis by precipitation with silver, this led to a doubling of the number of atoms in such compounds. Later, the French chemist Charles Gerhardt was to make the opposite assumption and calculate weights of atoms and radicals on the basis that all metal oxides were of the form M_2O![a]

The fifty-year neglect of Avogadro's 1811 paper is a particularly interesting example of scientific oversight. The facts upon which Avogadro based his system were available to Dalton, Gay-Lussac, and other contemporary chemists. André Ampère and, to some extent, Dumas were able to reach the same conclusion as Avogadro. The great majority of the nineteenth-century chemical establishment were not, however. The main reason for the neglect of Avogadro's work appears to be that its key concept, the diatomic gas, was philosophically unacceptable to most chemists of the time. Atoms were thought to combine into compounds only if they were of opposite electrical charge. Also, if atoms of the same element could bind to one another, why should this process stop at binary complexes?

Amedeo Avogadro is an example of the scientist who is able to believe the experimental evidence even when it contradicts assumptions so deeply embedded in the fabric of a discipline as to be almost subconscious. Avogadro's realization that Gay-Lussac's findings *required* the concept of diatomic gases when his contemporaries could not see this is similar to the ability of another lawyer turned chemist, Antoine Lavoisier, to discard the phlogiston theory when Priestley and Cavendish could not. Later, this iconoclastic ability would be exhibited by Linus Pauling in proposing that protein helices could contain non-integral numbers of amino acids (see Chapter 11).

---

a   For simplicity's sake, all empirical (elemental) formulas quoted below have been converted into values based on modern atomic weights ($C = 12$, $O = 16$, $N = 14$, $P = 31$, etc.).

Berzelius

Chemistry in the first half of the nineteenth century was dominated by the findings, teachings and personalities of three men: Jöns Jacob Berzelius, Justus Liebig and Friedrich Wöhler. Berzelius was the oldest of the three. Born in 1779 at Väversunda, Sweden, he graduated from Uppsala in 1802 with a medical degree and took an unpaid position in the College of Medicine in Stockholm. The discovery of the element cerium made his name, and in 1807 he became professor of medicine and pharmacy at the College of Medicine (from 1810, the Karolinska Institute). Berzelius continued to work on minerals, discovering the elements thorium and selenium, but also made important contributions to other branches of chemistry. His 1808 *Lärbok i kemien* (*Textbook of Chemistry*) was the most influential work of its kind since Lavoisier's *Elements of Chemistry*, and he befriended many prominent scientists during trips to Britain and continental Europe in 1812 and 1818, respectively. Every year from 1821 onwards, Berzelius produced a *Årsberättelse om Framstegen i Physik och Kemi*, which was translated into German as *Jahresbericht über die Fortschritte der physichen Wissenschaften* (*Annual Report on Progress in the Physical Sciences*). From the fringes of the scientific world, and with few disciples, Berzelius still managed to dominate opinion in chemistry.

Berzelius was principally interested in inorganic chemistry. Early in his career he discovered that salts could be decomposed into their component elements by electricity (electrolysis). Electrolysis of common salt (sodium chloride), for example, produced sodium metal at the negative pole and chlorine gas at the positive pole. Sodium was thus described as electropositive, and chlorine as electronegative. Berzelius eventually concluded that all compounds were salts of acids and bases. Calcium sulfate ($CaSO_4$), for example, was considered by Berzelius to consist of the base CaO plus the acid SO_3. In turn, CaO was a compound composed of the acid O and the base Ca.

Berzelius was above all a classifier – the Linnaeus of chemistry. He proposed the system now used to represent the elements by the first letter, or if necessary the first two letters, of its Latin name. He was also the first to use empirical formulas in which a compound is represented by the one- or two-letter symbols of its constituent elements followed by a number to indicate their relative amount. Berzelius proposed the phenomena of isomerism (see below) and catalysis (see Chapter 4) to bring together examples of different compounds exhibiting a similar property.

Liebig

Justus Liebig was born in Darmstadt in 1803. He left school without qualifications at the age of fourteen and was apprenticed to a pharmacist for several months before being forced to leave for lack of money (although he later claimed that he was dismissed for causing an explosion in his room). Fortunately, his father, a paint-maker, had supplied chemicals to Karl Wilhelm Kastner, professor of chemistry at the University of Bonn. In 1820, Kastner agreed to take on Liebig as an assistant. When Kastner moved to Erlangen a year later, he took Liebig with him.

An incident at Erlangen illustrates the pugnacious attitude that Liebig would later bring to scientific disputes. He joined other students from the Rhineland states in an illegal political organization, the Korps Rhenania. For his part in a violent confrontation with townspeople, Liebig was sentenced to three days in jail. According to the university records: 'this punishment was fully merited, for he was particularly active in this disturbance, he made scurrilous remarks about those in authority, and he knocked the hat from the head of not only police officer Scramm, but even of Councillor-at-Law Heim.' Even more scurrilously, it appears that Liebig may have betrayed colleagues in the Korps Rhenania to the university authorities in return for lenient treatment.

Whatever he thought of Liebig's character or political activities, Kastner was obviously impressed with his scientific abilities. On his recommendation, Grand-Duke Ludwig of Hessen-Darmstadt provided funds for Liebig to go to Paris for further chemical studies. Kastner also pulled some strings to ensure that Liebig was awarded the degree of Doctor of Philosophy from the University of Erlangen without having completed the requirements.

During his two years in the French capital, Liebig worked in the laboratory of Joseph Gay-Lussac and met many of the leading chemists of the day. His work on the salts of fulminic acid was presented at the Academy of Sciences, which occasioned a meeting with Alexander von Humboldt, the Prussian ambassador to France. Humboldt suggested to Grand-Duke Ludwig that Liebig be given an academic position upon his return to Germany. In 1824, at the age of twenty-one, Liebig was appointed *ausserordentlicher* (extraordinary) professor at Giessen.[b]

---

b   In nineteenth-century Germany, a student who had completed a *Habilitation*, consisting of a published paper, a public lecture and oral examination, was given the title of *Privatdozent* and expected to support himself by tutoring. Holders of university chairs were referred to as (*ordentlicher*) professor, but additional salaried positions, of lower status, existed for what were known as *ausserordentlicher* professors.

Liebig wished to establish a laboratory that would provide for large numbers of students the same training he had received from Gay-Lussac. To do this, he had to overcome a number of obstacles. The faculty at Giessen resented not being consulted in his appointment. The professor of chemistry, Wilhelm Zimmermann, was particularly hostile, apparently to the point of taking home equipment that had been assigned to Liebig. The building given Liebig for his teaching laboratory was the unheated guardhouse of a police barracks. His salary was small, and from this he had to buy laboratory equipment; consequently, he was for many years in debt.

Once the Giessen laboratory was operational, students of chemistry and pharmacy (and later other subjects) were instructed in the techniques of elemental analysis and then set to work analyzing compounds of unknown composition. Liebig provided this account of the operation of his laboratory:

> There was no actual instruction. Every morning I received from each individual a report on what he had done the previous day, as well as his views about what he was engaged on. I approved or criticised. Everyone was obliged to follow his own course. In the association and constant intercourse with each other and by each participating in the work of all, everyone learned from the others. Twice a week in winter I gave a sort of review of the more important questions of the day. We worked from break of day till nightfall. Dissipations and amusements were not to be had at Giessen. The only complaint which was continually repeated was that of the attendant, who could not get the workers out of the laboratory in the evening when he wanted to clean it.[8]

With his students, Liebig set out to standardize the process of elemental analysis. His main innovation over the techniques of Gay-Lussac was to introduce 'traps' for carbon dioxide and water. The former was a solution of caustic potash (potassium hydroxide), the latter a container of calcium chloride. Following combustion, these traps were weighed to determine the amount of carbon dioxide and water produced. The major gain was in speed, not in accuracy – Liebig claimed to be able to do as many analyses in three months as Berzelius could do in five years. When Liebig was satisfied with the accuracy of his analytical methods, he set his students to work on characterizing as many animal and plant materials as he could lay his hands on.

In 1826, following Zimmermann's death by suicide, Liebig became *ordentlicher* professor of chemistry, with a modest increase in salary. Expansions of his laboratory in 1835 and 1839 resulted in separate rooms for chemistry and pharmacy, plus a lecture theater, allowing thirty students to be instructed. The addition of an annex in 1843 increased this number to forty-five. By the time he left Giessen in 1852, over seven hundred *Chemikers* had been trained in Liebig's laboratory. These included such major figures in late nineteenth-century chemistry as Edward Frankland, Charles Gerhardt, August Kekulé, John Stenhouse, Adolph Strecker, Alexander Williamson and Charles Wurtz.

In 1832, Liebig founded the *Annalen der Pharmacie* (*Annals of Pharmacy*); eight years later he renamed it the *Annalen der Chemie und Pharmacie* (*Annals of Chemistry and Pharmacy*). The *Annals* was one of the first scientific journals devoted to chemistry, and its editor ran it as a forum for his own view of the discipline.[c] By the mid-1830s, Liebig's influence over chemistry surpassed that of Berzelius. Berzelius had his network of correspondents, but Liebig had his army of *Chemikers*. Berzelius had the *Jahresberichte*, but Liebig had his journal. Together, the Giessen laboratory and the *Annals* gave Liebig an ability to dominate his field of research possibly unrivaled in the history of science – a power that Liebig was not above abusing. By mid-century, Liebig was believed to have a virtual veto on professorial appointments in German chemistry: Theodor Schwann blamed Liebig for his failure to be awarded the chair at Bonn (see Chapter 3), and August Kekulé for his failure at Zürich (see below). An American chemist who spent a year at Giessen in the 1840s wrote: 'it seems to be a settled point that no young man can be expected to know anything of chemistry unless he has studied with Liebig.'

Historians do not paint a pleasant picture of Justus von Liebig.[d] Vance Hall wrote that 'Hasty judgement, arrogance, misrepresentation of [others'] positions, and reluctance to acknowledge his own mistakes can be found throughout his life's work.' Satish Kapoor labeled Liebig 'the undisputed champion of . . . squalid German nationalism in scientific affairs'. Even the sympathetic William Brock had to acknowledge 'Liebig's feeling of self-importance, his belief in the "rightness" of his own approach and his touchiness in the face of opposition from any quarter.' To this should be added three specific instances that reflect very poorly upon Liebig's humanity: his tendency to claim sole credit for work done collaboratively with his great friend Wöhler; his denigration of the man to whom he owed his career, Karl Kastner; and his solicitation for the chair of chemistry at Giessen only four days after the suicide of Zimmermann.

---

c   Liebig took on his editorial duties solely for financial reasons. In 1831, he complained in a letter to Berzelius: 'I have, recently, been burdened with a heavy load, by joining [Philipp] Gieger as co-editor of his journal, all for the sake of the damned money involved. At the small university where I live, I am almost on the verge of starvation'.

d   Liebig was ennobled in 1845.

Wöhler

Friedrich Wöhler was born in 1800 in Escherscheim, near Frankfurt. He went to the University of Marburg in 1820, transferring after one year to Heidelberg out of admiration for the chemist Leopold Gmelin.[e] To his surprise, Gmelin advised him not to attend his lectures; as a result, Wöhler never attended any formal course of instruction in chemistry. After Wöhler graduated with a doctorate in medicine in 1823, Gmelin advised him to study chemistry with his own former mentor, Berzelius. Wöhler spent a year in Stockholm, and recorded his impressions of Berzelius' laboratory, which was not at the Karolinska Institute, but in his house:

> Adjoining the living-room, the laboratory consisted of two ordinary chambers with the simplest fittings; there was neither oven nor fume chamber, neither water nor gas supply. In one room stood two ordinary work-tables of deal; at one of these Berzelius had his working place, the other was assigned to me. On the walls were several cupboards with reagents which, however, were not provided very liberally, for when I wanted prussiate for my experiments I had to get it from Lübeck. In the middle of the room stood the mercury trough and glass-blower's table, the latter under one of the chimney-places provided with a curtain of oiled silk. The washing place consisted of a stone cistern having a tap with a pot under it. In the other room were the balances and other instruments, beside a small work-bench and lathe. In the kitchen, where the food was prepared by the severe old Anna, cook and factotum of the master who was still a bachelor, stood a small furnace and the ever-heated sandbath.[8]

During his time with Berzelius, Wöhler analyzed cyanic acid. At the same time, Liebig had discovered a compound with quite different chemical properties, fulminic acid, which nonetheless gave an identical elemental analysis, CHNO. When Wöhler returned from Sweden, he arranged to meet Liebig at the house of a mutual friend in Frankfurt. On examining each other's data, Wöhler and Liebig agreed that two different compounds could indeed have the same composition. This was the first instance of the phenomenon that Berzelius would later term 'isomerism' (see below). It was also the beginning of a life-long friendship between Wöhler and Liebig.

Wöhler intended to set up shop as a *Privatdozent* in Heidelberg, but found that he had been nominated for and appointed to a teaching position at the

---

e   Gmelin was then studying the intestinal transformation of food. However, in 1829 he would make the important proposal that the chemical action of plants mainly involves 'desoxidation' (reduction) whereas that of animals involves oxidation.

Gewerbeschule (Technical School) in Berlin. There, Wöhler made his first important contributions to chemistry by performing the first isolations of the metals aluminum[f] and beryllium – and an even greater contribution to biology by performing the first organic synthesis (see Chapter 3). He moved to Kassel in 1831, and five years later narrowly beat his friend Liebig to the chair of chemistry at Göttingen. Wöhler was an excellent linguist who translated Berzelius' works, including the *Textbook of Chemistry* and the *Jahresberichte*, into German. He was also a noted teacher; when he asked Liebig to provide laboratory space for one of his students, Liebig replied: 'Those are rather stupid fellows who go from Göttingen to Giessen for the sake of chemistry.'

August Wilhelm Hofmann, one of the most distinguished graduates of the Giessen laboratory, has left this reflection on Liebig and Wöhler:

> Liebig, fiery and impetuous, seizing a new thought with enthusiasm, and giving to it the reins of his fancy, tenacious of his convictions, but open to the recognition of error, sincerely grateful, indeed, when made conscious of it, – Wöhler, calm and deliberate, entering upon a fresh problem after full reflection, guarding himself against each rash conclusion, and only after the most rigorous testing, by which every chance of error seemed to be excluded, giving expression to his opinion, – but both following the path of inquiry in their several ways, and both animated by the intense love of truth! Liebig, irritable and quick to take offence, hot-tempered, hardly master of his emotions, which not infrequently found vent in bitter words, involving him in long and painful quarrels, – Wöhler, unimpassioned, meeting even the most malignant provocation with an immovable equanimity, disarming the bitterest opponent by the sobriety of his speech, a firm enemy to strife and contention, – and yet both men penetrated by the same unswerving sense of rectitude! Can we marvel that between two such natures, so differently ordered, and yet so complementary, there should ripen a friendship which both should reckon as the greatest gain of their lives?[9]

Wöhler's attempts to moderate the combative personality of his friend were largely in vain; Liebig managed to alienate, at one time or another and usually permanently, all the leading chemists of his day. In each case, a scientific issue escalated into a personal feud: with Dumas, over the substitution theory; with Berzelius, over *Animal Chemistry*; with Gerrit Mulder, over the

---

f   Twenty years later, Wöhler's method of producing aluminum was used as an industrial process. The first bar of metal was used to strike a medal with the name of Wöhler on one side and a likeness of the emperor on the other. Napoleon III presented this medal to Wöhler on the occasion of his induction into the Legion of Honour.

protein radical; with Schwann, over the nature of fermentation; and with Gerhardt, over the composition of mellon.

Berzelius, Liebig and Wöhler were involved in almost all the major developments in 'animal chemistry' during the first decades of the nineteenth century. To a remarkable extent, progress in chemistry was defined by these three men, and depended upon the relationships between them. Berzelius was the elder statesman, Liebig and Wöhler the *enfants terribles*. Wöhler was caught in the middle when the new ideas of Liebig came into conflict with the orthodoxy of Berzelius.

The Theory of Substitution

According to the theory of Lavoisier and Guyton de Morveau, organic compounds differed from inorganic ones in having more complex bases, or 'radicals'. In 1832, Wöhler and Liebig, working directly together for the first and only time, extended the radical theory from inorganic to organic chemistry. Starting with oil of bitter almonds (benzaldehyde), they were able to produce and analyze a wide variety of derivatives, including benzoic acid, benzoyl chloride, benzamide and ethyl benzoate – what Timothy Lipman called 'a veritable romp through organic chemistry'. All of these substances appeared to contain an identical 'compound base' that Wöhler and Liebig called the benzoyl radical (C_7H_5O). It had been realized since the time of Lavoisier that all chemical reactions were rearrangements of elements; what Wöhler and Liebig were now proposing was that discrete groups of elements were conserved during a variety of different reactions.

Berzelius applauded this work, attaching a note to Wöhler and Liebig's paper describing it as 'the dawning of a new day in vegetable chemistry'. He claimed that this study 'proves that there are ternary composed atoms (of the first order) and the radical of benzoic acid is the first example proved with certainty, of a ternary body possessing the properties of an element'. However, Berzelius later came to believe that the radical conserved in these reactions was a hydrocarbon one to which oxygen became attached by an electrochemical mechanism.

A serious challenge to the radical and electrochemical theories soon arose from a most unlikely direction. In the early 1830s, a royal soirée at the Tuileries was disrupted by irritating candle fumes. The director of the Sèvres porcelain factory was by tradition chemical adviser to the royal household. King Charles X ordered the incumbent, Alexandre Brogniart, to investigate the source of the fumes. Brogniart was actually a geologist, but fortunately was able to delegate the king's problem to the leading French chemist of the day – his son-in-law, Jean-Baptiste Dumas.

Dumas was born in Alais in 1800, and was thus an exact contemporary of Wöhler. At the age of sixteen he set out on foot for Geneva, where he had relatives. He trained as a pharmacist and became involved in a variety of chemical studies. In 1822, a chance meeting with Alexander von Humboldt, the man who would later recommend Liebig for the position at Giessen, led to Dumas's decision to move to Paris. There he was befriended by a number of eminent scientists, including the former Lavoisier protégé, Pierre Laplace, and was appointed to positions at the École Polytechnique and the Atheneum. In 1832, he replaced Gay-Lussac as professor of chemistry at the Sorbonne. Dumas was responsible for a number of innovations, including a method of estimating the atomic weights of non-gaseous elements by measurement of their vapor densities. He also devised an improved technique of nitrogen analysis,[g] and established that a series of alcohols differed in their numbers of methylene (CH_2) groups.

When Dumas examined King Charles's offending candles, he soon discovered that they had been bleached with chlorine. Further study showed that hydrogen atoms in compounds such as candle wax could be replaced with chlorine atoms, producing substances whose chemical properties did not differ significantly from those of the original substance. Dumas's most spectacular example of this phenomenon came in 1839, when he showed that three hydrogen atoms in acetic acid could be replaced with chlorine atoms, producing a new compound, trichloroacetic acid, whose physical and chemical properties were virtually identical to those of the parent compound:

> Chlorinated vinegar is still an acid, like ordinary vinegar; its acid power has not changed . . . Here then is a new organic acid in which a very large amount of chlorine is present, but which shows none of the typical reactions of chlorine; in which the hydrogen has disappeared, and has been replaced by chlorine; but which has, as a result of this remarkable substitution, suffered only a slight alteration in its physical properties. *All the essential properties of the substance have remained intact* [emphasis in original].[10]

From such observations, Dumas proposed the theory of substitution, which stated that the properties of a compound depended more upon the number and relative positions of its constituent atoms than upon their nature. The idea that hydrogen and chlorine were interchangeable elements was

---

g   The Dumas procedure for nitrogen analysis continued to be used until as recently as the 1950s.

completely inconsistent with the electrochemical theory of Berzelius, in which hydrogen was electropositive and chlorine was electronegative, as Dumas realized:

> These electrochemical conceptions, this special polarity which has been assigned to the elementary atoms, – do they really rest on such evident facts that they are to be accepted as articles of faith? Or if we regard them only as hypotheses, do they possess the property of adapting themselves to facts; are they capable of explaining them; can we assume them with such a complete certainty that in chemical investigations they appear as useful guides? We must admit that such is not the case.[9]

The theory of substitution had been foreshadowed by Wöhler and Liebig's studies on the benzoyl radical, which showed that chlorine could replace hydrogen in benzaldehyde. Dumas, however, took the idea one step further by proposing substitution *within* the radical. Liebig had no particular problem with this idea. He and Dumas co-authored a magisterial 'Note on the present state of organic chemistry' in 1837, and the following year Dumas was invited to become an editor of the *Annals*. However, Liebig parted company with Dumas when the latter proposed that even carbon could be substituted without changing the properties of the compound.

In 1840, a famous spoof of the substitution theory appeared in the *Annals*. The author of this article, one 'S. C. H. Windler' of Paris, described an experiment in which the hydrogen atoms of manganous acetate $(Mn[C_2H_4O_2]_2)$ were substituted with chlorine. Monsieur Windler then goes on to substitute with chlorine the oxygen, carbon and even the manganese, producing a compound of the formula $Cl_8Cl_6Cl_6$, which still had the properties of manganous acetate! 'S. C. H. Windler' was in fact Friedrich Wöhler. He was surprised when Liebig published his letter, which he had intended as a private joke for Berzelius. Had he known that it was going to be published in the 'Annals', Wöhler later told Liebig, he would have signed the letter, which was written in French, as 'Professor Ch. Arlatan'!

However, the substitution theory was here to stay, as even Liebig was soon forced to admit. The realization that some atoms were more easily replaced than others spawned the 'nucleus' theory of Dumas's student Auguste Laurent, according to which compounds contained a core of strongly bound atoms and an outer shell of exchangeable atoms. The chemical similarity between the acetic and trichloroacetic acids influenced chemists to classify compounds into 'types'. These type theories were to lead eventually to the valence theory of Kekulé and Archibald Couper.

Animal Chemistry

In 1842, the results of Liebig's analyses of organic compounds were summarized in his book, *Die organische Chemie in ihrer Anwendung auf Physiologie und Pathologie* (translated into English as *Animal Chemistry, or Organic Chemistry in its Application to Physiology and Pathology*). Perhaps to make further amends for his wavering on the theory of substitution, Liebig dedicated the German edition of *Animal Chemistry* to Berzelius. The aim of the book was nothing less than to explain physiological processes by accounting for every atom entering and leaving the body. To do this, Liebig had made heroic efforts, including measuring the total amount of carbon in the food and excrement of 855 soldiers from a Hessian army company. For Liebig realized, more clearly than anyone to that time, that ingestion and excretion must be balanced:

> The carbon of the carbonic acid given off, with that of the urine, the nitrogen of the urine, and the hydrogen given off as ammonia and water; these elements, taken together, must be exactly equal in weight to the carbon, nitrogen, and hydrogen of the metamorphosed tissues, and since these last are exactly replaced by the food, to the carbon, nitrogen and hydrogen of the food. Were this not the case, the weight of the animal could not possibly remain unchanged.[11]

Liebig believed that not only the overall metabolism of the organism must be in balance, but also that specific physiological processes could be expressed as balanced chemical equations. His calculations convinced him that plants could synthesize organic compounds from their component elements, but animals could only 'metamorphose' them. In the animal body, 'nitrogenized' substances (the 'plastic elements of nutrition') were converted into blood and organized tissues, whereas 'non-nitrogenized' ones such as sugars, starches and alcohol (the 'elements of respiration') were used to support respiration and to produce animal heat. Liebig also believed that bile was re-absorbed from the gut into the blood and used for respiration.

A typical example of this approach to physiology was Liebig's explanation of the medicinal effects of caffeine. Liebig found that by adding some oxygen and water, it was possible to convert caffeine to the bile acid taurine:

1 'atom' caffeine ($C_8N_2H_5O_2$) + 9 water (HO) + 9 oxygen (O) = 2 taurine ($C_4NH_7O_{10}$)

By contributing to the secretion of bile, therefore, caffeine could aid in the digestion of food. To justify other parts of his metabolic scheme, however,

Liebig had to construct equations of such monstrous proportions as to be almost a *reductio ad absurdum* of the approach proposed by Lavoisier. For example, to explain why the urine of herbivores contained ammonia, urea and hippuric or benzoic acid, but no uric acid, Liebig proposed that blood, for which he had calculated the empirical formula $C_{48}N_6H_{39}O_{15}$, underwent the following reaction:

$$5 (C_{48}N_6H_{39}O_{15}) + O_9 = 6 \text{ hippuric acid } (C_{18}NH_8O_5) + 9 \text{ urea } (C_2N_2H_4O_2)$$
$$+ 3 \text{ choleic acid } (C_{38}NH_{33}O_{11}) + 3 \text{ ammonia } (NH_3) + 3 \text{ water } (HO)$$

Even in the 1840s, it must have been clear that a great number of different things could be made with 240 carbon atoms. It is difficult to avoid the conclusion that the equations were constructed to justify Liebig's pre-existing physiological theories, rather than the theories arising from the equations.

The publication of *Animal Chemistry* created a sensation. To many readers, Liebig's approach had at a stroke reduced physiology to chemistry. The eminent physiologist Johannes Müller delayed publication of his *Handbook of Physiology* in order to incorporate discussion of *Animal Chemistry*. On reading it, he realized that 'physiology based on comparative anatomy would give way to physiology based on chemistry and physics'. The English philosopher John Stuart Mill wrote in 1843: 'The recent speculation of Liebig in organic chemistry shows some of the most remarkable examples since Newton of the explanation of laws of causation subsisting among complex phenomena, by resolving them into simple and more general laws.' As Frederic Holmes put it, 'The calculations seemed able to reduce innumerable complex transformations to order, uniformity and accuracy.'

The chemists were not convinced, though. One showed that calculations similar to those used by Liebig could 'prove' that the oxidation of protein produced the poison prussic acid and the explosive fulminic acid! Mulder compared Liebig's metabolic equations to determining what takes place in a house by measuring the food that enters the door and the smoke that comes out of the chimney. Berzelius was outraged that Liebig would present hypothetical schemes as physiological reality, exploding in a letter to Wöhler, 'My God, what drivel!'. In his *Jahresbericht* for 1843, Berzelius wrote: 'This easy kind of physiological chemistry is created at the writing-table, and is the more dangerous, the more genius goes into its execution, because most readers will not be able to distinguish what is true from mere possibilities and probabilities.' Wöhler softened the language when he translated the *Jahresbericht* into German, but not enough for Liebig, who, as Holmes noted, 'was rarely able to preserve a distinction between intellectual disagreements and personal attacks'. In the *Annals* of May 1844, Liebig accused Berzelius of failing to realize that animal chemistry had advanced in the past thirty years.

The two men never reconciled. However, by the time the second edition of *Animal Chemistry* appeared in 1846, the equations describing physiological processes had disappeared, replaced with a warning about the dangers of writing chemical schemes without proof of the existence of such reactions. Typically of Liebig, this sermon was illustrated with examples from the work of others!

Perhaps the most important aspect of Liebig's approach in *Animal Chemistry* was the assumption that at least some food molecules could be converted into, not just assimilated by, animal tissues. In 1842 this was a controversial view and was explicitly denied by, among others, Dumas.

The Protein Radical

The first protein isolated was probably gluten, which was identified as a 'glue-like' constituent of wheat flour by Iacopo Bartolomea Beccari in work published in 1745. Later in the eighteenth century, a variety of substances from animal tissues were described as 'albuminous' because of their apparent similarity to egg white (*album ovi*). This group included fibrin, which was obtained as the insoluble residue of washed blood clots; serum albumin, a substance precipitated by heating the serum remaining when clots were removed from blood; gelatin, obtained by boiling skin, tendon or cartilage; and casein, obtained in a like manner from milk.

The first inkling that these substances were related came from the work of Lavoisier's colleagues Fourcroy and Berthollet in the 1780s. Berthollet showed in 1785 that the combustion of many animal tissues produced 'volatile alkali' (ammonia). Fourcroy refined this analysis by showing that the albuminous substances contained nitrogen, and that similar substances were also present in plants; he reported in 1789 that: 'Albumen [*sic*] extracted from plants and dried gives on distillation ammonium carbonate, foetid red oil, hydrogen gas and carbonic acid; it leaves in the retort a light charcoal, difficult to incinerate, and of which I have not yet been able to obtain enough ashes to examine.' Nonetheless, the presence of nitrogen was thereafter considered to be characteristic of animal tissues, and the term 'animalization' was coined to describe the process by which nitrogen–free plant substances were converted to animal tissues.

Berzelius believed that vegetable substances were oxides of radicals containing carbon and hydrogen, whereas in animal substances the radicals contained carbon, hydrogen and nitrogen. By the 1830s, it was known that 'albuminoid' substances also contained sulfur, which could be detected by its ability to form a salt with lead or silver. The Dutch physician–chemist Gerrit Jan Mulder found in 1838 that fibrin, ovalbumin and serum albumin all had similar contents of phosphorus and sulfur, except that serum albumin had twice as much sulfur as the other two proteins. Heating in dilute alkali removed the sulfur and phosphorus, and left a 'radical' with the empirical formula $C_{40}H_{62}N_{10}O_{12}$. Mulder was one of the many correspondents of Berzelius, who suggested the name *proteine* for this radical, 'because it appears to be the primary or principal substance of animal nutrition which the plants prepare for the herbivores, and which these then supply to the carnivores'. Mulder believed that proteins consisted of the protein radical (Pr) combined with phosphorus and sulfur in different proportions: ovalbumin and fibrin were $Pr_{10}SP$, serum albumin was $Pr_{10}S_2P$ and the lens protein crystallin $Pr_{15}S$.[h]

The protein radical theory was initially accepted by Liebig, who in 1841 hailed Mulder's 'elaborate and conscientious investigations . . . which opened up a universe of new discoveries'. However, other analytical data did not support Mulder's contention of a consistent protein radical. In 1842, Dumas showed that globin, an albuminous constituent of blood corpuscles, had more nitrogen and less carbon than serum albumin. The idea of a protein radical was contrary to the spirit, if not the letter, of the substitution theory, which by 1845 Liebig no longer opposed. He put a visiting student to work on the composition of proteins. The resulting paper in the *Annals* stated bluntly that: 'the empirical formulae advanced for Herr Mulder for protein-like bodies are in no way acceptable, as they do not agree with the results of analysis.' Typically of Liebig, a scientific issue rapidly became personal. Appended to the article was a note requesting that Mulder supply full details of his experimental procedures – virtually a public accusation of fraud or incompetence. Mulder responded with a pamphlet entitled *Liebig's Question to Mulder Tested by Morality and Science.* Liebig wrote in 1847 that 'the so-called proteine theory' was 'supported by observations both erroneous in themselves and misinterpreted to their significance'.

---

h   These formulas corresponded to molecular masses of approximately 9000 – much lower than the actual weights of these proteins, but much higher than many scientists were prepared to accept even as late as the 1920s (see Chapter 7).

On this occasion, Liebig was correct – there was no protein radical. Indeed, it was already becoming clear that proteins contained a number of smaller components. The earliest reported isolation of a compound of the class that later became known as the amino acids was in 1806, when Louis Nicolas Vauquelin and Pierre Robiquet, both former assistants to Fourcroy, purified a crystalline material from the sap of asparagus plants. Although they did not report an analysis, Vauquelin and Robiquet claimed that this material, later known as asparagine, contained carbon, hydrogen, oxygen and nitrogen. Another amino acid, cysteine, was isolated from kidney stones by William Hyde Wollaston in 1810. Dry distillation of this substance produced ammonia and carbon dioxide. An elemental analysis of cysteine was performed in 1820 by William Prout; his reported values (30% C, 5% H, 11.7% N, 53.3% O) were, except for his misidentification of sulfur as oxygen, identical to the modern values (30% C, 5% H, 11.7% N, 26.6% O, 26.7% S).

The first amino acids to be identified as constituents of proteins were glycollol (later renamed glycine), a sweet-tasting crystalline substance produced by boiling gelatin in sulfuric acid, and leucine, a white precipitate produced by acid treatment of fibrin. The next amino acid isolated was tyrosine, which Liebig obtained, along with leucine, in 1847 by fusing casein with caustic potash (potassium hydroxide). Two years later, an associate of Liebig, Friedrich Bopp, showed that tyrosine and leucine were also present in albumin and fibrin. In this study, Bopp pioneered the use of hydrochloric acid, rather than sulfuric acid, for the hydrolysis of protein – still the method of choice today. It was becoming clear that substances such as tyrosine, leucine and glycollol were common constituents of proteins, but Bopp hinted that the proportions of these substances may vary: 'casein, albumin and fibrin distinguish themselves not with reference to the quality of the breakdown products, whether there may be small difference in this quantity, this is at present not ascertainable; the known analyses seem to point to such differences.'

The two strands in the analysis of proteins, elemental formulas and amino acid composition, started to come together in the late 1840s. It was a time of intense interest in organic taxonomy, and many systems for classifying organic compounds into 'types' (families) were being proposed. In 1848, the French chemists Auguste Laurent and Charles Gerhardt proposed the existence of a series of compounds with the general formula $C_nH_{2n+1}NO_2$. Of these, the C_1, C_4 and C_5 compounds were unknown. The C_2 compound was glycollol, the C_3 one was sarcosine and the C_6 one was leucine. Two years later Adolph Strecker synthesized a compound he named 'alanine' and proposed that it, not sarcosine, was the C_3 compound of the Laurent and Gerhardt series.

The fact that most of the first-identified amino acids fitted into this pattern of elemental composition was purely fortuitous – it turns out that there are no elemental ratios that are constant for all amino acids. Tyrosine, which had been shown in 1848 to have the empirical formula $C_9H_{11}NO_3$, clearly did not fit into the Laurent and Gerhardt series, and neither would most of the amino acids yet to be discovered. However, the presence of at least one nitrogen and at least two oxygen atoms per molecule indicated that the components of proteins always contain an amino (NH_2) group and a carboxylate (COOH) group. The term 'amino acid' was used for compounds containing these groups as early as 1845, but was not immediately applied to the hydrolysis products of proteins.

By mid-century, therefore, it was becoming clear that proteins contained smaller units that had certain chemical characteristics in common, but differed, at least, in size. As the idea of a protein radical had been discredited, it appeared likely that the differences in properties between proteins was because they differed in their relative contents of their amino acid constituents.

Valence

By around 1840, it was clear that the empirical formula was a woefully inadequate means of classifying organic compounds. On the one hand, two compounds such as the cyanic acid of Wöhler and the fulminic acid of Liebig could share the same empirical formula but have quite different properties. In the *Jahresbericht* for 1832, Berzelius listed several examples of this phenomenon: 'Since it is necessary for specific ideas to have definite and consequently as far as possible selected terms, I have proposed to call substances of similar composition and dissimilar properties *isomeric*, from the Greek ισομερης (composed of equal parts).'

On the other hand, two compounds such as acetic acid and trichloroacetic acid could have different empirical formulas but similar properties. This phenomenon was known as isomorphism. Clearly, characterizing compounds solely in terms of their elemental compositions was unsatisfactory.

If two compounds could have the same empirical formula but different properties, then their common atoms must be arranged in different ways. The radical theory of Wöhler and Liebig indicated that atoms occurred in compounds in certain characteristic groupings that could survive intact through chemical transformations. This suggested that isomers may have the same atoms organized into different radicals. Conversely, isomorphs could result from *different* atoms organized into *similar* groupings. A major challenge for eighteenth-century chemists was to determine the nature of these submolecular atomic groupings. As the number of organic compounds proliferated, the task of determining how their atoms were organized into molecules became more and more pressing. By 1844, Laurent wrote:

> When we consider the number of organic substances that have been discovered in the last ten years or so, and the increasing rate at which chemists discover new ones every day; when we see that a hundred new compounds have been obtained by the action of chlorine on a single hydrocarbon (naphthalene), and that from each of these a large number of others can be prepared; and finally, when we realize that there is no system of classifying this huge number of bodies, then we may very well wonder – with some apprehension – whether in a few years' time it will be possible to find our way at all through the maze of organic chemistry.[10]

With nothing much to go on, such radical theories proliferated – the 'etherin' theory of Dumas, the 'nucleus' theory of Laurent, the 'type' theories of Dumas and Gerhardt, the 'copulae' of Berzelius, the 'constitutional' theory of Hermann Kolbe. Each of these theories proposed a different way of dividing molecules up into their constituent radicals. In 1861, Kekulé's *Textbook of Organic Chemistry* listed nineteen different formulas for acetic acid!

As noted above, the situation was further complicated by inconsistencies in the use of atomic weights. In 1839, for example, Dumas used the formula $C_8H_8O_4$ for acetic acid. The number of carbon and oxygen atoms in this compound was overestimated by a factor of two as a result of Dumas using Dalton's values for the atomic weights of these elements. The numbers of all three types of atoms were then overestimated by another factor of two because of Berzelius's miscalculation of the atomic weight of silver! It was clearly much more difficult to determine the way atoms were arranged in molecules if there was no agreement on how many atoms of different elements were present.

Another problem for theoretical chemistry was that it was by no means clear how atoms were held together in molecules. The electrochemical theory of Berzelius, which stated that all interactions between atoms and groups of atoms were electrical in nature, had become untenable after Dumas's demonstration of substitution. Some other type of force, vaguely referred to as 'affinity', was believed to be responsible for interatomic interactions, but it was by no means clear which atoms had affinity and under what circumstances. The problem of atomic weight intruded here, too. Ever since

Dalton's theory of multiple proportions and Gay-Lussac's law of combining volumes, it had been clear that elements combined into compounds in characteristic whole-number ratios. By mid-century, it was recognized that elements had specific 'saturation capacities' with respect to other elements. Carbon, for example, could combine with a maximum of two 'equivalents' of oxygen. This suggested that affinity between atoms obeyed certain numerical rules. However, if it could not be decided whether carbon had an atomic weight of 6 or 12, or whether oxygen had an atomic weight of 8 or 16, saturation capacities would not be much help in solving molecular arrangement.

The solution to organic chemistry's two big problems – atomic weights and molecular structure – both came in the year 1858. A means of calculating atomic weights had, of course, been available since the publication of Avogadro's paper in 1811. The breakthrough required was to gain general acceptance of the concept of molecular gases. Most credit for achieving this feat is due to a countryman of Avogadro's, Stanislao Cannizzaro.

Cannizzaro is probably the only scientist whose most famous work was a course outline. In 1858, he published a paper entitled *Sunto di un corso di filosophia chimica* (Sketch of a course of chemical philosophy), which appears to be a distillation of his own experience in teaching chemistry. Cannizzaro's course was based squarely upon the ideas of Avogadro; the 'Sketch' begins with the words:

> I believe that the progress of science made in these last years has confirmed the hypothesis of Avogadro, of Ampère, and of Dumas on the similar constitution of substances in the gaseous state; that is, that equal volumes of these substances, whether simple or compound, contain an equal number of molecules; not however an equal number of atoms, since the molecules of the different substances, or those of the same substance in its different states, may contain a different number of atoms, whether of the same or diverse nature.[12]

Included in the 'Sketch' was a table of atomic and molecular weights that were are quite similar to those of Avogadro, except for the atomic weight of phosphorus, which Cannizzaro somehow managed to double. All the atomic weights listed were whole numbers, except for that of chlorine, 35.5.

Cannizzaro also tackled the issue of 'saturation capacities' and appeared to be well on his way towards a theory of valence:

To express the various saturation capacities of the different radicals, I compare them to that of hydrogen or of the halogens [e.g., chlorine] according as they are electro-positive or electro-negative. An atom of hydrogen is saturated by one of a halogen, and *vice versa*. I express this by saying that the first is a monatomic electro-positive radical, and the second a monatomic electro-negative radical: thus, potassium, sodium, lithium, silver, and the mercurous and cuprous radicals are monatomic electro-positive radicals. The bi-atomic radicals are those which, not being divisible, are equivalent to 2 of hydrogen or to 2 of chlorine; among the electro-positive radicals are the metallic radicals of the mercuric and cupric salts, of the salts of zinc, lead, magnesium, calcium, etc., and amongst the electro-negative we have oxygen, sulphur, selenium, and tellurium, i.e., the amphidic substances. There are, besides, radicals which are equivalent to three or more atoms of hydrogen or of chlorine, but I postpone the study of these until later.[12]

Had Cannizzaro considered a radical equivalent to four atoms of hydrogen – namely, the element carbon – he might have been able to solve the problem of organic structure as well as help solve the problem of atomic weights. The time was clearly ripe to establish the relationship between saturation capacity and molecular structure; that same year, 1858, two such hypotheses were proposed.

According to legend, the idea of valence came to Friedrich August Kekulé as he drowsed on the top deck of a London bus. Kekulé was born in 1829, and between 1848 and 1851 studied analytical chemistry in Liebig's home town of Darmstadt and at Giessen, where he worked in the Liebig laboratory. He then studied in Paris and worked in Switzerland before becoming, in 1853, an assistant to John Stenhouse, another former *Liebig-Schüler*, at St Bartholomew's Hospital in London. The two years he spent in London were very influential for Kekulé, as he was able to have discussions with a number of leading theoretical chemists, including Edward Frankland and Alexander Williamson. In 1856, having failed to obtain the chair in chemistry at the polytechnic school in Zürich, Kekulé became a *Privatdozent* in Heidelberg.

Frankland, who had performed extensive studies on organometallic compounds, noted in 1852 that 'even a superficial observer is impressed with the general symmetry of their construction'. Nitrogen, phosphorus, antimony and arsenic tended to combine with other elements in a ratio of 1:3 or 1:5. Thus it appeared that 'the combining power of the attracting element, if I may be allowed the term, is always satisfied by the same number of atoms'.

Williamson, in an 1850 study on the formation of ethers, had proposed a 'water type' of molecule, which had the general form of two radicals (R groups) attached to a single oxygen atom. According to Williamson, all alcohols and ethers conformed to this structure. Kekulé's studies showed that sulfur could play the same role as oxygen in organic molecules. From this, Kekulé realized in 1854 that oxygen and sulfur were always bound to two other atoms or groups of atoms; these elements were 'diatomic'. The idea of a key central atom to which other atoms were attached either singly or in radicals led him to realize that certain elements always seemed to be associated with a characteristic number of attached atoms. By 1857, this concept had been applied to a number of organic and inorganic elements: hydrogen, chlorine, bromine and potassium were considered monobasic, or monoatomic: oxygen and sulfur were considered dibasic, or diatomic; nitrogen, phosphorus and arsenic were considered tribasic, or triatomic.

The following year, Kekulé wrote: 'it is striking that the amount of carbon which the chemist has known as the least possible, as the *atom*, always combines with four atoms of a monoatomic, or two atoms of a diatomic, element; that generally, the sum of the chemical unities of the elements which are bound to one atom of carbon is equal to 4. This leads to the view that carbon is *tetra-atomic* (or tetrabasic).'[2]

This concept of a characteristic number of atoms with which an element could bind – which became known as valency – explained why chlorine could replace hydrogen in acetic acid: both were monoatomic. It also explained why elements occurred in compounds in certain characteristic ratios: oxygen with two hydrogens in water (H_2O), nitrogen with three hydrogens in ammonia (NH_3), carbon with four hydrogens in methane (CH_4). As carbon was tetra-atomic and oxygen was diatomic, both criteria were satisfied in carbon dioxide (CO_2).

Kekulé also realized that carbon atoms must be able to bind to one another: 'For substances which contain more atoms of carbon, it must be assumed that at least part of the atoms are held just by the affinity of carbon and that the carbon atoms themselves are joined together.' Later, Kekulé was to propose the existence of 'chain' and 'network' polymers based on these carbon–carbon bonds (see Chapter 5).

By the late 1850s, although not yet in public, Kekulé was representing organic compounds with 'sausage formulas' in which the atoms were represented by circles or ovals. In these formulas, the atoms were grouped in such a way as to explain their reactivities (Figure 2.1). However, it appears that Kekulé regarded his sausage formulas as diagrammatic representations rather than having any physical reality.

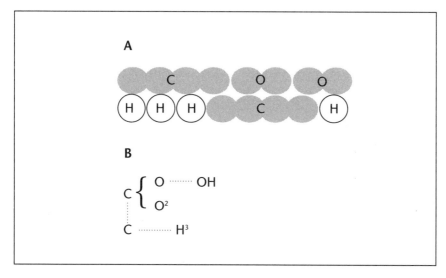

Figure 2.1: Early structural formulas of acetic acid. (A) August Kekulé's 'sausage' formula of 1859. (B) Archibald Couper's formula of 1858. Because Couper used an atomic weight of 8 for oxygen rather than 16, his structure of acetic acid contains twice as many oxygen atoms as Kekulé's

A valence theory was also published in 1858 by the Scottish chemist, Archibald Scott Couper. A former student of the humanities, Couper was working in Paris in the laboratory of Charles Wurtz and had been a chemist for only two years when he wrote a paper entitled 'On a new chemical theory'. Couper's paper, like that of Kekulé, proposed that elements had characteristic valences ('elective affinities') – two for oxygen, four for carbon. He also suggested the existence of carbon–carbon bonds in compounds containing more than one carbon atom.

However, Couper went much further than Kekulé in proposing structural formulas for organic molecules. He drew structures for alcohols, acids and ethers with dotted lines representing the 'affinities' between pairs of atoms (Figure 2.1). Couper soon replaced the dotted lines with solid ones, and it was this way of representing molecular structure, rather than Kekulé's 'sausage formulas', that became standard. By the mid-1860s, such structural formulas were in common use to depict organic molecules, and were increasingly regarded as representing the actual arrangement of atoms.

The events of 1858 established the reputation of Kekulé, but destroyed the career of Couper. The same year that he proposed the quadrivalency of carbon, Kekulé became professor of pure chemistry at the University of Ghent in Belgium.[i] Couper blamed Wurtz for the delay that resulted in Kekulé's paper being published before his own, and as a result was dismissed from Wurtz's laboratory. He returned to Scotland and died in 1892, four years before Kekulé. Couper's brilliant work was forgotten until 1909, when it was rediscovered and publicized by Richard Anschütz, Kekulé's successor at Bonn.[j]

It should be emphasized that the two breakthroughs of 1858 – Cannizzaro's advocacy of Avogadro's atomic weights and the valence theories of Kekulé and Couper – were interrelated. The idea of valence bonding depended upon elements combining in characteristic ratios, and the identification of such ratios required agreed-upon atomic weights. Similarly, the difficulty in accepting Avogadro's hypothesis – the idea of affinity between two atoms of the same element – was a crucial feature of the valence theories of Kekulé and Couper, both of whom proposed the existence of carbon–carbon bonds. The Avogadro/Cannizzaro theory and the Kekulé/Couper theory lived or died together.

Acceptance of these views was facilitated by the international chemistry congress that was held in the German resort city of Karlsruhe two years later. The idea of a conference to discuss the problems of organic chemistry originated with Kekulé, and was endorsed by a constellation of prominent chemists, including Cannizzaro, Dumas, Frankland, Liebig, Louis Pasteur, Williamson and Wöhler. The invitation sent out stated that the purposes of the conference were to reach a consensus on the definitions of the terms atom, molecule, equivalent, atomicity and basicity; to determine the nature of equivalents and their formulas; and to initiate a uniform chemical notation and nomenclature.

On the surface, little was achieved at the Karlsruhe conference, which was plagued by poor organization and fundamental disagreements on every issue. No consensus was achieved and no report was published. However, the conference provided an ideal forum for the dissemination of the views of

---

i   In 1867, following his discovery of the ring structure of benzene, Kekulé became professor of chem-
    istry at the University of Bonn.

j   Sir James Irvine wrote: 'No finer example of the brotherhood of science could be found than the
    efforts made by Anschütz, then at the height of his fame, to do justice to an obscure stranger to
    whom he owed nothing, not even national sympathy.' Without taking anything away from the
    admirable Anschütz, the 'brotherhood of science' appears to be one of those ideals that are honored
    more in the breach than in the observance.

Cannizzaro, who gave a talk supporting the molecular theory of Avogadro and distributed reprints of his 'Sketch'. Several influential chemists were persuaded to Cannizzaro's view, and by 1870 his values for atomic weight were being widely used. The tide had turned – the idea of bonds between atoms of the same element was no longer unthinkable, as it had been in the time of Avogadro, and the heuristic power of depicting molecular structure was irresistible.

Considering molecules as geometric assemblages of atoms provided an explanation for the phenomena of isomorphism and isomerism. Dumas's observation that the isomorphic compounds acetic acid and trichloroacetic acid have similar reactivities was not surprising, in view of the similarity of their structures (Figure 2.2). Likewise, the different chemical properties of the isomeric compounds ethanol and dimethyl ether, both of which were known to have the empirical formula C_2H_6O, could be explained by arranging these atoms into two quite different structural formulas.

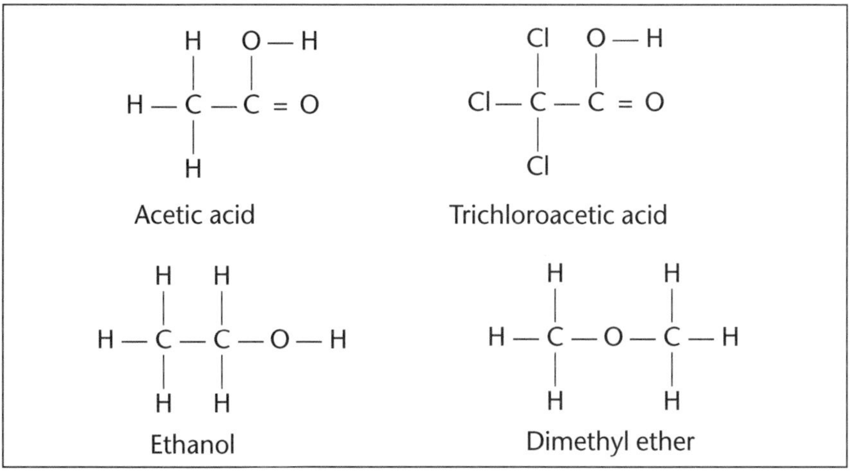

Figure 2.2: Isomorphs and isomers. Acetic acid and trichloroacetic acid are isomorphs – these compounds have different empirical (elemental) compositions but otherwise identical structural formulas. Ethanol and dimethyl ether are isomers – these compounds have identical elemental compositions (C_2H_6O) but different structural formulas. If the structures of isomers are known, the empirical formulas can generally be written in such a way as to distinguish between them; for example, C_2H_5OH (ethanol) and $(CH_3)_2O$ (dimethyl ether)

The establishment of a rational system of organic weights and the acceptance of the valence theory placed organic chemistry on a firm theoretical foundation. For physiological chemistry, however, a major problem remained. Because of isomerism, the structure of a molecule was not uniquely defined by its empirical formula. For this reason, long periods might elapse between the determination of an empirical formula and the determination of the corresponding structural formula. The amino acid cysteine was discovered in 1810 and a correct empirical formula ($C_3O_2H_7S$) reported in 1838; however, the structure was only determined in 1902. Yet cysteine contains only thirteen atoms; how would it ever be possible to determine the correct structure of a protein if, as Mulder reported, these substances contain a minimum of 1200 atoms?

Chapter 3
A Singular Inward Laboratory

'To Explain Life Completely Without Recourse to Life'

From the beginnings of recorded history, humans have observed, commented upon and attempted to explain the apparent difference between living beings and inanimate objects. Speculation about the special nature of life, like virtually everything else in science, can be traced back to Aristotle (or, no doubt, to Thales of Miletus). To prescientific societies, it must have seemed obvious that there was a qualitative difference between life and non-life, and the distinctions made by the earliest writers were also those debated in the nineteenth century: the difference between a living body and a corpse; the unique ability of living things to react to light, pain, threats and other stimuli; the existence of purposeful behavior in the animate world and random in the inanimate; the fact that only organic life can grow and multiply.

The idea that living things are inherently different from inanimate objects was first seriously challenged in the late seventeenth century, when the success of the Newtonian revolution in explaining the motions of celestial bodies, and the rationalism of philosophers such as René Descartes and Gottfried Leibniz, made it conceivable that life could be explained in terms of physics and chemistry. As scientific progress continued to throw up new findings, and the Industrial Revolution provided a ready source of mechanical analogies, the hope that living bodies could be explained as machines gained strength.

As described in Chapter 1, the last decades of the eighteenth century and the first decades of the nineteenth brought another set of challenges to traditional ideas of the nature of living matter. Respiration was revealed as a form of combustion; animal tissues were shown to consist almost entirely of the elements carbon, hydrogen, oxygen and nitrogen; certain vital processes were apparently mediated by substances similar to chemical catalysts. It is therefore not surprising that an active debate on the difference between animate and inanimate matter went on for the remainder of the nineteenth century. The terms of this debate were, on the surface, simple – could the operations of living systems be explained by the same laws that govern the behavior of inorganic substances?

The two answers to this question became known as 'vitalism' and 'materialism'. The aim of materialism has been elegantly summed up by Jean Rostand as 'to explain life completely without recourse to life'. Vitalism, in contrast, was the view that life can never be *completely* explained by chemical and physical laws. However, the distinction between the vitalist and the materialist philosophies was not set in stone. As George Hein put it: 'Since the "laws of chemistry and physics" and even the laws of mechanics are subject to change without notice, it is sometimes a little difficult to pin down the boundaries between mechanism and vitalism.' Even more 'difficult to pin down' is the extent to which biology has succumbed to a physical interpretation. As Henry Leicester observed: 'The term vitalism . . . signifies that sooner or later each scientist reached a level of speculation as to the mechanisms of the living organism at which he could no longer explain these mechanisms with the facts at his disposal.' Similarly, the biochemist Hans Krebs wrote in 1971: 'What has happened in recent decades is the constant enlarging of the areas approachable on a mechanistic basis and the shrinking of the areas remaining open to vitalistic concepts.'

In other words, every scientist should be a vitalist with respect to that portion of life phenomena which, to that point in time, remains inexplicable by physics and chemistry. In practice, however, there has never been a consensus on how much of biology can be explained in such a manner. Therefore, the materialist–vitalist debate was one in which the rules keep changing and the interpretation of these rules was never agreed upon. Clearly it would be an oversimplification to label any participant in this debate as a 'vitalist' or a 'materialist'; at any point in the history of biology, there existed a spectrum of opinions, and many biologists held beliefs about the nature of life that seem self-contradictory from a contemporary point of view. If, by the late nineteenth century, the materialist–vitalist debate had subsided, it was not because the former view had triumphed over the latter, but rather because the terms of the debate had changed so much that the original positions of both sides were irrelevant.

The Vital Force

It was in 1774, the year oxygen was discovered, that Friedrich Medicus coined the term *Lebenskraft* (vital force) to designate a physical entity that entered into the operation of living systems. Like Descartes, Medicus assumed that the soul was responsible for consciousness. However, certain 'vital' processes, such as digestion and respiration, were not under conscious control. These latter processes, Medicus believed, were manifestations of the *Lebenskraft*. By vital force, then, Medicus meant those processes that we now describe as subconscious or autonomic.

Medicus was attempting to find a middle ground between those who explained all life phenomena by reference to the soul, and those who considered living organisms as simply mechanical devices. However, his quite reasonable suggestion was soon taken to mean something entirely different – a physical force operating in living but not in non-living systems.

The chemical revolution was already completed and its conclusions widely accepted when, in 1796, Johannes Riel attempted the first systematization of vital and non-vital forces. Riel distinguished five types of force. These were the common 'physical force', which acted upon all objects, animate and inanimate; 'vital force', a property of living matter; 'vegetative force', which was responsible for the growth of plants; 'animal force', which was characterized by muscular movement; and finally the 'faculty of reasoning' which was specific to humans. Riel's distinction between 'vegetative force' and 'animal force' derived from the belief, common at the time, that plants and animals were composed of different elements; specifically, that only animal tissues contained nitrogen (see Chapter 2). Within the ambit of the 'animal force' he included both 'sensitive force' (irritability) and 'motive force' (conscious muscular movement). It was for the former that Medicus had coined the term *Lebenskraft*. Riel's vital force, however, lay somewhere between chemistry and physiology, as something that arose from organized matter. Three years later, Riel felt that the vital force could be dispensed with altogether:

> The vital force, which we consider to be the cause of these phenomena, is not something different from the organic matter; rather, the matter itself, as such, is the cause of these phenomena. Most of the animal phenomena can largely be explained on the basis of the general forces of matter. We therefore do not need any vital force as a unique primary force to explain them: we only use the word to designate concisely the concept of the physical, chemical, and mechanical forces of organic matter, through whose individuality and cooperation the animal phenomena are effected.[13]

Others, although impressed by the simple chemical nature of living substances, were not prepared to abandon the distinction between life and non-life. Antoine Fourcroy considered inorganic materials such as minerals to be the products of physical forces, but organic materials the products of a vital force. Georges Cuvier thought that the chemical difference between animate and inanimate matter was more subtle than that between rocks and animals: 'the very essence of life consists of a perpetual variation of proportions among a few elements. A little more or less of oxygen or nitrogen; that is, in the present state of science, the sole apparent cause of the innumerable products of organized bodies.' A different position was taken by Jöns Jacob Berzelius, who in the 1827 edition of his *Textbook of Chemistry* wrote: 'In living Nature the elements seem to obey entirely different laws than they do in the dead . . . The essence of the living body consequently is not founded in its inorganic elements, but in some other thing, which disposes the inorganic elements . . . to produce a certain result, specific and characteristic of each species.'

'The Ammonium Salt of Cyanic Acid is Urea'

A corollary to the idea that organic molecules have different properties from inorganic ones is that the former cannot arise spontaneously from the latter; if this were not the case, then life could be created in the laboratory. Therefore, it was assumed by the so-called 'chemical vitalists' that organic molecules could not be made artificially. A challenge to this view came in 1828, when Friedrich Wöhler synthesized urea. Wöhler found that cyanogen reacted with aqueous ammonia to produce not the expected salt, ammonium cyanate, but a white crystalline substance. The elemental composition of this material, $C_2N_4H_8O_2$, was that of ammonium cyanate, but its chemical properties were quite different. Wöhler noted that this elemental composition was also identical to that previously determined for urea by William Prout. Comparing the mysterious white crystals with urea that he isolated from his own urine, Wöhler found that their properties were identical.

Wöhler drew two conclusions from this surprising finding. First, it was 'noteworthy inasmuch as it furnishes an example of the artificial production of an organic, indeed, a so-called animal substance, from inorganic materials'. Second, he noted 'the like elementary and quantitative composition of compounds of very different properties', which had previously been observed in the case of fulminic acid and cyanic acid (see Chapter 2).

When Wöhler sent news of this experiment to his friend and mentor Berzelius, it was the aspect of organic synthesis that he emphasized: 'I can no longer, as it were, hold back my chemical urine; and I have to let out that I can make urea without needing a kidney, whether of man or dog; the ammonium salt of cyanic acid is urea.' However, the implications of this were qualified, perhaps in deference to Berzelius' beliefs:

This artificial formation of urea, can one regard it as an example of the formation of an organic substance from inorganic matter? It is striking that one must
have for the production of cyanic acid (and also of ammonia) always at the start
an organic substance, and a nature philosopher would say that the organic
[power] has not yet disappeared from either the animal carbon or the cyanic
compounds derived therefrom and an organic body may always be brought from
it.[14]

Wöhler was referring to the fact that the cyanic acid he used in the synthesis of urea was prepared from biological sources. Perhaps Wöhler's tact was
unnecessary, for Berzelius seemed much more interested in the fact that he
had identified two compounds of identical composition but different properties. Four years later, Berzelius proposed the name 'isomer' for compounds
of this type (see Chapter 2).

The chemical vitalists either ignored or rationalized Wöhler's synthesis of
urea. The 1827 edition of Edward Turner's *Elements of Chemistry* stated that
'a circumstance characteristic of organic products is the impracticability of
forming them artificially by direct union of their elements'. Turner felt no
need to revise this statement after Wöhler's paper appeared. As late as 1843,
Johannes Müller's *Handbook of Physiology* contained an explicit statement
of chemical vitalism: 'the mode in which the ultimate elements are combined
in organic bodies, as well as the energies by which the combination is effected,
are very peculiar; for although they may be reduced by analysis to their
ultimate elements, they cannot be regenerated by any chemical process'.
Müller, who was described by Ernst Haeckel as 'the most versatile and most
comprehensive biologist of our age', was aware of Wöhler's experiment, but
chose not to consider urea as a 'real' organic molecule: 'Wöhler's experiments afford the only trustworthy instance of the artificial formation of these
[organic] substances; as in his procuring urea and oxalic acid artificially.
Urea, however, can scarcely be considered an organic matter, being rather
an excrement than a component of the animal body.'

However, the leading French chemist of the day, Jean-Baptiste Dumas,
wrote: 'All chemists have applauded the brilliant discovery by Wöhler of
the artificial formation of urea.' Wöhler's good friend Justus Liebig loyally
hailed the synthesis of urea as an epoch-making discovery: '[The] extraordinary, and to some extent inexplicable, production of this substance without
the assistance of the vital functions, for which we are indebted to Wöhler,
must be considered one of the discoveries with which a new era in science
has commenced.' In their joint 1838 paper on the derivatives of uric acid,
it was apparently Liebig who inserted the following passage:

The philosophy of chemistry will draw the conclusion from this work that the production of all organic substances, as long as they no longer belong to the organism, must be considered not merely as possible within our laboratories but rather as certain. Sugar, salicin, morphine will be prepared artificially. We do not yet know the way to reach this end result, because the antecedents, from which these substances develop, are unknown to us, but we shall learn them.[15]

The significance of Wöhler's synthesis of urea has been extensively debated. Clearly it did not demolish the doctrine of vitalism, if only because no one finding could possibly disprove all forms of vitalistic beliefs. However, the synthesis of urea, and the ensuing syntheses of other organic substances, could be said to disprove the doctrine of chemical vitalism, the rearguard action of Müller notwithstanding. An equally important result of Wöhler's work was that it opened the way to the important organic chemistry technique of retrosynthesis – proving the structure of an organic molecule by resynthesizing it from its analytical products. By the time of Emil Fischer (see Chapter 4), retrosynthesis was a standard procedure of analytical chemistry.

'An Immaterial Agency, Which the Chemist Cannot Employ at Will'

Even Berzelius and Liebig, both sympathetic to Wöhler, had difficulty assimilating organic synthesis into animal chemistry. In *Animal Chemistry* (1848), Liebig wrote: '. . . in the animal body we recognize as the ultimate cause of all force only one cause, the chemical action which the elements of the food and the oxygen of the air mutually exercise on each other. The only known ultimate cause of vital force, either in animals or in plants, is a chemical process.'[a] Elsewhere in the same work, however, he stated: '. . . everything in the organism goes on under the influence of the vital force, an immaterial agency, which the chemist cannot employ at will.'

For Liebig, as for Müller, the vital force acted in opposition to the normal chemical forces of attraction. Thus the vital force was responsible for breaking the bonds that held food substances together, for recreating them as animal tissues, and for growth and motion. Disease occurred when the vital force became weaker than chemical forces, and death ensued when it was overcome completely. However, the vital force had no relationship to

a   It may have been because of this that the extreme vitalist Charles Caldwell wrote: 'if there be in existence a more unscrupulous system of materialism than that contained in Liebig's "Animal Chemistry", I know not where it is to be found.'

consciousness, which Liebig attributed to 'an agency entirely distinct from the vital force, with which it has nothing in common'.

Perhaps because of his feud with Pasteur over the nature of fermentation (see Chapter 4), Liebig became more of a vitalist with age. In the fourth edition of his *Familiar Letters on Chemistry*, published in 1859, he wrote:

> If any one assured us that the palace of the king, with its entire internal arrangement of statues, and pictures, started into existence by an accidental effort of a natural force, which caused the elements to group themselves into the form of a house, – because the mortar of the building is a chemical compound of carbonic acid and lime, which any novice in chemistry can prepare . . . we should meet such an assertion with a smile of contempt, for we know how a house is made.[16]

Berzelius, in contrast, appears to have become more of a materialist in his later years. In a letter written in 1831, he flatly rejected the idea of a vital force: 'To suppose that the elements are imbued with other fundamental forces in organic Nature than in the inorganic is an absurdity . . . the fact that we cannot rightly understand the conditions prevailing in organic Nature gives us no sufficient reason to adopt other forces.' This appears to be a major change from Berzelius's position four years earlier that organic nature involves 'some other thing, which disposes the inorganic elements . . .' (see above). Could it have been Wöhler's synthesis of urea that changed his mentor's attitude to the vital force? Be that as it may, by the time Berzelius prepared the fifth edition of the *Textbook of Chemistry* in 1847, he was prepared to go public with his materialism:

> It is evident that if one is to understand . . . by the effect of the vital force, something other than the characteristic conditions, cooperating in various ways, under which the usual natural forces are put into action in organic Nature and that, if it is thought of as a specific chemical power in living Nature, this inclination is a mistake.[17]

The increasing implausibility of chemical vitalism in the years after 1828 removed the debate on vitalism from the question of organic synthesis. If organic molecules could be created artificially, then the seat of the vital force could not be at the molecular level. Rather, vital properties must arise at some higher level of biological structure, such as the way in which organic molecules were organized within living tissues. Two plausible vital agents, cells and protoplasm, were suggested in the middle years of the nineteenth century.

'Omnis Cellula e Cellula'

Theodor Schwann was born in 1810 in Neuss, Germany. In 1829 he entered the University of Bonn, where he studied with and assisted Johannes Müller. Following a period in Würzburg, Schwann moved to the University of Berlin to rejoin Müller, who had been appointed professor of anatomy and physiology. Following his graduation with an MD degree in 1834, Schwann became an assistant in Müller's laboratory and a full-time researcher.

Unlike his mentor, Schwann rejected the notion that different laws operated in animate and inanimate systems:

> Never was I able to conceive of the existence of a simple force which would itself change its mode of action in order to realize an idea, without however possessing the characteristics of intelligent beings. I have always preferred to seek in the Creator rather than in the created the cause of the finality to which the whole of nature evidently bears witness; and I have also always regarded as illusory the explanation of vital phenomena as conceived of by the vitalist school. I laid down as a principle that these phenomena must be explained in the same way as those of inert nature.[18]

Schwann was a brilliant experimenter who, in a research career lasting only five years, determined the relationship between length and tension in muscle, discovered the gastric enzyme pepsin and was a co-discoverer of the role of yeast in fermentation (see Chapter 4). However, his most important contribution to biology was the theory that all life is based on cells.

Cells had been recognized as the structural units of plants by Robert Hooke in 1665. Because animal cells lack a cell wall, however, it was much more difficult to discern the cellular structure of their tissues. The recognition that animals, like plants, were composed of cells, was made possible by a major innovation in microscopy in the 1830s.

The resolving power of the earliest microscopes was limited by two forms of image distortion: chromatic aberration and spherical aberration. A theoretical solution to the problem of chromatic aberration was found in the 1730s, although not fully incorporated into commercial instruments until around 1800. A solution to the problem of spherical aberration was found in 1830 by Joseph Lister, the father of the famous surgeon of the same name. By the late 1830s, microscopes with Lister optics were widely available. Solving the problems of aberration resulted in a tremendous improvement in the capability of microscopes; by the 1840s, objects less than 1 μm in diameter could be resolved.[b] By 1870, a resolution of one-third of a micrometer, close to the physical limit of light microscopy, would be achieved. Using the new Lister instruments, microscopists of the 1830s were able to advance from the tissue level to the cellular and subcellular levels. This produced two major breakthroughs: the cell theory and the recognition of yeasts as living organisms (see Chapter 4). Not until the advent of the electron microscope in the years following World War II would such rapid progress in ultrastructural analysis occur.

In developing his cell theory, Schwann was also drawing upon a philosophical tradition of viewing organisms as machines that derived from Descartes, and gained more metaphorical power with the onset of the Industrial Revolution. Riel had noted in his 1796 article on the *Lebenskraft* that 'Not only the whole body or its coarse components, but even its smallest parts are machines; everything, down to the smallest fibre, dissolves itself into nothing but purposefully formed bodies'. In 1827, the French zoologist Henri Milne-Edwards compared higher organisms to factories, in which different areas performed different functions for the purposes of the whole. According to these views, the body was not a single machine, but a collection of tiny machines each with specialized functions.

Schwann may have been a brilliant scientist with early access to a revolutionary piece of technology, but he would probably never have conceived of the cell theory but for a fortuitous meeting with the eccentric botanist Schleiden. Jacob Mathias Schleiden studied law at the University of Heidelberg and established a legal practice in Hamburg. Following a period of depression that culminated in an unsuccessful suicide attempt, he studied natural science in Göttingen and Berlin. In Berlin, he encountered the Scottish botanist Robert Brown, who had discovered the plant cell nucleus in 1833, and worked in the laboratory of Johannes Müller, where he met Schwann.

In an 1838 paper entitled 'Contributions to phytogenesis', Schleiden concluded that the nucleus of plant cells was responsible for the formation of new cells. He believed that the new cell arose from the growth of a 'vesicle' upon a pre-existing nucleus, the nuclear membrane giving rise to the new cell wall. The nucleus of the new cell then formed by aggregation of 'mucous granules' in the intracellular fluid. As they did not appear to be formed of cells, Schleiden did not believe that this form of growth could occur in animals.

---

b   1 µm (micrometer) = 10^{-6} m.

Schwann later recalled: 'One day, when I was dining with Mr Schleiden, this illustrious botanist pointed out to me the important role that the nucleus plays in the development of plant cells. I at once recalled having seen a similar organ in the cells of the notochord, and in the same instance I grasped the extreme importance that my discovery would have if I succeeded in showing that this nucleus plays the same role in the cells of the notochord as does the nucleus of plants in the development of plant cells.'[19]

The two men immediately went to Schwann's laboratory in the anatomical institute, where Schleiden confirmed that the frog notochord contains nuclei. Schwann's subsequent studies convinced him that the notochord and branchial cartilage both contain 'cavities enclosed by a membrane', the membrane thickening during growth, indicating that at least certain animal tissues consisted of cells: 'Cells presented themselves in the animal body having a nucleus, which in its position with regard to the cell, its form and modifications, accorded with the cytoblast [nucleus] of vegetable cells, a thickening of the cell-wall took place, and the formation of young cells within the parent-cell from a similar cytoblast, and the growth of these without vascular connexion was proved.'[20]

Anatomists of the time believed that different animal tissues were composed of different structural elements, including globules, granules, fibers, tubes and bladders. To Schwann, however, these were merely 'natural-history ideas'. Inspired by Schleiden, he decided to see whether these different structural units of animal tissues could all represent products of the same formative unit, the cell. Examination of various animal tissues persuaded Schwann that this was so: 'The elementary parts of all tissues are formed of cells in an analogous, though very diversified manner, so that it may be asserted, *that there is one universal principle of development for the elementary parts of organisms, however different, and that this principle is the formation of cells*' [emphasis in original].

Schwann's microscopic investigations, and the conclusions he drew from them, were described in his 1839 book, *Mikroskopischer Untersuchungen über die Übereinstimmung in der Struktur und dem Wachstum der Thiere und Pflanzen* (*Microscopical Researches into the Accordance in the Structure and Growth of Animals and Plants*). The first section of this work was a description of the structures of cartilage and the spinal cord. In the second section, Schwann divided animal tissues into five types, based on the degree of association of their constituent cells. In the third and final section, Schwann laid out his theory of cell formation and cell function. In this third section, Schwann attempted to fuse his recognition of membrane-bounded animal cells and a pronounced philosophical materialism into a new theory of biological organization:

> We set out, therefore, with the supposition that an organized body is not produced by a fundamental power which is guided in its operation by a definite idea, but is developed, according to blind laws, by powers which, like those of inorganic nature, are established by the very existence of matter. As the elementary materials of organic nature are not different from those of the inorganic kingdom, the source of the organic phenomena can only reside in another combination of these materials, whether it be in a peculiar mode of union of the elementary atoms to form atoms of the second order [molecules], or in the arrangement of these conglomerate molecules when forming either the separate morphological elementary parts of organisms, or an entire organism.[20]

Schwann divided the 'fundamental powers' of cells into two types: plastic phenomena, or cell formation; and 'metabolic phenomena', the ability to induce chemical changes within the cell or in the surrounding 'cytoblastema' (extracellular) fluid. Unlike plant cells, those of animals did not usually arise from the nuclei of pre-existing cells. Instead, animal cells mostly formed spontaneously from 'conglomerate molecules' contained within the 'cytoblastema' fluid. The 'metabolic phenomena' of cells were manifested by chemical differences between the cell contents and the extracellular fluid: 'The cytoblastema, in which the cells are formed, contains the elements of the materials of which the cell is composed, but in other combinations; it is not a mere solution of cell-material, but it contains only certain organic substances in solution.' The cell fluid, therefore, was a chemically transformed version of the extracellular fluid. A similar transformation of the extracellular fluid occurred in vinous fermentation, where sugar was converted to alcohol by the action of yeast cells. The source of this 'metabolic power' of the cell apparently lay in the nucleus and membrane: 'In the cells themselves again, it appears to be the solid parts, the cell-membrane and the nucleus, which produce the change . . . It may therefore, on the whole, be said that the solid component particles of the cells possess the power of chemically altering the substances in contact with them.'[20]

The cell membrane could not only effect chemical transformations on cytoblastema fluid, but also had the ability to admit some components of the cytoblastema while excluding others. This suggestion was remarkably prescient; even more remarkably, Schwann hinted that this ability of the membrane may have an electrical basis. However, he believed that the 'cell contents' that lay between the nucleus and membrane were metabolically inert – subject to, rather than agents of, chemical change.

In the final pages of the *Mikroskopischer Untersuchungen*, Schwann speculated upon the kind of matter that could be responsible for the plastic and metabolic phenomena of cells. He noted that organic bodies are intermediate between solids and liquids. Like solids, they have definite shapes; like liquids, they are capable of imbibition (being infiltrated by water). Cell formation, he thought, was somewhat akin to crystallization, although cells grew not by apposition (surface deposition) but by intussusception (deposition within). If substances capable of imbibition could exist in crystalline form, their properties might be consistent with the behavior of cells.

The contributions of Schleiden and Schwann to the cell theory were somewhat complementary, although that of Schwann was far greater. Schleiden's mechanism of cell formation in plants turned out to be incorrect. However, it did enable Schwann to extend the concept of cell-based life to all animal tissues. Schwann's mechanism of cell formation was even further from the mark than Schleiden's; the latter at least recognized that cells arise from other cells. However, Schwann's concept of cellular metabolism as chemical transformation of the cytoblastema, and in particular the role of the membrane in this process, was both insightful and influential.

It was his extreme materialist outlook that led Schwann to believe that cells arose by spontaneous precipitation of cytoblastema. He had earlier been involved in the debate over spontaneous generation – the question of whether life could arise spontaneously from dead organic matter. Schwann's studies in the mid-1830s showed that 'infusoria' (micro-organisms) did not arise in meat extracts that had been boiled, as long as the air in contact with the extracts had also been heated. Heating air did not affect its ability to support life, however, leading Schwann to suppose that unheated air contains organisms that can multiply in a suitable nutrient medium. Studies by other researchers, notably Louis Pasteur, confirmed that putrefaction failed to occur under sterile conditions (see Chapter 4). However, the growing unpopularity of the doctrine of spontaneous generation at mid-century made Schwann's idea of spontaneous formation of cells from cytoblastema fluid less attractive.

It was another student of Müller's, Rudolf Virchow, who revised the cell theory to remove this increasingly untenable feature. In his 1858 book *Cellularpathologie* (*Cellular Pathology*), Virchow distanced himself from the extremes of both the holist and reductionist positions. As an opponent of spontaneous generation, Virchow did not believe that cells could arise from extracellular fluid any more than flies could arise from decaying carcasses, frogs from mud or mice from grain:

Even in pathology we can now go so far as to establish, as a general principle, *that no development of any kind begins* de novo, *and consequently as to reject the theory of equivocal* [spontaneous] *generation just as much in the history of the development of individual parts as we do in that of entire organisms* . . . Where a cell arises, there a cell must have previously existed (*omnis cellula e cellula*), just as an animal can spring only from an animal, a plant only from a plant [emphasis in original].[21]

The idea that all cells arise from other cells made the Schleiden–Schwann theory consistent with a repudiation of spontaneous generation, but could hardly be said to represent a material explanation of life. The vital phenomena had merely been transferred from the organism as a whole to its constituent cells.

'We Should Constitute Physiology on a Chemico-physical Foundation'

Although Johannes Müller encouraged an independence of mind in his trainees and Germany had become the center of the materialist philosophy, it is difficult to understand why the views on vitalism of so many Müller students were diametrically opposed to his own. Staunch materialists though they were, Schwann and Virchow were in fact rather restrained in their views compared with the 'organic physics' group that arose in Müller's laboratory in the 1840s. The founder and leader of this group was Emil DuBois-Reymond, who wrote in 1841: 'I am gradually returning to Dutrochet's view "The more one advances in the knowledge of physiology the more one will have reasons for ceasing to believe that the phenomena of life are essentially different from physical phenomena".' By 1847, DuBois-Reymond had recruited his fellow Müller students, Ernst von Brücke and Hermann von Helmholtz, and also Carl Ludwig, a former student of Ludwig Fick, and now *ausserordentlicher* professor at Marburg: 'We four imagined that we should constitute physiology on a chemico-physical foundation, and give it equal rank with physics.'

With the arguable exception of Brücke, the members of the 'group of 1847' went on to have distinguished scientific careers. DuBois-Reymond discovered the action potential, the electrical depolarization that causes nerve transmission, and thus became the founder of electrophysiology. Helmholtz is mainly remembered as one of the founders of physical chemistry and the proposer of the law of conservation of energy, but he also performed the first measurement of the velocity of nerve impulses. Ludwig became one of the leading physiologists of his generation, making many contributions to the physiology of nerves and heart. However, their program of putting physiology on a physical basis was impossibly premature in the mid-nineteenth century. As Fick wrote in 1874: 'the absolute dominance of the mechanistic-mathematical orientation in physiology has proved to be an Icarus flight, and unhappily in this case, the wax of the wings was not melted by too great a nearness to the sun of knowledge but through the heating which the unaccustomed effort brought with it.'[22]

The organic physics program had considerable propaganda value, however. Liebig no doubt had the Berlin physiologists in mind when, in his 1856 *Critique of Contemporary Materialism*, he described materialists as 'total strangers to all investigations connected with chemical and physical forces . . . amateurs . . . ignorant and presumptuous dreamers'. DuBois-Reymond, in turn, referred to Liebig as *Gottes-Geissel* (the scourge of God). But new findings were undermining Liebig's position. Pasteur's demonstration that organic matter kept under sterile conditions did not undergo putrefaction (see Chapter 4) contradicted Liebig's idea that the vital force opposed the normal chemical forces of affinity. Claude Bernard's demonstration that the liver could synthesize the glucose polymer glycogen (see below) was inconsistent with Liebig's belief that only plants could make complex organic molecules.

'A Living Jelly'

As noted above, the cell theory of Schwann emphasized the importance of the 'solid' parts of the cell, the nucleus and membrane, in the generation and growth of organisms. Another school of thought held that the soluble contents of the cell might be responsible for the phenomenon of generation. The major impetus behind the protoplasm theory, however, was to find an explanation for the characteristically vital property of movement.

In 1835, the French zoologist Félix Dujardin observed that protozoa placed in water extruded a clear jelly-like material. Dujardin believed that this material was responsible for the locomotion of these organisms, and suggested the name 'sarcode' to emphasize the similarity with muscle: 'I propose to give this name to what other observers have called a living jelly – this glutinous, transparent substance, insoluble in water, contracting into globular masses, attaching itself to dissecting-needles and allowing itself to be drawn out like mucus; lastly, occurring in all the lower animals interposed between the other elements of structure.'[19]

Jan Evangelista Purkyně (Purkinje), professor of physiology at the University of Breslau, was also struck by the presence in many organisms of a gelatinous substance, for which in 1839 he proposed the name 'protoplasm'. Purkyně, however, emphasized the generative function of protoplasm, stating that it constituted the embryonic tissue of plants and animals, as well as the living part of adult animal tissues. The Swiss botanist Carl von Naegeli had noticed the presence in plant cells of a material he described as 'a homogeneous, thickish colourless matter' for which he proposed the name *Schleim*. Not surprisingly, the name 'protoplasm' was preferred to 'slime' for the generative material of plant cells.

By 1850, when the German botanist Ferdinand Cohn suggested that the generative protoplasm may be the same thing as the contractile sarcode, protoplasm had became a popular concept – not least because it provided a scientifically respectable 'explanation' of vital phenomena. For example, Franz Unger in 1852 described protoplasm as 'this marvellous substance, this self-moving wheel'. Ernst Haeckel wrote that protoplasm is 'the active substrate of all vital motions and of all vital activities: nutrition, growth, motion and irritability'.

The proponents of protoplasm as the seat of vital activity felt free to dispense with other cellular components. Max Schultze's studies on the fusion of muscle cells into fibers led him to suggest in 1861 that some animals cells lacked membranes, and to define cells as 'a small naked clump of protoplasm with a nucleus'. That same year, Ernst von Brücke denied the necessity for a nucleus. Brücke also emphasized that protoplasm was a form of matter that could not be properly described by the physical terms 'solid' or 'liquid', and that a special organization of this material was necessary for the manifestation of vital phenomena: 'To the living cell, apart from the molecular structure of the organic compounds which it contains, there must be attributed yet another structure complicated in a different manner, and it is this to which we give the name organisation.'

The idea of organization as a fundamental characteristic of life became a popular concept. In part, this was probably because it provided a way out of the increasingly sterile vitalist–materialist debate by suggesting that vital phenomena were not the products of a special form of matter, but rather of ordinary matter organized in a special way. However, as pointed out by Eduard Glas, describing living systems as an organized form of inanimate matter was not so very different from Liebig's belief that the vital force opposed the normal chemical forces of affinity. Another weakness in Brücke's argument was noted by Joseph Henry in 1866:

Although we cannot perhaps positively say in the present state of science that this directing principle will not manifest itself when all the necessary conditions are present, yet in the ordinary phenomena of life which are everywhere exhibited around us, organization is derived from vitality, and not vitality from organization. That the vital or directing principle is not a physical power which performs work, or that it cannot be classed with heat or chemical action is evident from the fact that it may be indefinitely extended – from a single acorn a whole forest of oaks may result.[23]

Henry's criticism pointed to the fact that the debate of the nature of life was changing – increasingly it would be less about the nature of living matter than about its generation; less about chemistry than about genetics.

'The Physical Basis of Life'

The concept of life as organized protoplasm reached its zenith in 1868 with Thomas Henry Huxley's lecture 'On the physical basis of life'. Huxley was born in 1825 in Ealing, England, and graduated with a degree in medicine from the University of London in 1845. Like Charles Darwin, Huxley was converted into a naturalist by a long sea voyage; in Huxley's case, a four-year surveying expedition in the Torres Strait (Australia) aboard HMS *Rattlesnake*. In 1854, following the publication of a series of papers on marine invertebrates, he was appointed lecturer in natural history at the Government School of Mines in London.

Huxley now became more interested in the comparative morphology of the vertebrates, and on the basis of his studies developed a profound skepticism concerning the transmutation of species. He became acquainted with Darwin, who sent Huxley one of three pre-publication copies of his great work *The Origin of Species* (1859). Huxley was immediately convinced by Darwin's arguments for evolution by natural selection; when Samuel Wilberforce, bishop of Oxford, denounced natural selection at the 1860 meeting of the British Association for the Advancement of Science, held in Oxford, Huxley rose to Darwin's defence. Wilberforce had been well briefed on the subject by anti-evolutionist scientists, but he was no match for Huxley's knowledge of taxonomy and rhetorical skills. His tenacious defence of evolutionary theory on this occasion earned Huxley the nickname of 'Darwin's bulldog'.

The high and low points of Thomas Huxley's career both came in 1868. That summer, he announced to the British Association that he had found in sediment samples from the North Atlantic a new form of life. *Bathybius haeckelii*, according to Huxley, was composed of undifferentiated protoplasm (*Urschleim*) and formed great rafts on the ocean floor. Marine biologists immediately challenged the view that Huxley's samples contained any living organisms, and it was eventually found that '*Bathybius*' was actually a precipitate of calcium sulfate in the alcohol used to preserve the sediment samples.

In November of that year, Huxley delivered the most famous of his public addresses. 'On the physical basis of life' was one of a series of Sunday evening talks given in the Hopetoun Rooms, Edinburgh, on non-theological topics. Public interest in science was high in Victorian Britain, and Huxley had a reputation not only for his controversial views on evolution and religion (he later coined the term 'agnostic' to describe his religious beliefs) but also as a brilliant speaker. A contemporary account of the lecture reported that 'the audience seemed almost to cease to breathe, so perfect was the stillness'.

The physical basis of life, according to Huxley, was protoplasm. This protoplasm consisted of the elements carbon, hydrogen, oxygen and nitrogen, and constituted the chemical basis of all living matter:

> For example, this present lecture, whatever its intellectual worth to you, has a certain physical value to me, which is, conceivably, expressible by the number of grains of protoplasm and other bodily substances wasted in maintaining my vital processes during its delivery . . . By-and-by, I shall probably have recourse to the substance commonly called mutton . . . Now this mutton was once the living protoplasm, more or less modified, of another animal – a sheep. As I shall eat it, it is the same matter altered, not only by death, but by exposure to sundry artificial operations in the process of cooking.

> But these changes, whatever be their extent, have not rendered it incompetent to resume its old functions, as matter of life. A singular inward laboratory, which I possess, will dissolve a certain portion of the modified protoplasm, the solution so formed will pass into my veins; and the subtle influences to which it will then be subjected will convert the dead protoplasm into living protoplasm and transubstantiate sheep into man.

> Nor is this all. If digestion were a thing to be trifled with, I might sup upon lobster, and the matter of life of the crustacean would undergo the same metamorphosis into humanity. And were I to return to my own place by sea, and undergo shipwreck, the crustacea might, and probably would, return the compliment, and demonstrate our common nature by turning my protoplasm into living lobster.[24]

Huxley noted that the passage of an electrical spark through a mixture of hydrogen and oxygen produced a compound, water, with different chemical properties from those of its component gases. Likewise, this water could in plants be combined with carbonic acid and ammonia to make something different again, protoplasm:

> We do not assume that a something called 'aquosity' entered into and took possession of the oxide of hydrogen as soon as it was formed, and then guided the aqueous particles to their places in the facets of the crystal, or amongst the leaflets of the hoar-frost. On the contrary, we live in the hope and in the faith that, by the advance of molecular physics, we shall by-and-by be able to see our way as clearly from the constituents of water to the properties of water, as we are now able to deduce the operations of a watch from the form of its parts and the manner in which they are put together.

> Is the case in any way changed when carbonic acid, water, and ammonia disappear, and in their place, under the influence of pre-existing living protoplasm, an equivalent weight of the matter of life makes its appearance?

> What justification is there, then, for the assumption of the existence in the living matter of a something which has no representative or correlative in the not living matter which gave rise to it? What better philosophical status has 'vitality' than 'aquosity'?

> If the properties of water may be properly said to result from the nature and disposition of its component molecules, I can find no intelligible ground for refusing to say that the properties of protoplasm result from the nature and dispositions of its molecules.[24]

Indeed, Huxley went even further: if the human body consists of protoplasm, then all the vital properties, including consciousness, should be regarded as 'the result of the molecular forces of the protoplasm'. This bold version of the protoplasm theory thereby leaped the mind-body gap of Cartesian dualism.

'On the physical basis of life' caused almost as great a sensation as *The Origin of Species*. The February 1869 issue of the *Fortnightly Review*, in which the lecture appeared, went to seven editions. The full text was also published in the New York *World* under the headline 'New Theory of Life'. The interpretation of Huxley's lecture was that the protoplasm theory had, by identifying the chemical entity responsible for all life phenomena, abolished vitalism from biology.

However, the protoplasm theory did not reduce vital phenomena to a 'physical basis', but rather relocated them to a subcellular level. Huxley's theory did not explain why a sheep was alive and a piece of mutton dead, or why the same protoplasm could become a lobster or a human. In the early twentieth century, the concept of protoplasm would be swallowed whole by a new physical basis of life – the colloids.[c] Around the standard of 'bio-colloidology', vitalism would make its last stand (see Chapter 7).

Huxley's great achievement was to be part of a generation of naturalists who achieved the secularization of biology. A chemistry that could characterize organic molecules only in terms of their elemental content was incapable of providing a chemical explanation of vital phenomena. However, men like Huxley, Darwin and DuBois-Reymond could at least point the way to a new biology in which divine creation and vital forces would no longer be necessary.

'Nothing That Lives is Alive in Every Part'

An alternative view of protoplasm was articulated by Lionel Beale, professor of physiology and morbid anatomy (pathology) at King's College, London. From his histological studies, Beale had shown that some tissues stained red with the dye carmine, whereas others did not. In books published in 1870 and 1872, Beale proposed that carmine-staining tissues were enriched in 'germinal matter' (later called 'bioplasm'), which was capable of growth and assimilation, whereas non-staining tissues were composed of 'formed matter', dead germinal matter. Germinal matter was a clear, structureless fluid that was abundant in the embryo and in the nervous system. It also occurred in tissues rich in blood vessels, from which it assimilated a nutrient medium that Beale called 'pabulum'. Formed matter was more structured and solid, and was abundant in tissues such as hair, nail and tooth.

Because all organisms contained both germinal matter and formed matter, Beale summarized his theory in the phrase: 'Nothing that lives is alive in every part'. Disease resulted from disruptions in the formation of germinal matter. A lack of pabulum caused hardening, shrinking or wasting of tissues; an excess caused cancer, inflammation, pneumonia and other 'morbid growths'.

Beale's germinal matter differed from Huxley's protoplasm in two important respects. First, bioplasm was irreversibly converted to formed matter

---

c   In 1923, for example, Edmund Wilson wrote an article on the importance of colloids in biology entitled 'The physical basis of life'.

in death, whereas protoplasm could apparently be boiled, roasted or dissolved and still be protoplasm. Second, protoplasm was an organic substance subject to ordinary physical and chemical laws, whereas germinal matter exhibited 'some force or power of a nature different from any form or mode of energy yet discovered'. Therefore, Huxley was a 'materialist', while Beale was a 'vitalist'.

From the standpoint of contemporary science, however, Beale's vitalism appears more acceptable than Huxley's materialism. Beale's germinal matter and formed matter can be seen as metaphors for stem cells and fully differentiated cells, or even for DNA and protein. In retrospect, Beale's views appear as a conceptual – although not historical – link between the dragon-slaying materialism of Thomas Huxley and the subtle organic philosophy of Claude Bernard.

The Legislative Force

Claude Bernard was born in 1813 in St-Julien, France, and educated in the humanities. He went to Paris in 1834 to seek fame as a writer, but was persuaded to seek the security of a profession first. A mediocre student, Bernard qualified for the faculty of medicine only with great difficulty. He graduated in 1843, having spent several years working with the great digestive physiologist François Magendie, but failed to obtain a teaching position in Paris and was forced to become a country doctor. Marriage to a wealthy woman allowed Bernard to devote himself to research in physiology, and in 1848 he succeeded Magendie as professor of physiology at the Collège de France. Following Magendie's lead, Bernard made important contributions to the physiology of digestion and nerve action, including studies on the role of the pancreas in the digestion of fats.

However, Bernard's most acclaimed work was on the glucose-regulating function of the liver. This work arose from Bernard's chance 1848 discovery of sugar in the blood of an animal that had been starved for several days. Further investigation showed that the sugar concentration in the blood is kept constant by uptake and release of glucose from the liver. For this latter study, he received in 1853 a doctorate in zoology from the Sorbonne. Bernard followed up his thesis work on the glycogenic function of the liver by showing in 1855 that sugar was released from a perfused liver even after the death of the animal. He attributed the glucose storage capacity of the liver to the presence of an insoluble 'animal starch', although this glucose polymer, which he named glycogen, was not isolated in pure form until 1857.

Recognition followed quickly: Bernard was elected to the Academy of Sciences in 1854 (having been unsuccessful four years earlier), and in 1855 became professor of medicine at the Collège de France. Perhaps most gratifying of all to the former writer, in 1869 Bernard became one of the 'immortals' of the Académie Française.

In the latter part of his career, Bernard retreated from experimental physiology to concentrate more on what one might nowadays describe as theoretical biology and philosophy. This led to the proposal of concepts such as 'intermediary metabolism' and the *milieu intérieur* (internal environment). Bernard's view of biology was described in his last published work, the 1878 book *Leçons sur les phénomènes de la vie communs aux animaux et aux végétaux* (*Lectures on the Phenomena of Life Common to Plants and Animals*).

Like Virchow, Bernard dissociated himself from both vitalism and materialism. The vital force was for Bernard a 'metaphysical concept' that had no place in science, which deals with the 'determinism of phenomena'. Materialism, however, was inadequate because physical forces alone do not produce biological organization. Bernard proposed instead a dualistic view that distinguished between two 'orders of phenomena' in living beings: 'organic creation', which included both the synthesis of organic molecules and their assembly into organized structures; and 'organic destruction', which included processes such as combustion and fermentation. Both were essential for life, as the products of one became the precursors of the other, but only organic creation was quintessentially vital.

The difference between the liver and the laboratory was that the former organized chemical reactions in a particular way: 'Vital properties are in reality only the physicochemical properties of organized matter.' Life was therefore the manifestation of what Bernard called the 'legislative force':

> Observation teaches us only this: it shows us an *organic plan*, but not an *active intervention* by a vital principle. The only *vital force* that we would accept would only be a sort of legislative force, but in no way executive.

> To summarize our thoughts, we would say metaphorically, the vital force directs the phenomena it does not produce; physical agents produce the phenomena that they do not direct.[25]

The legislative force was a set of laws for the creation and maintenance of the organism, 'a preestablished design for each being and each organ'. It was because of these laws that the organism could direct its chemistry towards the production of highly organized structures, the generation of progeny, the assimilation of nutrients and all the other 'vital' phenomena. The rules for making a particular species had been established at the time of its creation and were recapitulated at every generation: 'they derive by atavism from organisms which the living being continues and repeats, and they can be followed back to the very origin of living beings'.

Bernard's legislative force was an extraordinarily original and profound concept. It obviated the need for a vital force by stipulating that living organisms and inanimate objects both exhibit similar chemical reactions, but in the former these reactions are timed and concerted in a particular way. Bernard's differentiation between the legislative and executive forces represents the earliest antecedent of the distinction between genotype and phenotype, which was to become a central concept of twentieth-century biology (see Chapter 6). His idea of a 'preestablished design' anticipates Erwin Schrödinger's 'miniature blueprint' for the development of the organism (see Chapter 10). Bernard's powerful metaphor would reach its apotheosis in the 1950s when it was realized that the legislative force resided in DNA, and the executive force in proteins.

Standing as it does roughly equidistant from Lavoisier and the genetic code, Bernard's *Phenomena of Life* represents a watershed in the attempt of biological scientists to understand the nature of living systems. In the hundred years prior to the publication of *Phenomena of Life*, natural philosophers had concentrated on characterizing the components of living systems and debating whether these systems could be understood without recourse to vital forces. In the hundred years that followed, biological scientists would increasingly identify the central mystery of life to be the transmission and expression of hereditary information.

Chapter 4
The Catalytic Force

'Slumbering Affinities'

The rise of the enzyme theory in the nineteenth century is to a remarkable extent the story of alcoholic fermentation. This humble process, familiar since antiquity, which had already led to the proposal of the law of conservation of matter (see Chapter 1), was now to have profound effects on the vitalist–materialist debate and lead, more than any other single process, to the new science of biochemistry.

In 1833, Anselme Payen and Jean Persoz found that the addition of alcohol to an aqueous extract of germinating barley (malt) precipitated a substance that could dissolve the internal contents of starch grains. Payen was the director of a sugar factory in Paris and the discoverer of cellulose; Persoz had earlier discovered another carbohydrate, dextrin. Both men were former students of Louis Nicolas Vauquelin, the discoverer of asparagine (see Chapter 2) and himself a student of Antoine François Fourcroy. Because the extract of barley separated the soluble dextrin from the insoluble coats, they proposed the name 'diastase'. This substance could be purified by serial precipitation with alcohol, and was inactivated by boiling. Its activity on starch seemed to be to convert it into a 'gum' and fermentable sugar. Payen and Persoz concluded: 'There remains before us a vast field of researches to be traversed as to the existence of diastase in the various parts of the vegetable organization, its atomic weight, its elementary composition, its combinations, and the products of its markedly specific reaction on vegetables that contain starch.'

The discovery of diastase appears to have inspired Friedrich Wöhler to attempt to convert amygdalin, a compound found in bitter almonds, to benzaldehyde ('bitter almond oil'). An aqueous extract of almonds did, indeed, have this effect, which Wöhler attributed to the 'albumen' present. For the active principle, which was produced as an emulsion of bitter almonds, Justus Liebig and Wöhler proposed in 1837 the name 'emulsin'.

That same year, Theodor Schwann found that the ability of gastric fluid to digest food was not solely due to the presence of hydrochloric acid, as had previously been believed. A solution of hydrochloric acid of the same strength as that found in the stomach was only one-tenth as effective at curdling milk as the gastric fluid itself. The other activity was precipitated by alcohol and destroyed by heating: 'By all these reactions the digestive principle is characterized as an individual substance, to which I have given the name *pepsin*.'

Substances such as diastase, emulsin and pepsin were termed 'ferments' because of the obvious analogy between the reactions they produced and the process of alcoholic fermentation. From these early studies, the general features of these ferments were already becoming clear. They were probably proteins ('albumen'); they converted complex biological compounds to simpler ones; they were precipitated by alcohol and inactivated by heating.

When news of these developments reached Stockholm, the arch-classifier Jöns Jacob Berzelius saw an analogy between such activities and certain chemical agents that speeded up reactions but were not consumed in them. In particular, Gottlieb Sigismund Kirchhof, an apothecary in St Petersburg, had shown in 1811 that starch was decomposed into 'dextrin' and 'saccharose' by sulfuric acid, but the acid was not used up in the reaction. No doubt struck by the fact that sulfuric acid and diastase had the same effect upon starch, Berzelius proposed in 1836 that such phenomena represented the action of a hitherto unrecognized force:

> This new force, which was unknown until now, is common to organic and inorganic nature. I do not believe that it is a force entirely independent of the electrochemical affinities of matter; I believe, on the contrary, that it is only a new manifestation, but since we cannot see the connection and mutual dependence, it will be easier to designate it by a different name. I will call this force *catalytic force* [emphasis in original].[2]

The key characteristic of the catalytic force was that its agents promoted chemical reactions without being altered by these reactions. Berzelius was also insightful enough to sense that the catalytic force allows reactions to occur at lower temperatures than in its absence: 'The catalytic force appears to consist intrinsically in this: that bodies through their mere presence, and not through their affinity, may awaken affinities slumbering at this temperature'.

As in the case of Antoine Lavoisier's analogy between respiration and combustion, the theory of catalysis is an example of a scientific concept that arose from both chemistry and biology. As Malcolm Dixon noted: 'It is important to realize that it was very largely the action of enzymes that gave rise to the idea of catalysis, not the converse as is often assumed.'

Sugar Fungus

In the early nineteenth century, three forms of fermentation were recognized; vinous fermentation, acetous fermentation, and putrefaction. The last was usually associated with animal matter and involved the production of ammonia. Because fermentation processes occurred naturally, they were regarded as chemical decompositions. Lavoisier, for example, found that a sugar solution would not ferment unless he added a little 'yeast of beer', but nonetheless regarded vinous fermentation as a chemical process. Joseph Gay-Lussac believed that fermentation was a process of oxidation initiated by exposure to air.

In 1837, however, the presence of what appeared to be living cells in fermenting solutions was reported independently by Charles Cagniard de la Tour in Paris, Theodor Schwann in Berlin and Friedrich Kützing in Nordhausen. These simultaneous discoveries can be directly attributed to the increased resolving power of the achromatic compound microscope (see Chapter 3); what had previously appeared as amorphous material was now clearly visible as small individual cells. During a night spent in a brewery (surely one of the lesser sacrifices made in the name of science), Cagniard conducted a series of observations on fermenting wort in the course of which he saw yeast cells dividing. In subsequent studies, he was able to show that neither drying nor freezing (with the newly available reagent solid carbon dioxide) abolished the ability of yeast to cause fermentation. Cagniard stated that yeast consisted of 'a mass of little globular bodies able to reproduce themselves, consequently organized, and not a substance simply organic or chemical, as one supposed'. Because of their lack of movement, he classified yeast cells as plants rather than animals – a belief that would prevail until the end of the century.

Schwann was studying spontaneous generation, not fermentation. He showed that boiled extracts of animal tissues did not putrefy when exposed to air that had also been heated. Anticipating the possible criticism that heating air destroyed its ability to sustain life, Schwann tried to show that this air was still capable of supporting fermentation. When he incubated a boiled solution of sugar and yeast with heated air, Schwann found, to his surprise, that no fermentation occurred. Microscopic investigation of fermenting solutions revealed the presence of 'infusoria' that Schwann described as *Zuckerpilz* (sugar fungus) because of their similarities in form and growth to those of fungi.

The microscopical evidence seemed to prove irrefutably that fermenting wort contained the cells of a living organism. But were these cells the cause of the fermentation? Schwann believed so, citing three lines of evidence:

> Now, that these fungi are the cause of fermentation, follows, first from the constancy of their occurrence during the process; secondly, from the cessation of fermentation under any influence by which they are known to be destroyed, especially boiling heat, arsenate of potash, etc.; and thirdly, because the principle which excites the process of fermentation must be a substance which is again generated and increased by the process itself, a phenomenon which is met with only in living organisms.[26]

Schwann's third criterion is the strongest. The agent of fermentation exhibited the process of extensibility – it could be divided indefinitely and still convert sugar into alcohol. As described in Chapter 3, the extensibility of living systems was used by Joseph Henry in 1866 to refute the materialist views of Ernst von Brücke. Likewise, Lionel Beale had noted in 1861 that 'everything that lives – every so-called living machine – grows of itself, builds itself up and multiplies, while every non-living machine *is made, does not grow, and does not produce machines like itself*' [emphasis in original]. A century after Schwann, Oswald Avery would use the same criterion of extensibility in support of his view that the pneumococcal transforming principle was DNA (see Chapter 10).

Kützing had made similar observations, but was only motivated to publish them when he encountered the works of Cagniard and Schwann. He also observed a smaller organism associated with the production of acetic acid, probably the bacterium *Acetobacter*. Kützing concluded: 'organic life = fermentation'.

The claim that fermentation was a vital rather than a chemical process was a radical one in 1837. In Paris, the Academy of Sciences established a committee under Pierre Turpin to investigate Cagniard's claims. The committee confirmed the original findings, concluding that fermentation was 'a purely physiological process'. Turpin published his own study on fermentation in 1838, a German-language abstract of which appeared in Liebig's *Annals of Pharmacy* the following year. One might have thought that a vital explanation of vinous fermentation would have suited Liebig, but he viewed all forms of fermentation as spontaneous decompositions resulting from the cessation of action of the vital force (see Chapter 3). The Turpin abstract was followed by a paper by Wöhler and Liebig purporting to describe their own observations on fermentation:

Brewer's yeast, dispersed in water, is resolved by this instrument [a Pistorius microscope] into infinitesimally small spherelets, hardly 1/400 of a line in diameter, and into fine threads which unmistakably are a kind of albumin. If these spherelets are placed into sugar water, they can be seen to be animal eggs; they swell up, they burst; small animals develop from them which multiply with inconceivable rapidity, in the most unparalleled manner. In appearance, these animals diverge from all those hitherto described, 600 species; their shape is that of a Beindorf's retort (without condenser). The delivery tube is a kind of proboscis, the interior of which is lined with fine bristles, approximately 1/2000 line in length; teeth and eyes cannot be detected, but otherwise one can discern with clarity a stomach, the intestinal canal, the anus (a rose-colored point), and the organs for the secretion of urine. It can be observed that the animals absorb the sugar from the solution from the very moment in which they escape from the eggs; the sugar can very clearly be seen to reach the stomach. It is digested instantaneously, and presently and with the greatest certainty, this digestion can be recognized as such, by virtue of the discharge of excrements. In a word, these infusoria feed on sugar, discharge spirits from the intestinal canal, and carbonic acid from the urinary apparatus. When full, the bladder assumes the form of a champagne bottle; empty, it is a little knob.[27]

Nineteenth-century science was rather more robustly personal than its modern counterpart. In the *Jahresbericht* for 1839, for example, Berzelius described Charles Gerhardt's system for organic classification with the words: 'He fills up 37 pages with this drivel'; Mathias Schleiden wrote of one of his fellow cytologists: 'To enter upon Raspail's work appears to me incompatible with the dignity of science'; Liebig and Wöhler, as described in Chapter 2, satirized Jean-Baptiste Dumas as 'S. C. H. Windler' and 'Professor Ch. Arlatan'; Hermann Kolbe described Louis Pasteur in a review of 1872 as being 'not quite of sound mind'.

Schwann's bitter reaction to the satire on fermentation was therefore a reflection more of his life-long propensity to anxiety and depression than of the article itself. Coupled with his unsuccessful candidacy for a chair at the University of Bonn (for which he probably also blamed Liebig), it led to Schwann leaving Germany for Belgium. He became professor of anatomy at the University of Louvain in 1839, moving to Liège nine years later. In Belgium, Schwann performed some studies in digestive physiology, but his brilliant research career was essentially over. While still working on a great work that would cover all aspects of biology, from atoms to the brain, Theodor Schwann died in 1882.

Liebig followed up the satire on fermentation with a more serious article in

1839. In this he proposed that the agents of fermentation were not living cells but rather plant proteins; in the presence of oxygen, these proteins underwent a reaction with sugar that decomposed both:

> The atoms of a putrefying body, the ferment, are in a ceaseless movement; they change their positions in forming new combinations. These moving atoms are in contact with the atoms of sugar, whose elements are held together as sugar by a weak force. The movement of the atoms of the ferment cannot be without effects on the atoms of the elements of the substance mixed with it; either their motion is abolished or the atoms of the latter move too. The sugar atoms suffer a displacement; they arrange themselves in such a way that they hang together more firmly, so that they no longer follow the impact, that is to form alcohol and carbon dioxide.[28]

Liebig's view was heavily influenced by Berzelius' theory of catalysis; like Berzelius, he saw a clear analogy between catalytic processes in the organic and inorganic domains and compared the effects of 'ferments' on chemical reactions to the action of heat. Unlike Berzelius, however, Liebig believed that the ferment (plant protein) was decomposed in the course of the reaction. Berzelius disagreed both with Liebig and with the vital theory of fermentation, holding that this process represented the action of an organic catalyst.

The Asymmetric Force

Louis Pasteur was born in 1822 in Dôle, in the Jura region of France, moving five years later to Arbois. He was the son of a former sergeant in the army of Napoleon, who imbued in young Louis a sense of national pride and patriotic duty. Pasteur obtained a baccalaureate in arts from the *Collège royale* in Besançon in 1840, and a degree in science two years later. Little promise of scientific brilliance was apparent at this stage of Pasteur's career, as he was rated 'passable' in physics and 'mediocre' in chemistry. He moved to Paris for a year at preparatory school, and attended the lectures of Jean-Baptiste Dumas at the Sorbonne. Pasteur became an admirer of the great chemist, who in turn later became an important patron of the younger man. In 1843, Pasteur entered the *École normale supérieure*, graduating with his doctorate in 1847.

Pasteur's instructor in crystallography at the *École normale* was the eminent physicist Jean-Baptiste Biot. In 1815, Biot had discovered a curious phenomenon – that shining a beam of polarized light through solutions of some organic molecules resulted in a rotation in the plane of polarization of the light. One substance that had this optical activity was tartaric acid, one of the 'vegetable acids' originally isolated by Carl Wilhelm Scheele. The physical basis of optical rotation was completely unknown. In 1820, however, a substance that became known as paratartaric acid was identified as a by-product of the industrial preparation of tartaric acid. Paratartaric acid had the same elemental composition as tartaric acid, but was optically inactive, causing no rotation of polarized light.

In 1844, Pasteur read a paper by Eilhard Mitscherlich, a former student of Berzelius, which noted that the sodium (and also the ammonium) salts of tartaric acid and paratartaric acid not only had the same elemental composition, but also exhibited the same crystalline form, optical refraction and specific weight. Surely the physical and chemical properties of compounds were wholly determined by the nature and arrangement of their atoms. How then could two substances identical in these respects have different optical activities? The young Pasteur decided this would be a worthwhile topic for investigation.

Examining Mitscherlich's salts of the tartaric and paratartaric acids, Pasteur made a novel observation. These crystals had the form of cubes with some of the corners cut off. The small oblique faces at the corners of the cubes were referred to as hemihedral faces. Pasteur noticed that in tartaric acid all the crystals had the hemihedral faces located on the same corners of the cube, but in paratartaric acid the crystals were of two types, which were mirror images of one another (Figure 4.1). He suspected that the occurrence of these different crystal forms was somehow related to optical activity. To

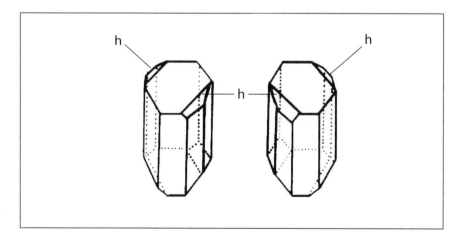

Figure 4.1: Crystal morphologies of the tartrates. Paratartaric acid crystallizes in two mirror-image forms, distinguishable by the location of the hemihedral faces (h). Reproduced from Kottler, D. B. (1978) *Studies in the History of Biology* **2**, 69

test this hypothesis, Pasteur in 1848 laboriously separated out the two mirror-image forms of paratartaric acid crystals. In one of the truly definitive experiments in the history of science, he showed that one of the hemihedral forms rotated polarized light to the right (dextrorotation), like tartaric acid, while the other rotated polarized light to the left (levorotation). The optical inactivity of paratartaric acid was therefore due to the presence of two different isomeric substances, one of which was identical to the tartaric acid found in plants.

News of this finding produced a summons to the *Collège de France*, where Biot asked for a personal demonstration of the separation of the two crystalline forms of paratartaric acid. Biot himself dissolved the two sets of crystals and placed the solutions in the polarimeter. When he saw that the two solutions had different optical activities, the old man was moved to tears.

Like many great scientists, Pasteur was quick to draw generalizations from novel observations. From his studies on the tartrates, he inferred two such principles: the 'law of hemihedral correlation', and the 'asymmetric force'. The law of hemihedral correlation stated that the crystals of all optically active compounds have hemihedral faces, and that there exists a correlation between the symmetry of these faces and the direction of optical rotation. Despite Pasteur's ingenuity in coaxing crystals to develop hemihedral faces, numerous 'exceptions' to this law were found. In 1849, Biot informed him that amyl alcohol was optically active. Pasteur succeeded in isolating crystals of both the dextrotatory and levorotatory type. Both of these, however, crystallized without hemihedral faces. By 1855, the law of hemihedral correlation was effectively abandoned.

The concept of the asymmetric force arose from the fact that optically active substances such as tartaric acid seemed to be exclusively of biological origin, whereas chemical process appeared never to produce such substances. In 1853, Pasteur found that the dextrorotatory (vegetable) and levorotatory forms of tartaric acid reacted quite differently with optically active bases. He proposed that this kind of reaction specificity results from an 'asymmetric force' operating only in living organisms, and that the action of this force was responsible for the optical activity of many biological molecules:

> In order to account for the exclusive formation of molecules of a single order of asymmetry it therefore suffices to admit that *at the moment of their grouping the elementary atoms are subjected to an asymmetric influence*, and as all organic molecules which have arisen in analogous circumstances are identical, *this influence must be universal*. It would embrace the entire terrestrial globe. To it would be

due the molecular asymmetry of organic natural products of vegetable organisms, products which we rediscover among animals almost without alteration and where they play a mysterious role of which we do not yet have the slightest idea [emphases in original].[29]

For Pasteur, then, a distinguishing characteristic of living systems was the presence of one optical isomer rather than the other. In fact, this was the only chemical difference between the molecules found in organic and inorganic systems: 'the molecular asymmetry of natural organic products' represented 'perhaps the only well-marked line of demarcation that can at present be drawn between the chemistry of dead matter and the chemistry of living matter'.

The asymmetric force acted at the genesis of an organism, causing its matter to be asymmetric. Thereafter, this asymmetry was maintained by the fact that the presence of an asymmetric environment ensured that only optically active matter would be produced. The idea that animate systems differed from inanimate ones in the operation of the asymmetric force was to colour Pasteur's subsequent studies on fermentation and putrefaction.

'Life Without Air'

In 1848, Pasteur became the professor of physics at the *Lycée* in Dijon, then moved to the University of Strasbourg as deputy professor of chemistry. In 1854, still only thirty-two years old, Pasteur became professor of chemistry and dean of the faculty of sciences at Lille. It was there, at the heart of the French sugar-beet fermentation industry, that he first became interested in fermentation processes.

Amyl alcohol was a by-product of alcoholic fermentation. According to Pasteur's idea of the asymmetric force, the production of optically active substances was the hallmark of life. Proponents of the chemical nature of fermentation, such as Liebig, argued that amyl alcohol merely preserved the optical activity of the sugar molecule from which it was produced. Pasteur, however, had found that optical activity was easily lost in chemical transformations, and therefore believed that the production of amyl alcohol in fermentation must represent the action of the asymmetric force – in other words, fermentation must be a vital, rather than a chemical, process.

The legend that Pasteur became interested in fermentation because a local distiller begged his assistance in identifying the reason for his low alcohol yields may, however, have some basis in fact, because Pasteur's first paper on this subject dealt with lactic fermentation rather than the alcoholic variety. Lactic acid was occasionally produced during the fermentation of sugar, together with or instead of the production of alcohol, and represented a major problem to brewers. In a paper presented in Lille in 1857, Pasteur showed that: 'just as an alcoholic ferment exists, namely, brewers' yeast, which is found wherever sugar breaks down into alcohol and carbonic acid, so too there is a special ferment, a lactic yeast, always present when sugar becomes lactic acid'. To prove this, Pasteur used techniques that would revolutionize microbiology. He took some of the grey deposit that characterized lactic acid fermentation, and added it to a solution containing cane sugar, chalk and an extract of brewers' yeast. On incubation at 30–35°C, a fermentation resulted, accompanied by the dissolution of the chalk and the appearance of a crystalline matter. These crystals, Pasteur showed, were of calcium lactate. The 'lactic yeast' that mediated this transformation of sugar to lactic acid appeared quite different under the microscope from brewers' yeast, being composed of 'little globules or very short segmented filaments'.

With this memoir, Pasteur came down firmly on the side of those, such as Schwann, Cagniard and Turpin, who believed that fermentation was a vital process, and against those, such as Liebig and Felix Hoppe-Seyler, who believed that it was a chemical process: 'it is my opinion . . . that fermentation appears to be correlative to life and to the organization of globules, and not to their death or putrefaction.'

Just after presenting this memoir on the lactic fermentation, Pasteur returned to the *École normale* as director of scientific studies and administration. In 1860, he published a detailed study on alcoholic fermentation. This included an important innovation – the growth of yeast in a solution containing only sugar, ammonia tartrate, and the ash obtained by incinerating yeast. The sugar served as a source of carbon, the salt as a source of nitrogen, and the yeast ash as a source of minerals. The elimination from the growth medium of the yeast extract that had been used for the 'lactic yeast' study meant, to Pasteur, that Liebig was wrong in believing that organic matter was necessary for fermentation. Pasteur also showed that fermentation did not convert all the sugar utilized to carbonic acid and alcohol, as one would expect of a catalyst, but also produced lesser amounts of other organic molecules such as succinic acid, cellulose and 'fatty matters'. He concluded:

> My present and most fixed opinion regarding the nature of alcoholic fermentation is this: The chemical act of fermentation is essentially a phenomenon correlative with a vital act, beginning and ending with the latter. I believe that there is never any alcoholic fermentation without there being simultaneously

the organization, development, multiplication of the globules, or the pursued, continued life of globules, that are already formed . . . I profess the same view on the subject of lactic fermentation, butyric fermentation, the fermentation of tartaric acid, and many other fermentations properly designated as such.[13]

The following year, Pasteur backed up this claim by isolating the micro-organism responsible for butyric acid fermentation. This could be grown in a medium containing only sugar, ammonia and phosphates. Under the microscope, these infusoria were 'small cylindrical rods, rounded at the extremities, usually straight, isolated or united into chains of two, three, four, and sometimes more joints'. The 'butyric ferment' not only grew in the absence of oxygen but was killed by it, leading Pasteur to his famous definition of fermentation as 'life without air'. In 1863, he reported that the organism responsible for the fermentation of tartaric acid was also killed by oxygen.

When he investigated the effect of atmospheric air on brewers' yeast, he discovered that these organisms grew more rapidly in the presence of oxygen, but did not produce any alcohol. The explanation of this observation, which became known as the 'Pasteur effect', was that yeast could use oxygen to break down sugar completely to carbonic acid and water, but in its absence only partially to alcohol and carbonic acid.

Pasteur also studied the role of micro-organisms in the third form of fermentation – putrefaction. Many scientists of the vitalist persuasion, including Liebig, believed that putrefaction was due to a spontaneous breakdown of animal tissues once the chemical forces of affinity were no longer held in check by the vital force. It was also widely believed – and seemingly supported by observation – that infusoria and other organisms associated with decay were spontaneously generated in dead tissues. In this way, Pasteur was drawn into the acrimonious debate about the spontaneous generation of life.

Careful experiments conducted by Pasteur showed that the 'combustion' of organic substances does not occur in the absence of micro-organisms. In a prize-winning essay of 1861, he described a variety of experiments that showed airborne micro-organisms to be responsible for the putrefaction of organic solutions. The most definitive of these was the demonstration that a boiled sugar solution in a swan-necked flask left open to the air did not undergo putrefaction, apparently because the airborne organisms became trapped in the bend of the neck. It could still be argued, however, that boiling organic solutions destroyed their ability to undergo spontaneous decomposition. Pasteur managed to disprove this by showing in 1863 that urine and blood, drawn sterile from the body but otherwise untreated, did not take up any significant amount of oxygen over a three-year period. Only if micro-organisms were present would fermentation or putrefaction occur, and only in this case would oxygen uptake occur:

> Dead substances that ferment or putrefy do not yield solely to forces of a purely physical or chemical nature. It will be necessary to banish from science the whole of that collection of preconceived opinions which consist in assuming that a certain class of organic substances – the nitrogenous plastic substances – may acquire, by the hypothetical influence of direct oxidation, an occult power, char-acterised by an internal agitation, communicable to organic substances supposed to have little stability . . . the slow combustion which takes place in dead organic substances, when they are exposed to the air has, in most cases, an equally intimate connection with the presence of the lowest forms of life.[30]

Throughout the rest of the 1860s, Pasteur's debates with the proponents of spontaneous generation overshadowed his studies on fermentation, in which little novel was accomplished. In 1867, Pasteur succeeded his friend and mentor Dumas as professor of chemistry at the Sorbonne. The following year, he suffered a stroke that temporarily paralyzed the left side of his body and from which he never fully recovered. For the remainder of his life, Pasteur concentrated on practical applications of his earlier scientific breakthroughs. These included the development of treatments for diseases of silkworm and the development of vaccines for anthrax, rabies and fowl cholera.

In 1870, however, the controversy over the vital nature of fermentation was rekindled by a paper from Liebig. In this, Liebig appeared willing to reach a compromise between the vital and the chemical views of fermentation. He now accepted that the globules present in fermenting were growing yeast cells, not protein aggregates (or miniature distillation vessels). The role of the yeast was to produce the 'albuminous substance' that decomposed the sugar:

> There seems to be no doubt as to the part which the vegetable organism plays in the phenomenon of fermentation. It is through it alone that an albuminous substance and sugar are enabled to unite and form this particular combination, this unstable form under which alone, as a component part of the mycoderm [yeast], they manifest an action on sugar. Should the mycoderm cease to grow, the bond which unites the constituent parts of the cellular contents is loosened, and it is through the motion produced therein that the cells of yeast bring about a disarrangement or separation of the elements of the sugar into other organic molecules.[31]

However, Liebig still questioned some aspects of Pasteur's work, including the crucial growth of yeast in chemically defined media.

Pasteur was a scientist who typically reacted to any criticism of his work with a blizzard of new studies supporting his original position and a propaganda offensive using all the channels at his disposal. His opponents were not so much defeated by the scientific evidence as buried under an avalanche of restatements of Pasteur's arguments, public demonstrations and commission reports. Liebig may have been willing to split the difference, but Pasteur, now at the height of his influence, was in no mood to compromise with 'the celebrated chemist of Munich'.[a]

In his reply to Liebig in 1871, Pasteur proposed that the question of whether yeast could grow in protein-free medium be referred to a commission of the Academy of Sciences. In the presence of such a commission, he offered to prepare 'as great a weight of ferment as Liebig could reasonably demand'. The chemist of Munich did not take up this challenge, and died two years later.

By this point in the debate over fermentation, the terms 'vitalist' and 'materialist' had become almost meaningless. Liebig, who at the same time fulminated against materialists like Emil DuBois-Reymond (see Chapter 3), nonetheless believed that fermentation was a purely chemical process; whereas Pasteur, whom no-one would consider, based on his career as a whole, a vitalist, perpetuated the view that fermentation was an intrinsic property of living cells.

Organized and Unorganized Ferments

By the time of his death, it appears that Liebig had realized that the true nature of alcoholic fermentation lay somewhere between the vital process of Pasteur and his own earlier view of it as a spontaneous process of decay. In his essay of 1870, he wrote:

It may be that the physiological process stands in no other relation to the fermentation process than the following: a substance is produced in the living cells

---

a   Pasteur was also incensed by the bombardment of Paris during the Franco-Prussian War of 1870–1. In January 1871, he wrote to the dean of the faculty of medicine at the University of Bonn, which three years earlier had awarded him an honorary degree: 'I am obeying a cry of my conscience in asking you to erase my name from your faculty archives and to take back the diploma as a sign of the indignation, which has been inspired in a French scientist by the barbarity and hypocrisy of him, who in order to satisfy a criminal pride insists upon the massacre of two great nations.'[32]

which, through an operation similar to that of emulsin on salicilin and amygdalin, leads to the decomposition of sugar and other organic molecules; the physiological process would in this case be necessary to produce this substance but would stand in no further relation to the fermentation.[33]

This view was probably inspired by the writings of the German wine-maker and former *Liebig-Schüler*, Moritz Traube. Starting in 1858, Traube promoted the view that ferments were specific protein molecules that catalyzed oxidation-reduction reactions. Two years later, the French organic chemist Marcellin Berthelot isolated from yeast cells a ferment that could 'invert' (hydrolyze) sucrose. Berthelot distinguished between 'soluble ferments', such as diastase and his new 'invertase', and 'insoluble ferments', such as yeast. The former were able to produce their catalytic effects in the absence of life, but the actions of the latter required the organized structures of living tissues.

Berthelot was an extreme materialist, who in 1860 wrote in connection with the fermentation controversy: 'to banish life from all explanations concerning organic chemistry, this is the objective of our studies.' The death of Liebig left Berthelot as the chief opponent of the vital theory of fermentation.

In 1861, Ernst von Brücke proposed that protoplasm was a form of organized cellular matter (see Chapter 3). It was therefore natural to associate insoluble ferments with protoplasm. As a result, catalytic activities were divided into two categories: 'organized ferments', which performed complex chemical transformations within the cell; and 'unorganized ferments', which were secreted from the cell to perform the preliminary digestion of food molecules. This view was consistent with the activities of the ferments then known. For example, the unorganized ferment diastase performed the extracellular breakdown of starch to glucose; the glucose was then taken up by the cell and converted to alcohol and carbonic acid by the intracellular organized ferment. Willi Kühne, a former student of Claude Bernard, suggested in 1876 a term for the unorganized type of ferment that would eventually be used to describe both classes of activity – enzyme ('in yeast'). Kühne, professor of physiology at Heidelberg, was a supporter of the Pastorian view of fermentation as a vital process.

However, it was also possible that the distinction between organized and unorganized ferments was merely an operational one. Perhaps the only difference between these classes of catalyst was that one was present within the cells, the other outside. As Traube wrote in 1874: 'the activity of this ferment only appears to be tied to the cell in so far as no means have been found up to the present of isolating the ferment from the cell without destroying it.' Such studies were in fact attempted by, among others, Pasteur, who tried to extract an 'alcoholase' from yeast by grinding. All such efforts were, however, unsuccessful (presumably because the grinding procedures used caused mechanical degradation of the ferments responsible). Chemical extraction was no better – boiling in glycerine or sodium hydroxide was, at least in retrospect, unlikely to preserve biological activity.

The division of metabolic activities into those catalyzed by extracellular/ soluble/unorganized ferments and those catalyzed by intracellular/insoluble/organized ferments marked the beginning of a twenty-year truce in the debate on the nature of fermentation. During that period, much effort was expended in trying to incorporate catalytic activity into the concept of protoplasm. Eduard Pflüger proposed in 1875 that protoplasm contained unstable chemical groups (aldehydes) that allowed it to participate in certain reactions. The presence of such groups distinguished 'living' cellular proteins from 'dead' extracted proteins. A similar view was later expressed by Max Verworn, who suggested that protoplasm consisted of a simple nucleus to which were attached numerous side-chains containing different chemical activities. Verworn used the term 'biogen' for this giant multi-functional molecule. Biogens, which were related to proteins, contained labile chemical groups that were capable of undergoing spontaneous decomposition. However, these labile groups could also catalyze rearrangements of atoms in other molecules. The 'living protein' and 'biogen' mechanisms therefore drew heavily upon Liebig's view of fermentation, but attempted to put it on a firmer chemical basis and extend it to respiration.

The Asymmetric Carbon Atom

In 1874, an explanation of the optical activity that had been studied by Biot, Mitscherlich and Pasteur was independently proposed by Jacobus Henricus van't Hoff and Joseph Achille Le Bel. Although their approaches were somewhat different, both men concluded that the presence of two optical isomers resulted from the two possible arrangements of four different radicals around a tetrahedral carbon atom.

Van't Hoff was born in Rotterdam in 1852 and studied at the Polytechnic School in Delft. He then worked with August Kekulé in Bonn and Charles Wurtz in Paris before returning to the Netherlands in 1874 to start work for his doctorate at the University of Utrecht. His paper on the tetrahedral carbon atom was van't Hoff's first scientific work, published when he was only twenty-two. Le Bel was not much older. He was born in 1847 and studied at the *École polytechnique*. Le Bel then worked in Wurtz's laboratory where he made the acquaintance of van't Hoff, but the two men apparently did not discuss their ideas on the relationship between carbon tetravalency and optical activity.

In his 1874 paper, van't Hoff noted that, if the four valences of carbon were all in the same plane, there were three different ways in which four different groups could be attached to the same carbon atom (Figure 4.2). These three conformations were not superimposable, and thus could be expected to rotate polarized light differently. However, optical isomers came in pairs, not trios. Van't Hoff proposed instead that the four carbon valences formed a tetrahedron:

> The theory is brought into accord with the facts if we consider the affinities of the carbon atom directed toward the corners of a tetrahedron of which the carbon

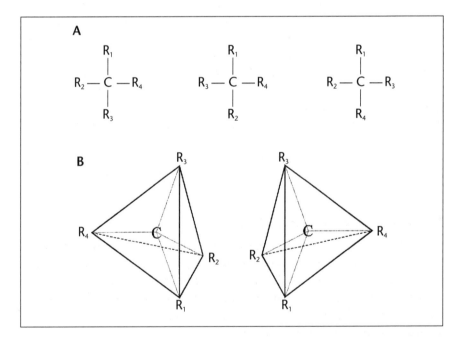

Figure 4.2: Jacobus van't Hoff's 1874 model of the asymmetric carbon atom. (A) The three possible ways in which four different groups (R_1–R_4) can be arranged around a tetravalent carbon atom in a planar molecule. (B) The two possible ways in which four different groups can be arranged around a tetravalent carbon atom that occupies the center of a tetrahedron. These structures correspond to the two optical isomers of a compound containing an asymmetric carbon atom. Adapted from van't Hoff, J. H. (1874) *Archives néerlandaises des sciences exactes et naturelles* **9**, 454

atom itself occupies the center . . . *When the four affinities of the carbon atom are satisfied by four univalent groups differing among themselves, two and not more than two different tetrahedrons are obtained, one of which is the reflected image of the other, they cannot be superposed; that is, we have here to deal with two structural formulas isomeric in space* [emphasis in original].[34]

Carbon atoms of this type, which van't Hoff termed 'asymmetric', occurred in all compounds known to exhibit optical activity, including tartaric acid, malic acid, aspartic acid and glucose.

Le Bel's reasoning was more abstractly mathematical, but reached the same conclusion: 'if it happens not only that a single substitution furnished but one derivative, but also that two or even three substitutions give only one and the same chemical isomer, we are obliged to admit that four atoms A occupy the angles of a regular tetrahedron, whose planes of symmetry are identical with the whole molecule MA_4.'

The recognition of the asymmetrically bonded carbon atom as the basis of optical activity was a dividend from the valence theory and the use of structural formulas. In a single elegant step, the observations of Pasteur were reduced to chemistry. Had Biot still been alive, he might have shed another tear.

In 1860, Pasteur had speculated on the molecular asymmetry that caused optical activity:

Are the atoms of the dextro [right] acid arranged in the form of a right-handed spiral, or are they situated at the corners of an irregular tetrahedron, or do they have some other asymmetric grouping? This we do not know. But without doubt the atoms possess an asymmetric arrangement like that of an object and its reflected image. Quite as certain is it that the atoms of the laevo [left] acid possess exactly the opposite grouping.[34]

It would be easy to read into this statement more than Pasteur intended. His 'irregular tetrahedron' was not an asymmetric carbon atom, but a solid with non-identical sides. Pasteur believed that optical activity resulted from a shape imposed on the molecule, in the sense that a spiral staircase is asymmetric whereas a straight one is not. Thus he proposed that asymmetric substances could exist in four forms: left-handed, right-handed, a racemic mixture of both, and 'the substance which is neither right nor left nor formed by the combination of the right and the left'. In contrast, the asymmetric carbon atom of van't Hoff and Le Bel could not be 'untwisted' to make a non-asymmetric molecule.

Of more lasting significance was Pasteur's suggestion of helicity as a possible basis of molecular asymmetry. In the 1950s, it would be recognized that both the DNA molecule and the protein α-helix were right-handed spiral structures.

'Like Lock and Key'

The elucidation of the relationship between the asymmetric carbon atom and the action of enzymes was one of several great contributions to physiological chemistry by the organic chemist Emil Fischer. In this way, Pasteur's 'asymmetric force' and Berzelius' 'catalytic force' both were subsumed into a new paradigm of enzyme action that provided one of the underpinnings of twentieth-century biochemistry.

Emil Fischer was born in 1852 in Euskirchen, near Cologne. On leaving school in 1869 he wished to study science, but his father insisted that he join the family lumber business. Eventually concluding that young Emil was 'too stupid to be a businessman', his father sent him in 1871 to the University of Bonn. The following year, Fischer transferred to the University of Strasbourg to study chemistry under Adolf von Baeyer. On graduating in 1874, he moved with Baeyer to Munich, where he became *ausserordentlicher* professor in 1879. Fischer became professor of chemistry at Erlangen in 1882, and at Würzburg in 1885. In 1892, propelled by his brilliant work on the purines and carbohydrates, he reached the pinnacle of academic chemistry in Germany by being appointed professor of chemistry and director of the chemical institute at the University of Berlin.

In 1884, while still at Erlangen, Fischer began his studies on the carbohydrates. At this time, three monosaccharides were known – glucose, galactose and fructose – all of which had the empirical formula $C_6H_{12}O_6$. To be considered pure, a compound had to be recrystallizable to constant melting point and optical rotation. Impure sugars, however, formed syrups that resisted crystallization. Fischer used phenylhydrazine, a powerful new reagent that he had discovered, to make osazone derivatives of the monosaccharides which, unlike the parent compounds, had definite melting points. It was still difficult work, though; Fischer wrote to Baeyer in 1889: 'Unfortunately, the experimental difficulties in this group are so great, that a single experiment takes more time in weeks than other classes of compounds take in hours . . .'

By 1888, Fischer was able to synthesize not only glucose and fructose but also a novel six-carbon sugar, mannose. Mannose differed from glucose only in the orientation around one asymmetric carbon atom. Fischer's synthesis of mannose thus provided experimental vindication for the 'stereochemical' theory of van't Hoff and Le Bel: 'In the sugar group, dextrose [glucose] and mannose are the first examples of two isomers, which have the same structure and can be converted into each other. For the explanation of this form of isomerism, we draw entirely upon the principles of the Le Bel–van't Hoff theory.' He also introduced a notation for designating pairs of stereoisomers and a means of representing them in two dimensions.

Monosaccharides with the structural formula of glucose (aldohexoses) have four asymmetric carbon atoms, and therefore should exist in sixteen (2^4) stereoisomeric forms. By 1891, in one of the classic feats of organic chemistry, Fischer had synthesized all of these compounds (Figure 4.3). Even

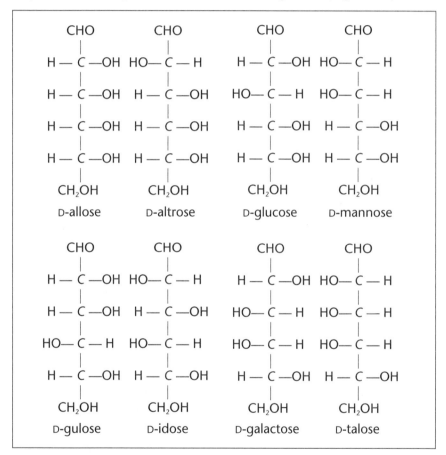

Figure 4.3: The family of D-aldohexoses (Fischer projections). Because aldohexoses (six-carbon sugars with an aldehyde [CHO] group) contain four asymmetric carbon atoms (shown in italic), there are sixteen (2^4) stereoisomeric forms: the eight shown, plus their L isomers, which differ in the orientation around the carbon atom closest to the CH$_2$OH group

more remarkably, from his studies on their reactions he was able to construct a formal proof of the relative stereochemistry of each of the sixteen aldohexoses. Both the Fischer scholar Frieder Lichtenthaler and the historian Joseph Fruton regard Fischer's work on sugar stereochemistry as the high point of his scientific career.

Since reading Pasteur's book *Études sur la bière* (*Studies on Beer*) in the winter of 1876–7, Fischer had used fermentation as a means of characterizing new sugars. It was this approach that allowed Fischer to discover the stereospecificity of enzymes in 1894. Studying the activities of invertase and emulsin against methyl derivatives of glucose, he found that the former enzyme hydrolyzed the α stereoisomer of methyl-glucoside but not the β isomer, whereas the latter hydrolyzed the β but not the α isomer. Fischer realized that enzymes could discriminate between the mirror-image isomers of their substrates because the interaction was three-dimensional, like a key fitting a lock:

> The restricted action of the enzymes on glucosides may therefore be explained by the assumption that only in the case of similar geometrical structure can the molecules so closely approach each other as to initiate a chemical reaction. To use a picture I would like to say that enzyme and glucoside have to fit together like lock and key [*Schloss und Schlüssel*] in order to exert a chemical effect on each other. The finding that the activity of enzymes is limited by molecular geometry to so marked a degree, should be of some use in physiological research.[35]

Fischer's demonstration of the stereospecificity of enzyme action, together with his earlier discovery of chemical reactions that generate asymmetric products, led him to conclude that there was nothing specifically 'vital' about stereospecific reactions; Pasteur's 'asymmetric force' was merely the inevitable consequence of a reaction between two asymmetric substances. The optical activity of natural plant carbohydrates resulted from their production by the optically active enzymes of the plant cell.

When he demonstrated in 1860 that the two isomeric forms of tartaric acid could be separated by forming salts with optically active bases, Pasteur appeared to realize that stereospecificity could be a significant factor in the reactions of optically active compounds: 'The molecular asymmetry of a substance obtrudes itself on chemistry as a powerful modifier of chemical affinities. Towards the two tartaric acids, quinine does not behave like potash, simply because it is asymmetric and potash is not. Molecular asymmetry exhibits itself henceforth as a property capable by itself, in virtue of its being asymmetry, of modifying chemical affinities.'

Pasteur believed that the production of optically active substances was due to a physical force that could, under the appropriate conditions, be replicated in the laboratory, and in fact expended a considerable effort trying to do this. Fischer, however, associated enzyme stereospecificity with Rudolf Virchow's dictum on the fundamental properties of living systems: just as cells only arise from pre-existing cells, optically active molecules only arise from pre-existing optically active molecules. According to Fischer, therefore, stereospecificity was not the product of a physical force, but merely the inevitable consequence of interaction between optically active molecules; as such, enzymes could react with only one of the possible stereoisomeric forms of their substrate, and could produce only one of the possible stereoisomeric forms of their products. Fischer's lock and key analogy allowed Pasteur's 'asymmetric force' to be subsumed into the new enzyme theory of life.

The idea of stereospecific reactions was also applied to many other forms of molecular interaction in biology, including antibody-antigen recognition and the replication of the gene. Through Paul Ehrlich and Karl Landsteiner, the lock and key analogy was to eventually lead to Linus Pauling's concept of molecular complementarity (see Chapter 9).

Zymase

The final obstacle to the general acceptance that life is based on the action of enzymes was overcome in 1897 when the Buchner brothers discovered cell-free fermentation. Eduard Buchner was born in Munich in 1860 and studied chemistry at the *Technische Hochschule* (polytechnic). Because of financial problems, he had to abandon his studies in 1880 and spent the next four years working in canning factories. With the help of his older brother Hans, Eduard Buchner was able to study chemistry under Fischer's mentor, Baeyer, and botany under Carl von Naegeli. The latter interested Buchner in the study of alcoholic fermentation. Following the completion of his doctorate with Theodor Curtius in 1888, Buchner became an assistant to Baeyer, and *Privatdozent* three years later. In 1893, he moved to the University of Kiel to become head of the section for analytical chemistry, and three years later was appointed *ausserordentlicher* professor of analytical pharmaceutical chemistry at Tübingen.

The two men who were to usher in the enzyme theory, Fischer and Buchner, were both students of Baeyer, but otherwise quite different. Fischer was widely recognized as the outstanding organic chemist of his time and achieved the top position in German chemistry by the age of forty; Buchner was regarded by Baeyer as having 'no talent for chemistry' and while still in Munich was passed over for promotion in favor of a younger man. Fischer was 'indifferent to literature, music, art or other cultural distractions from his professional work'; Buchner was a skilled mountaineer with many first ascents to his credit. Fischer also appears to have been politically astute, achieving the exalted position of *Exzellenz* (Excellency) in 1910; Buchner was arrested on his wedding night for creating a disturbance in the streets of Tübingen, and lost his chance for a prized academic position by rebuking a powerful bureaucrat who had kept him waiting. However, both Fischer and Buchner were to achieve the distinction of winning the Nobel prize for chemistry. Although Fischer's achievements ranged much more widely over the terrain of chemistry, both men deserve equal credit for making possible the twentieth-century science of biochemistry. As will be seen in Chapter 7, both were also fervent German nationalists whose support for the Central Powers in World War I was to cost them dearly.

Eduard Buchner was overshadowed not only by Emil Fischer but even by his own brother. Hans Buchner was a student of Carl Ludwig who in 1894 became professor at the Institute of Hygiene in Munich. He was a bacteriologist interested in the mechanisms of disease and in the body's natural defense mechanisms against infection. It was known that animals infected with bacterial pathogens produced substances that were capable of neutralizing bacterial toxins. Hans Buchner believed that both the toxins and 'antitoxins' were bacterial proteins. If antitoxins could be purified from bacterial cells, these proteins could be used to protect humans against infection.

The discovery of cell-free fermentation was a fluke. Hans Buchner wanted to find a way of isolating active proteins from bacterial cells. On consulting his chemist brother, he was advised that chemical extraction would be too harsh, but that mechanical methods might be successful. To experiment with extraction methods, Hans Buchner used yeast – then, as now, readily available in Munich. His assistant, Martin Hahn, came up with the process of grinding the yeast cells with a mixture of kieselguhr and quartz sand, then expressing the fluid with a hydraulic press. The use of kieselguhr, which made the grinding mixture less abrasive, was the crucial difference between this study and earlier attempts to extract enzymes from yeast cells. The Hahn extraction process produced excellent yields of intracellular fluid – approximately 300 ml of 'press fluid' from 1 kg of yeast. However, Hahn noticed that the amount of coagulable protein in the extract decreased rapidly with time. Hans Buchner recommended that he try various preservatives, including the standard technique of adding high concentrations of sugar. At this point, in the fall of 1896, Eduard Buchner arrived in his brother's laboratory, on vacation from Tübingen. Seeing streams of bubbles emerging from the press juice to which sugar had been added, Eduard immediately recognized that the solution was fermenting. Hans Buchner thought that the press juice must contain a 'living plasmatic fluid', but found, to his great surprise, that the addition of antiseptics such as arsenic and toluene had no effect on the fermentation.

In view of the fact that the discovery of cell-free fermentation was made in Hans Buchner's laboratory, and that the key technical innovation was made by Hahn, it seems a little surprising that the credit for this discovery appears to have gone solely to Eduard Buchner. However, it was his name alone that appeared on the 1897 paper that announced the remarkable finding to the world:

> The following conclusions may be drawn with respect to the theory of fermentation. In the first place it has been demonstrated that for the production of the fermentation process no such complicated apparatus is necessary as is represented by the yeast cell. It is much more likely that the agent of the juice which is active in fermentation is a soluble substance, doubtless an albuminoid substance; this may be designated as zymase . . . It is possible that the fermentation of the sugar by the zymase takes place inside the yeast cells; it is more probable though that the yeast cells secrete this albuminoid substance into the sugar solution, where it causes the fermentation. If so, the process in alcoholic fermentation is perhaps to be regarded as a physiological act only insofar as it is living cells that secrete the zymase.[36]

Interestingly, Buchner portrayed zymase as a secreted enzyme, or 'unorganized ferment'. This suggestion was unlikely to be taken seriously – zymase catalyzed a far more substantial chemical transformation, involving the cleavage of carbon–carbon bonds and the rearrangement of oxygen atoms, than any soluble enzyme previously demonstrated. Also, alcoholic fermentation was the archetypical activity of organized ferments.

Supporters of Buchner's work concluded that the extraction of zymase was proof that complex intracellular processes were catalyzed by the same kind of soluble proteins that mediated extracellular hydrolyses. His opponents thought that the extraction process produced fragments of protoplasm that retained some of the 'vital' activity. Among the latter group was Richard Neumeister, a former student of Willi Kühne, who pointed out, 'It is more likely that the activity of the press juice is due not to one substance but to a number of different proteins that even after their removal from the living cell persist in the metabolic activities characteristic of the protoplasm.' Neumeister was correct in thinking that zymase contained more than one enzymatic activity – twelve different enzymes are required to convert glucose to ethanol. His main point, that zymase might be pieces of protoplasm, was, for Buchner, semantic. If the main characteristic of protoplasm was its state of organization, how could an extract of ground-up cells be described as protoplasm? And if fermentation was a 'vital' activity of protoplasm, why was the activity of zymase not affected by chemical treatments that killed cells?

The demonstration of cell-free fermentation in fact was the beginning of the end of the protoplasm theory. In its place arose a new theory of living processes in which enzymes played a central role – and a new science, biochemistry, which was initially based on the ability to extract and assay the activity of enzymes. It would take a couple of decades before the protoplasm theory was relegated to the dusty back shelves of science, but as early as 1901 Franz Hofmeister wrote: 'we may be almost certain that sooner or later a particular specific ferment will be discovered for every vital reaction.' This sentiment was echoed by Fischer the following year: 'Of the chemical aids in the living organism the ferments – mostly referred to nowadays as enzymes – are so pre-eminent that they may justifiably be claimed to be involved in most of the chemical transformations in the living cell.' Twenty years before zymase, Claude Bernard's telling metaphor had divided vital processes into the legislative and the executive. The enzyme theory of life supposed that enzymes were the mediators of Bernard's 'executive force'.

The discovery of zymase showed that both Liebig and Pasteur had been partly correct about the nature of fermentation. As Buchner stated in his Nobel lecture of 1907:

> The difference between the vitalistic view and the enzyme theory have been reconciled. Neither the physiologists nor the chemists can be considered the victors; nobody is ultimately the loser; for the views expressed in both directions of research have fully justified elements. The difference between enzymes and micro-organisms is clearly revealed when the latter are represented as the products of the former, which we must consider as complicated but inanimate chemical substances.[37]

Fermentation was neither a chemical decomposition, as Liebig had believed, nor a process that required living cells, as Pasteur believed. Vindicated was the intermediate position of Traube and Berthelot, which viewed fermentation as the action of intracellular enzymes. Yet perhaps most credit for the enzyme theory of life should go to Jöns Jacob Berzelius. In his *Jahresbericht* for 1836, Berzelius had written:

We have well-grounded reason to conjecture . . . that in the living plants and animals thousands of catalytic processes go on between the tissues and the fluids, and produce the amount of dissimilar chemical syntheses for whose formation from the common raw material, the plant sap or the blood, we could never see acceptable cause, which perhaps in the future we shall discover in the catalytic force of the organized tissue of which the organs of the living body consist.[38]

Chapter 5
Building Stones of Protoplasm

Nuclein

As described in Chapter 2, the carbohydrates, fats and proteins had all been recognized as distinct classes of organic compounds by around 1820. Another fifty years were to pass before Miescher identified the fourth major class of biological molecule.

Friedrich Miescher was born in Basel in 1844 into an eminent Swiss-German scientific family. His father had been one of the first students of the physiologist Johannes Müller, and his uncle, Wilhelm His, was professor of anatomy and physiology at the University of Basel from 1857 to 1872. Miescher studied medicine in Basel, taking one semester in Göttingen to learn 'practical chemistry' with Friedrich Wöhler. Following his graduation in 1868, a hearing problem discouraged him from practicing medicine, and he took his uncle's advice to become a chemist.

Miescher went to the laboratory of Felix Hoppe-Seyler at the University of Tübingen, which was located in an old castle overlooking the Neckar river. His intention was to study the chemistry of the cell nucleus, for which he needed a source of tissue rich in nuclei and poor in cytoplasm. Miescher chose lymph, as open infections were common in those days. It was a good scientific choice, if unpleasant work; as Miescher was later to say: 'he who has not learned to surmount difficulties in his youth cannot later cope with them'.

Miescher washed the lymphocytes out of the bandages with sodium sulfate, defatted them with warm alcohol and dissolved them in dilute hydrochloric acid. This precipitated the nuclei, from which he removed any accompanying cytoplasmic protein by digestion with an extract of pig's stomach, a source of the proteolytic (protein-digesting) enzyme pepsin. When the purified nuclei were dissolved in dilute alkali and then neutralized, a flocculent precipitate was obtained. This material, which Miescher termed 'nuclein', contained 14% nitrogen, 6% phosphate and 2% sulfur, in addition to carbon, hydrogen and oxygen. The nitrogen content was similar to that of proteins, and the phosphate content to that of lecithin, a recently dis-covered phospholipid. Since the nitrogen and phosphate appeared to be present in the same substance, however, Miescher concluded that: 'It is more likely that we here have a substance *sui generis*, not comparable with any other group at present known'.

In fact, the elemental composition suggests that the first nuclein prepara-tions must have contained a mixture of different components. From the harsh chemical methods used in the isolation, these must have been exten-sively degraded. Nonetheless, Miescher's 1869 study was the first known preparation of the material later to become known as deoxyribonucleic acid (DNA).

Phosphorus-containing organic compounds were almost unknown at this time, although Gerrit Mulder had reported the presence of phosphorus in proteins, particularly casein, as early as 1837 (see Chapter 2). Although lecithin had been isolated in Hoppe-Seyler's own laboratory, he was skep-tical about Miescher's results. When Miescher left Tübingen in 1869, he gave Hoppe-Seyler a manuscript on the nuclein work for publication in the journal *Medizinische-chemische Untersuchungen* (*Studies in Medical Chem-istry*), of which he was editor. However, Hoppe-Seyler kept the manuscript until Miescher's findings could be verified, so the paper did not appear until 1871, accompanied by a series of papers on nucleins from other sources. One of these papers concerned nuclein from yeast – probably the first description of the substance later known as ribonucleic acid (RNA).

After leaving Tübingen, Miescher furthered his scientific training by spending a year in the Leipzig laboratory of Carl Ludwig, previously a member of the 'organic physics' group of Emil DuBois-Reymond (see Chapter 3). In 1871, Miescher returned to Basel as *Privatdozent* in physiol-ogy. His inaugural lecture was not on the discovery that would make him famous, but on the respiratory physiology he had learned in Leipzig. One year later Wilhelm His did his nephew another career favor by moving to Leipzig himself; Miescher became professor of physiology in his place.

Realizing that a splendid source of material was available in his home town, Miescher decided to resume work on nuclein. Basel lies at the headwaters of the Rhine, to which millions of salmon return each year to spawn. The male salmon contain large quantities of sperm, each of which is about 90% nucleus. The ready availability of salmon sperm gave Miescher access to an unlimited supply of nuclein.

By 1874, Miescher had calculated an elemental composition of $C_{29}H_{49}N_9P_3O_{22}$ for salmon sperm nuclein. This material appears to have been relatively pure, but the relatively high content of phosphorus indicates that the mineral acids used during its isolation had removed some of the purine bases. The empirical formula of 1874 corresponds to a minimum molecular weight of 968, but Miescher observed that nuclein did not pass through a parchment filter, indicating that it may be a much larger molecule. His work on salmon also persuaded Miescher that the sperm head was a complex of nuclein and a basic protein he called 'protamine'.

Miescher may have been fortunate in having a ready source of experimental material in Basel, but he was less fortunate in his research facilities. His extractions were performed in a corner of the general chemistry laboratory and the subsequent analyses in the corridor of another building. The only support staff available was a quarter share of an anatomy *Diener* (assistant). To make things worse, the isolation of nuclein was a long and uncomfortable process. As Miescher described in a letter to a friend: 'When nucleic acid is to be prepared I go at five o'clock in the morning to the laboratory and work in an unheated room. No solution can stand for more than five minutes, no precipitate more than an hour before being placed under absolute alcohol. Often it goes on until late in the night. Only in this way do I finally get products of constant phosphorus proportion.'[39]

In 1885, Miescher was able to move into a new research laboratory. This was named the 'Vesalianum' in honour of the sixteenth-century anatomist, Andreas Vesalius, who had worked briefly in Basel. By this time, however, Miescher's health was broken. In 1895, at the age of fifty-one, he died of tuberculosis. Shortly before his death, his former mentor Ludwig wrote to Miescher: 'as men work on the cell in the course of the following centuries, your name will be gratefully remembered as the pioneer of this field'. A characteristically eloquent epitaph on the career of Miescher was written by Erwin Chargaff in 1978: 'what a time takes to its heart dies with the time . . . it augured well for the permanent value of Miescher's work that it found so little echo during his lifetime.'

Kossel

The work of Friedrich Miescher had shown that a major constituent of the cell nucleus was an acidic molecule containing nitrogen and phosphorus. The further characterization of this material was largely the work of another student of Hoppe-Seyler, Albrecht Kossel.

Karl Martin Leonhard Albrecht Kossel was born in 1853 in the Hanseatic port of Rostock. In 1872, he entered the medical school of the University of Strasbourg, where he was taught by Hoppe-Seyler, who had moved there earlier that year. Kossel transferred to the University of Rostock to complete his studies, graduating in 1877. He then returned to Strasbourg to work on nuclein with Hoppe-Seyler, in 1881 becoming *Privatdozent*. Two years later, Kossel moved to Berlin to become director of the chemistry division at the Institute of Physiology, becoming *ausserordentlicher* professor in 1887.

Kossel's initial studies (1879–80) were performed on nuclein from yeast, which was not associated with protein. He showed that this material contained the nitrogenous bases xanthine and hypoxanthine, which had been described earlier in the century and were known to be related to uric acid, one of the 'vegetable acids' isolated by Carl Wilhelm Scheele in the late eighteenth century (see Chapter 2). Another such substance, guanine, had earlier been isolated from sperm nuclei. In 1885–6, Kossel discovered a fourth nitrogenous compound, which he named adenine, in extracts of yeast and animal nuclein.

The structures of these compounds were determined by Emil Fischer between 1881 and 1898. In a brilliant chemical *tour de force*, Fischer showed that guanine, hypoxanthine, xanthine and adenine, as well as caffeine and uric acid, were all derivatives of a hypothetical parent molecule he called 'purine' (Figure 5.1). Using the new techniques of structural organic chemistry, of which he was the undisputed master, Fischer was able to deduce the structural formulas of all the purine compounds, and confirm their structures by retrosynthesis from products of known structure. It was for this work, together with his equally seminal studies on the carbohydrates (see Chapter 4), that Fischer was awarded the Nobel prize for chemistry in 1902.

In 1889, Richard Altmann showed that nuclein was a complex between protein and a phosphate-containing compound that he called *Nucleinsäure* (nucleic acid). This discovery made possible a proper chemical characterization of the nuclear components. In the early 1890s, Kossel analyzed the material remaining after the removal of the purine bases from animal nucleic acid with acid, and found two new nitrogenous compounds, which he named thymine and cytosine. These simpler compounds, consisting of a single ring, were of a class of molecules known as pyrimidines.

Another component of the nucleic acids was a sugar. In 1893, Kossel isolated a carbohydrate by acid hydrolysis of yeast nucleic acid and converted it to a crystalline form, as had Emil Fischer, with phenylhydrazine (see Chapter 4).

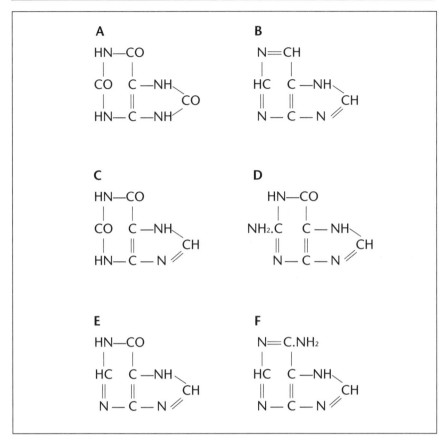

Figure 5.1: Emil Fischer's structures of the purine compounds. (A) Uric acid. (B) 'Purine'. (C) Xanthine. (D) Guanine. (E) Hypoxanthine (F) Adenine

From the observation that this sugar could be converted to furfural, Kossel concluded that it was a pentose (a five-carbon sugar). Hydrolysis of thymus nucleic acid produced a different five-carbon compound, levulinic acid ($C_5H_8O_3$), and formic acid (CH_2O_2). This suggested that the carbohydrate component of thymus nucleic acid was a hexose (six-carbon sugar).

In 1895, Kossel became professor of physiology at the University of Marburg. Here he enjoyed excellent research facilities, accommodated many foreign students, and spent what were probably the happiest years of his life. Kossel's discovery of the amino acid histidine signaled a change in his research interests; from this point onwards, he worked mainly on proteins. In 1901, having failed to obtain the position of director of the Institute of Physiology in Berlin made vacant by the death of Emil DuBois-Reymond, Kossel moved to Heidelberg to replace Willi Kühne as professor of physiology.

In a biographical sketch, Mary Ellen Jones wrote: 'Kossel enjoyed music and literature but they were allotted only a limited space in his life which was so filled with science and home that he had little energy left for pursuit of the arts.' Ernest Kennaway, who briefly worked with Kossel, recounted the story of how the Kossel family was setting off for a summer vacation when word came from the laboratory that a substance long refractory to crystallization had finally formed crystals; 'the trunks were unpacked, and the children returned to the nursery, while the new compound was analyzed'.

The Chemistry of Proteins

Two studies emanating from the Giessen laboratory of Justus Liebig suggested that proteins from different sources varied in composition. The elemental analyses of proteins performed in response to the protein-radical theory of Mulder showed that proteins from different sources differed in empirical formula (see Chapter 2). The studies of Friedrich Bopp suggested that proteins differed in their contents of individual amino acids. Thus arose from Liebig's laboratory an idea that was to be of central importance to the new science of physiological chemistry – and beyond it, to biochemistry and molecular biology – that the term 'protein' referred not to a single chemical entity, but rather to a class of compounds that had in common the presence of a particular type of subunit. This may seem a small thing, but it would be a hundred years before it was realized that the same was true of nucleic acids.

Few technical innovations in protein chemistry marked the latter half of the nineteenth century, and thus progress in elucidating the structure of proteins was slow. The first clear demonstration of differences in amino acid composition between proteins came from the analysis of plant proteins by Heinrich Ritthausen. In 1866, he isolated a novel amino acid from wheat gluten, and named it glutamic acid. Ritthausen was also the first to show that aspartic acid, previously known as a breakdown product of the amino acid asparagine, occurred in the hydrolysis products of proteins. His careful analyses of a range of plant proteins turned up large differences in the contents of aspartic and glutamic acids. Maize fibrin, for example, contained 1.4% aspartic acid and 10% glutamic acid, whereas broad bean legumin contained 3.5% aspartic acid and 1.5% glutamic acid: 'Differences like those in these figures, especially for glutamic acid (even if these still leave much to be desired in accuracy), can scarcely be seen as chance happenings; they depend on this, that the *protein bodies themselves* from which they are derived, *must differ from one another*' [emphases in original].[38]

The finding that individual proteins contained a variety of amino acids was quite common. In 1873, Heinrich Hlasiwetz and Josef Habermann showed that the hydrolysis of casein with stannous (tin) chloride in hydrochloric acid produced glutamic acid, aspartic acid, leucine and tyrosine. Also present in the products of hydrolysis was ammonia, which Hlasiwetz and Habermann realized was produced by the conversion of glutamine and asparagine to glutamic acid and aspartic acid, respectively. Application of their method to plant proteins did not result in quantitative data on amino acid composition, but Hlasiwetz and Habermann did feel able to conclude that differences exist:

> Only this we can state on the grounds of numerous already collected data that the different protein modifications yield different amounts of these products, and it appears to us now already more than a mere guess that the differences in the properties of the protein modifications are to be sought in a different proportion of the primary atom groups of which they are constituted.[38]

By 1900, thirteen amino acids had been isolated from the hydrolysis products of proteins: glycine (formerly known as glycollol), leucine, tyrosine, serine, glutamic acid, aspartic acid, phenylalanine, alanine, lysine, arginine, iodogorgoic acid, histidine and cysteine.

Colloids and Crystalloids

By the late nineteenth century, the chemical nature of proteins was clear. These substances consisted of simple subunits, the amino acids, each of which had the general structure $NH_2–CHR–COOH$. The 'side-chain' R could be an acidic group, as in the case of the glutamic and aspartic acids; a basic group, as in the case of lysine and arginine; an aromatic ring, as in the case of tyrosine and phenylalanine; or one of a variety of other chemical structures.

What of the physical nature of proteins? What sort of physical properties did these complex organic substances possess? This was an important physiological question in light of the fact that several of the known proteins, including gelatin (a major component of connective tissues and bones) and fibrin (the major component of blood clots) had important structural roles in the body. This question was addressed in 1861 by the Scottish chemist Thomas Graham.

Born in Glasgow in 1805, Graham became one of the leading chemists of his day, being elected as a Fellow of the Royal Society and subsequently winning its highest honor, the Copley medal. He held chairs in chemistry at Anderson's College in Glasgow and University College, London, before being appointed in 1855, like Isaac Newton before him, to the position of Warden and Master Worker of the Royal Mint in London.

In an 1861 paper entitled 'Liquid diffusion applied to analysis', Graham suggested that natural substances could be divided on the basis of their physical properties into two classes:

> It is proposed to designate substances of the [gelatin] class as *colloids*, and to speak of their peculiar form of aggregation as the *colloidal condition of matter*. Opposed to the colloidal is the crystalline condition. Substances affecting the latter form will be classed as *crystalloids*. The distinction is no doubt one of intimate molecular constitution [emphases in original].[40]

Colloids diffused slowly and formed amorphous solids; crystalloids diffused rapidly and formed solids with well-defined faces and angles: 'The hardness of the crystalloid, with its crystalline planes and angles, is replaced in the colloid by a degree of softness, with a more or less rounded outline. The water of crystallization is represented by the water of gelination.'

The protein gelatin (collagen) was Graham's archetype for the colloidal condition. Other proteins, however, were clearly crystalloids. Crystals of the blood pigment hemoglobin were well known by the late nineteenth century. In 1840, Friedrich Hünefeld wrote: 'I have occasionally seen in almost dried blood, placed between glass plates in a desiccator, rectangular crystalline structures, which under the microscope had sharp edges and were bright red.' By the time Wilhelm Preyer wrote his book *Die Blutkrystalle* (*Blood Crystals*) in 1871, hemoglobin crystals from over forty species had been described. A 1909 book entitled *The Crystallography of Hemoglobins*, by Edward Reichert and Amos Brown, reproduced 600 photomicrographs of hemoglobin crystals from a hundred different species. By this time, several other proteins, including egg white ovalbumin and a variety of plant globulins, had also been crystallized. Clearly at least some proteins behaved as crystalloids rather than as colloids.

The observation that proteins could exhibit either colloidal or crystalloid properties indicated to Graham that there was no absolute distinction between these two 'conditions of matter'. Rather, he believed that colloids were large groupings of smaller crystalloid units: 'It is difficult to avoid associating the inertness of colloids with their high equivalents, particularly where the high number appears to be attained by the repetition of a smaller number. The inquiry suggests itself whether the colloid molecule may not be constituted by the grouping together of a number of smaller crystalloid molecules, and whether the basis of colloidality may not really be this composite character of the molecule.'[40]

At the time these words were written, the concepts of valence and the structural formula were still very new; it is therefore not clear what Graham meant by the term 'colloid molecule'. However, the fact that Graham referred to '*the* molecule' suggests that he was not thinking of colloids as aggregates of many molecules but rather as single large entities.

It is also clear from the above quotations that Graham believed that proteins could exist as either colloids or crystalloids. However, his choice of the protein gelatin (collagen) as the archetype of the colloidal condition probably influenced later workers to think of proteins as colloids (see Chapter 7).

Indeed, there were reasons to believe that proteins were extremely large molecules. As described in Chapter 2, the elemental analyses of Gerrit Mulder had suggested that proteins had a minimum molecular mass of 9000. Although Mulder's concept of the protein radical had been discredited by Liebig, the use of elemental analyses to calculate minimum molecular masses was still valid. In 1885, Oscar Zinoffsky calculated the elemental composition of hemoglobin as $C_{712}H_{1130}N_{214}O_{245}S_2Fe$. Assuming that a single molecule must have at least one iron (Fe) atom, the minimum molecular mass of this protein was 16 710.

August Kekulé appears to have realized that the quadrivalency of carbon meant that the grouping of small crystalloid molecules into larger colloids could occur by valence bonds. In what seems to have been a throwaway line delivered during his inauguration as rector of the University of Bonn in 1877, Kekulé stated that: 'The hypothesis of chemical quantivalence further leads to the supposition that also a considerably large number of single molecules may, through polyvalent atoms, combine to *net-like*, and if we like to say so, *sponge-like masses*, in order thus to produce those *molecular masses* which resist diffusion, and which, according to Graham's proposition, are called *colloidal* ones' [emphasis in original].[41] Colloids for Kekulé were thus two- or three-dimensional lattices of small molecules.

Levene

Kossel had taken up the study of the nucleic acids in the twilight of the career of Miescher. In the 1890s, as Kossel turned his attention to proteins, the third of the great nucleic acid chemists entered the field. Phoebus Aaron Theodor Levene was born in Sagor, Russia, in 1869, his family settling soon thereafter in St Petersburg. He studied medicine at the Imperial Military Medical Academy, where he was taught physiology by Ivan Pavlov and chemistry by Alexander Borodin. Anti-Jewish pogroms led the Levene family to emigrate to the United States in 1891. 'Fedya' practiced medicine in New York until 1896, at the same time studying chemistry and working in a physiology laboratory. He started his work on nucleic acids in 1896, at the Pathological Institute of the New York State Hospitals. For the next four years, suffering intermittently from tuberculosis, Levene made several trips to Europe, during one of which he spent some time working with Kossel at Marburg. After two years working at the Saranac Laboratory for the Study of Tuberculosis, Levene spent the summer of 1902 with Fischer in Berlin. He then returned to the Pathological Institute, but three years later became an assistant at the new Rockefeller Institute for Medical Research in New York. Levene was appointed director of the division of chemistry in 1907. This division gradually expanded to include nine permanent staff members and several technicians, occupying two floors of the North Building of the Institute. Using the considerable resources available to him (and often more – he consistently overspent his budget), Levene was able to mount a large research program in organic chemistry, of which his studies on nucleic acids were only a small part.

Levene ran his laboratory like a benevolent autocrat. Apart from his own, the only authority he recognized was that of Emil Fischer. Although Levene loved to work with his hands, which as a result were stained yellow from picric acid, the mundane tasks were delegated to a series of *Diener* or 'lab-boys', whose duties included driving the 'Herr Doktor' to the opera. Almost every visiting scientist could be addressed in his native tongue, as Levene, in addition to speaking Russian and English, was fluent in French and German and spoke some Italian and Spanish.

When, in the final years of the nineteenth century, Levene began work on the nucleic acids, the state of knowledge was still primitive. Elemental analysis, such as Miescher's 1896 formula of $C_{40}H_{54}N_{14}O_{17} \cdot 2P_2O_5$ for salmon sperm nucleic acid, suggested a complex structure. It was well established that these substances consisted of four different types of compounds: purine bases, pyrimidine bases, sugars and phosphoric acid. Four different purine bases had been identified – guanine, adenine, xanthine and hypoxanthine – but it would soon become clear the latter two were not true constituents of nucleic acids. Three different pyrimidines had been found – thymine, cytosine and uracil – but their structures were not yet established.

Table 5.1: Compositions of the nucleic acids, *circa* 1900

Constituents	*Complex nucleic acids*		*Simple nucleic acids*	
	Thymus nucleic acid	*Yeast nucleic acid*	*Inosinic acid*	*Guanylic acid*
Phosphorus	Phosphoric acid	Phosphoric acid	Phosphoric acid	Phosphoric acid
Carbohydrate	Hexose?	Pentose?	Pentose?	Pentose?
Purine bases	Xanthine Hypoxanthine Guanine Adenine	Xanthine Hypoxanthine Guanine Adenine	Hypoxanthine	Guanine
Pyrimidine bases	Thymine Cytosine	Uracil Cytosine		

However, not all nucleic acids contained all of the above components (Table 5.1). The so-called simple nucleic acids, inosinic acid and guanylic acid, contained only one base: in the former, hypoxanthine; in the latter, guanine. The two 'complex' nucleic acids differed in their sugar and base contents: the nucleic acid of animal tissues and fish sperm ('thymus nucleic acid') appeared to contain a hexose sugar, and the bases adenine, guanine, cytosine and thymine; the nucleic acid of yeast and plants appeared to contain a pentose sugar and the bases adenine, guanine, cytosine and uracil. However, it was not known whether the nucleic acids were single molecules. On acid hydrolysis, the purine bases were cleaved much more easily than the pyrimidines, suggesting that they were less tightly associated with the sugar and phosphoric acid. It was also suspected that the phosphoric acid and the bases might not be part of the same molecule, but rather occur as a salt, as Miescher's 1896 formula implied.

The lack of certainty about the exact nature and relative amounts of the constituents of nucleic acids did not deter speculation about the structures of these molecules. In 1902, Thomas Osborne and Isaac Harris reported that a nucleic acid isolated from wheat, which they called *Triticonucleinsäure* (triticonucleic acid), appeared to be identical to that of yeast. Specifically, triticonucleic acid appeared to contain three moles of pentose, four of phosphoric acid, one of guanine, one of adenine, two of uracil, and another unidentified basic component. Their proposed structure was one of the first attempts to determine how the bases, sugars and phosphoric acids were put together into a nucleic acid molecule (Figure 5.2).

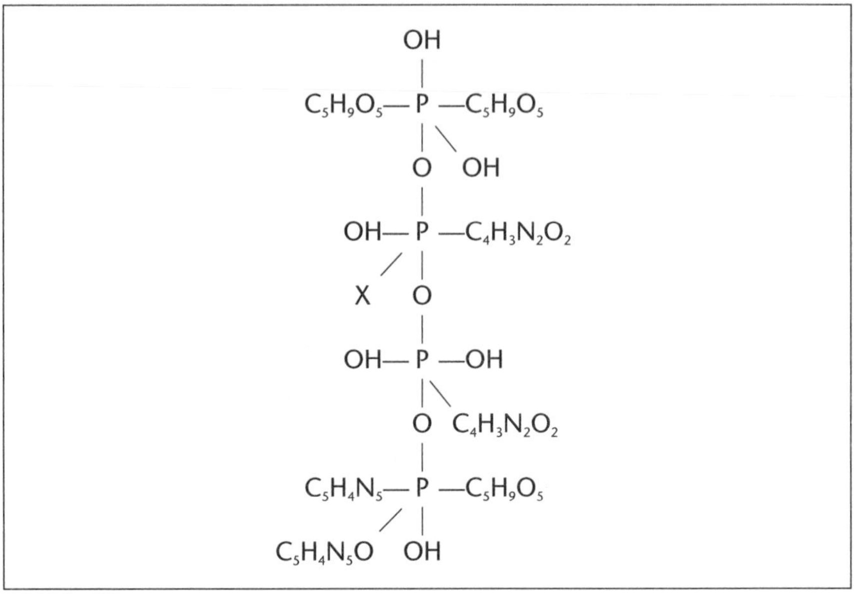

$$
\begin{array}{c}
OH \\
| \\
C_5H_9O_5\text{—} P \text{—}C_5H_9O_5 \\
| \ \backslash \\
O \quad OH \\
| \\
OH\text{—} P \text{—}C_4H_3N_2O_2 \\
/ \ | \\
X \quad O \\
| \\
OH\text{—} P \text{—}OH \\
| \backslash \\
O \quad C_4H_3N_2O_2 \\
| \\
C_5H_4N_5\text{—} P \text{—}C_5H_9O_5 \\
/ \ | \\
C_5H_4N_5O \quad OH
\end{array}
$$

Figure 5.2: Thomas Osborne's and Isaac Harris's 1902 structure of 'triticonucleic acid'. To the polyphosphate backbone are attached three molecules of pentose ($C_5H_9O_5$), one molecule of guanine ($C_5H_4N_5O$), one molecule of adenine ($C_5H_4N_5$), two molecules of uracil ($C_4H_3N_2O_2$) and 'an unidentified basic product' (X). Adapted from Osborne, T. B. and Harris, I. F. (1902) *Zeitschrift für physiologische Chemie* **36**, 120, with permission

Levene's initial investigations in nucleic acid chemistry were prompted by his belief that these substances played an important role in the development of tissues and regeneration. In 1899, he wrote that the 'nucleoproteids' were the key to understanding 'how the organism repairs its waste, and how we can successfully aid the organism in the most important of its tasks, when this power of restitution is for some reason or other diminished.'

If Levene was correct in this belief, then different tissues should contain different nucleic acids. He therefore began to prepare and analyze nucleic acid from a variety of different tissues and different organisms. In 1901, there appeared the first of a series of twelve papers entitled *Darstellung und Analyse einiger Nucleinsäuren* (Preparation and analysis of some nucleic acids). He developed a method for dissociating nucleic acid from proteins in tissue extracts and precipitating it in (relatively) pure form. By 1903, he had obtained elemental analyses of nucleic acids from mammalian pancreas, spleen, liver, brain and testicles, cod spermatozoa, yeast and bacteria (tubercle bacillus). Only the last had a significantly different elemental composition.[a]

---

a   The DNA of *Mycobacterium tuberculosis* is very rich in guanine and cytosine; this fact played an important role in Erwin Chargaff's demonstration that the bases do not occur in equimolar amounts (see Chapter 12).

As Levene realized, elemental analysis was a crude means of comparing complex molecules such as nucleic acids. As the bases had similar elemental compositions (adenine and guanine differed by only a single oxygen atom, for example), elemental analysis could not even determine which bases were present, far less their relative amounts. To determine whether different species and tissues contained distinct nucleic acid species required methods by which the bases could be independently measured. Levene therefore changed his approach; from now on, he would concentrate on separating and quantifying the bases.

Accurate measurement of nucleic acid bases stretched the limits of turn-of-the-century organic chemistry. The idea was to hydrolyze the nucleic acid, separate the bases by differential crystallization and weigh the individual crystalline precipitates. Every step presented formidable technical difficulties: the purines and pyrimidines required different hydrolysis conditions; uracil was difficult to crystallize; adenine and guanine could be oxidized to hypoxanthine and xanthine; cytosine and thymine could be converted to uracil. In an attempt to obtain satisfactory yields and purities, and to minimize artifacts, Levene was forced to tinker incessantly with his hydrolysis and precipitation conditions.

Between 1903 and 1906, Levene reported partial or (more rarely) complete base compositions for nucleic acids from a number of sources: mammalian pancreas, liver, mammary gland and kidney, fish sperm and yeast. These varied widely; whether the variations were due to tissue-specific differences or to experimental artifact was the big question.

Levene was looking for *differences* in base composition. Other workers, however, were interested in the relative amounts of the bases as a means of determining the structure of the nucleic acid molecule. In 1906, Hermann Steudel's studies on herring sperm and thymus nucleic acids led him to the conclusion that: 'The four nitrogen-containing components of nucleic acid, and only these four, appear in molecular proportion in the acid'. The following year, Walter Jones, a former student of Kossel at Marburg, used digestion with 'ferments' from pig spleen to liberate the purine bases of thymus nucleic acid. He concluded: 'The quantities of guanin [*sic*] and hypoxanthin [*sic*] (equivalent to adenin [*sic*]) formed by the action of the ferments are as nearly proportional to the molecular weights of the two bases as could reasonably be expected with the use of the methods at our disposal.'

Steudel's paper seems to have prompted Levene to re-examine his own assumptions about base compositions. Yet another change in experimental procedure produced the finding that the two purine bases of thymus nucleic acid were present in 'equimolecular' amounts. Soon thereafter, he concluded that all four bases were equally represented in both the thymus and yeast nucleic acids molecules.

By this time, Levene had abandoned his search for tissue-specific differences in nucleic acid content. In 1907, he had concluded that the proteins, not the nucleic acids, of the nucleus were responsible for its functions in heredity and development. However, he did not lose interest in the nucleic acids; instead, he concentrated on determining their structures.

The outstanding question in nucleic acid chemistry was the nature of their carbohydrate components. As noted above, the sugar of animal nucleic acid was thought to be a hexose, while that of yeast nucleic acid was thought to be a pentose. Eight different pentoses were known to exist: the D and L stereoisomers of arabinose, lyxose, ribose and xylose. Most of these had been proposed, at one time or another, to occur in yeast nucleic acid.

Levene's key insight was that it would be easier to isolate the mysterious sugar from simple nucleic acids than from complex ones. Guanylic acid and inosinic acid both contained only a single base, and in both cases it was one of the easily removed purines. Levene's initial attempts at isolating and characterizing the sugar from guanylic acid were inconclusive, but he had better luck with inosinic acid – the pentose obtained was D-ribose.

Even better, Levene's studies on inosinic acid showed that, depending upon the hydrolysis conditions used, one obtained either hypoxanthine–pentose or pentose–phosphate. This demonstrated that the order of attachment of the components of inosinic acid was hypoxanthine–pentose–phosphate. Levene coined the term *Mononucleotiden* (mononucleotides) for this novel type of compound. Further studies revealed that substance of the type base–sugar–phosphate could be produced by mild hydrolysis of both the thymus and yeast nucleic acids. This suggested that the complex nucleic acids were *Polynucleotiden* (polynucleotides).

Levene's first complete structure of a nucleic acid, the 1909 structure of yeast nucleic acid, is shown in Figure 5.3. The four mononucleotide units were joined via their phosphate groups to give a polyphosphate 'backbone'. The order of the bases within the molecule was presumably arbitrary. Levene also conceived of thymus nucleic acid as a polyphosphate.

The period 1908–9 had been a remarkably productive time for Levene – he

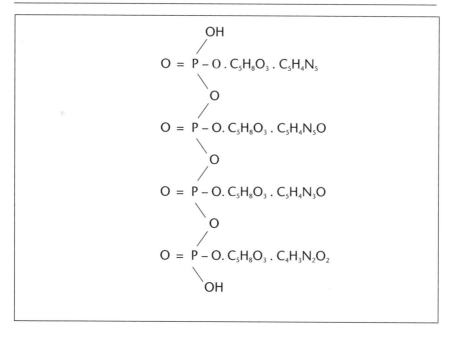

Figure 5.3: Phoebus Levene's 1909 structure of yeast nucleic acid. Like Osborne's and Harris's earlier structure of triticonucleic acid (Figure 5.2), it has a polyphosphate backbone. However, the bases are attached to the backbone via the pentose ($C_5H_8O_3$) groups, explaining why yeast nucleic acid can be dissociated into nucleotide (phosphate–sugar–base) units. Adapted from Levene, P. A. (1909) *Biochemische Zeitschrift* **17**, 122

had demonstrated that the simple nucleic acids were nucleotides and that the complex nucleic acids were polynucleotides, and identified the sugar of yeast nucleic acid, inosinic acid and guanylic acid as D-ribose. His decision to concentrate on determining the structures of nucleic acids rather than seeking tissue-specific differences had been spectacularly vindicated.

However, there was an anomaly in the chemistry of the nucleic acids that would confound Levene until the end of his career. Yeast nucleic acid was far more susceptible to hydrolysis under alkaline conditions. If both these compounds had the same polyphosphate backbone, surely they should be hydrolyzed to mononucleotides under similar conditions?

These early attempts to quantify the base content of the nucleic acids were mainly directed at determining whether xanthine and hypoxanthine were true constituents of the nucleic acids or artifacts of preparation, and whether all the bases were present in one molecule. However, an incidental conclusion of these studies – that adenine, guanine, cytosine and thymine were present in thymus nucleic acid in equal amounts – gave rise to the infamous tetranucleotide hypothesis of nucleic acid structure. In the following decades, it would be variously believed that nucleic acids contained equimolar amounts of the four bases, that they were small molecules, and that they contained a repeating sequence of bases. The general implication of the tetranucleotide hypothesis, in all its manifestations – that nucleic acids were 'simple' molecules, at least compared with proteins – would delay for half a century the recognition of DNA as the material of heredity.

The Peptide Bond

By the late 1890s, Emil Fischer had completed the work that would bring him the 1902 Nobel prize and had proposed the 'lock and key' mechanism of enzyme action. In 1893, he identified the chemistry of proteins as one of the key areas of research for the future: 'So long as we know little more about the chemical bearers of life, the proteins, than their percentage composition, so long as we cannot explain the most fundamental process of organic nature, the conversion of carbon dioxide into sugar by green plants, we must admit that physiological chemistry still remains in baby shoes'.[42]

In 1899, Fischer embarked upon an analysis of protein structure. He utilized a technical breakthrough made by Theodor Curtius, Eduard Buchner's doctoral supervisor from Munich. In the early 1880s, Curtius showed that amino acids could, by reaction with alcohol, be converted to esters. Fischer's studies on these amino acid esters showed that they differed in boiling point according to molecular weight, and therefore could be separated by fractional distillation. These compounds could be as useful in the analysis of amino acids as the osazone derivatives had been in his earlier studies on carbohydrates (see Chapter 4).

Fischer also found that the amino acid esters could easily be converted to cyclic compounds called diketopiperazines. These were formed when the amino (NH_2) group of one amino acid reacted with the carboxylate (COOH) group of another, and vice versa (Figure 5.4). These diketopiperazines were experimental artifacts, but Fischer realized that similar bonds between amino and carboxylate groups – amide bonds – could occur between amino acids in proteins. Hydrolysis of the protein silk fibroin under relatively mild conditions produced not only single amino acids but also a larger molecule that Fischer identified as glycyl-alanine; that is, glycine and alanine joined by an amide bond. Fischer thus conceived of proteins as chains of amino acids connected head to tail by amide bonds between their amino and carboxylate groups.

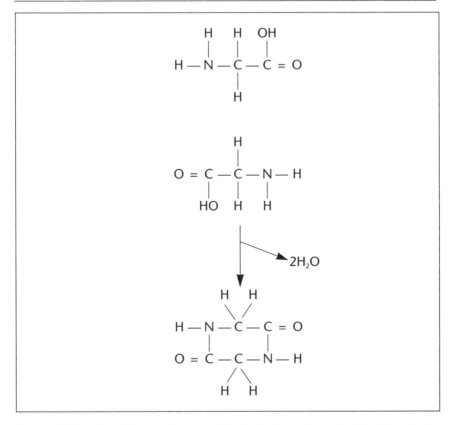

Figure 5.4: Reaction of two molecules of glycine to form the cyclic dipeptide diketopiperazine

Fischer presented his theory of protein structure at a meeting of the *Gesellschaft für deutscher Naturforscher und Ärzte* (German Society for Natural Science and Medicine) in Karlsbad in September, 1902: 'The speaker on this account makes the proposal, following the known division of the carbohydrates as disaccharides, trisaccharides, etc., to name the bodies of the type of glycylglycine dipeptides, and to designate anhydride-like combinations of a greater number of amino acids as tripeptides, tetrapeptides.'

For once Fischer had to take a back seat. The polypeptide structure of proteins had been presented earlier that day in a plenary lecture by Franz Hofmeister. Whereas Fischer was a recent convert to protein research, Hofmeister had been studying these compounds for his entire career, and was in 1902 probably the leading authority on the physiological chemistry of proteins.

Franz Hofmeister was born in Prague in 1850, and studied medicine at the German University of Prague. While still a student, he started work on peptones, the partially hydrolyzed proteins that were found in the intestinal tract. In 1881, Hofmeister became chief of a new institute of experimental pharmacology in Prague; four years later, he was professor of pharmacology. His studies on peptones led to the discovery of the Hofmeister series, representing the relative abilities of inorganic salts to precipitate proteins. One practical application of these studies was Hofmeister's 1889 demonstration that ovalbumin could be crystallized by precipitating the protein from solution with ammonium sulfate.

In 1896, Hofmeister replaced Felix Hoppe-Seyler as professor of physiological chemistry at the University of Strasbourg. Although Hofmeister was not the first choice of the Strasbourg authorities, this was nonetheless a prestigious appointment. Alsace had been annexed by Germany in the Franco-Prussian War of 1870–1, and the Imperial government was determined to use the University of Strasbourg as a showcase of German scientific leadership. Here Hofmeister commanded a talented group of investigators studying diverse aspects of protein function. One of his Strasbourg colleagues described Hofmeister as 'a man of wide-ranging spirit full of new ideas that he generously handed out to his co-workers'.

In his lecture at the Karlsbad conference, Hofmeister discussed various ways by which amino acids could be joined together in proteins. Attachment by C–C bonds was dismissed on the grounds that no enzyme able to cleave such bonds was known, yet trypsin could digest proteins. Ether and ester linkages were ruled out on the grounds that these required hydroxyl (OH) groups, and only one amino acid, serine, had such a group. On the other hand, the possibility of amino acids combining by amide bonds was consistent with the chemical evidence that proteins lacked free amino groups. Even more telling for Hofmeister was the observation that proteins reacted with the biuret reagent, which required the presence of two –CO–NH– groups.

Hofmeister concluded: 'On the basis of these given facts one may therefore consider the proteins as for the most part *arising by condensation of α-amino acids, whereby the linkage through the group –CO–NH–CH=* has to be regarded as the one regularly recurring' [emphasis in original]. An example of such a linkage between leucine and glutamic acid was shown in the published version of Hofmeister's lecture (Figure 5.5).

The historical significance of the Karlsbad conference is that two of the most influential figures in physiological chemistry publicly endorsed the same

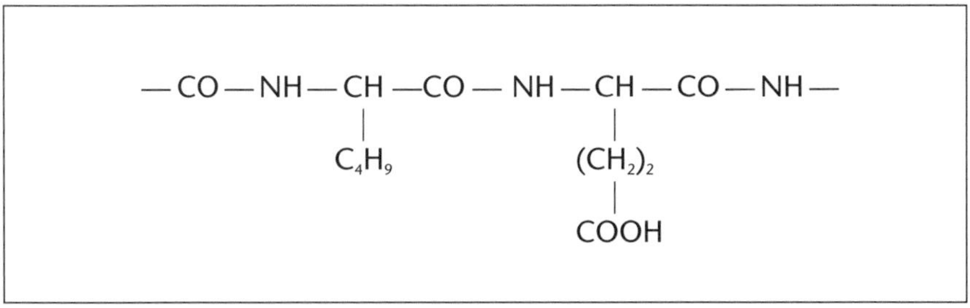

$$— CO — NH — CH — CO — NH — CH — CO — NH —$$

$$C_4H_9 \qquad\qquad (CH_2)_2$$

$$COOH$$

Figure 5.5: Franz Hofmeister's 1902 structure of proteins. Individual amino acids have condensed together by formation of an amide (–CO–NH–) bond between the carboxylate (COOH) group of one and the amino (NH₂) group of the next, as shown for glycine in Figure 5.4. The two amino acids shown in full are leucine and glutamic acid

structure for proteins. The polypeptide structure appeared to be based on strong experimental evidence, including Fischer's retrosynthesis of dipeptides. However, the polypeptide theory would almost immediately come under attack by proponents of the biological importance of colloids (see Chapter 7).

It is important to realize that polypeptides represented a novel type of chemical structure. The existence of polymers had been suggested by Berzelius to explain the existence of different compounds with identical empirical formulas (see Chapter 2). By 1900, it was believed that carbohydrates such as cellulose consisted of sugar molecules linked into a chain structure. Cellulose, therefore, was a repeating, or periodic, polymer. Polypeptides, however, were composed of different monomeric units, and therefore represented a new type of non-repeating polymer. Synthesizing a protein, whether the feat was attempted by an organic chemist or a living cell, was a more complicated process than simply attaching identical amino acids together in head-to-tail fashion. Yet there was no point in proposing a non-repeating polymeric structure for proteins if the biosynthesis of such molecules were not possible. Protein chemists would soon have to wrestle with the problem of explaining how cell chemistry could produce such complicated molecules as polypeptides.

The *Baustein* Hypothesis

During the latter part of the nineteenth century, much evidence had accumulated that the complex organic molecules of animals and plants were composed of smaller molecules linked together. As early as 1861, Graham had proposed that colloids, such as the proteins, were 'groupings' of the smaller crystalloids. The discovery of the tetravalent carbon atom showed that such groupings could occur by normal valence bonds, producing Kekulé's hypothetical 'net' and 'sponge' polymers. The polypeptide theory of Hofmeister and Fischer associated these two ideas by suggesting that proteins were amide-bonded polymers of amino acids.

In October 1911, Albrecht Kossel visited the USA and delivered lectures to the Harvey Society of New York and the Herter Foundation of Baltimore. He used these lectures to describe a view of biological molecules that had arisen from his forty years of work on the nucleic acids and proteins. The central concept was that the complex molecules of the cell were composed of *Bausteine*, or building blocks – simple organic molecules that could be assembled in different combinations:

> So long as one considers the mass of living substance as a whole, an analysis of its activity can scarcely be undertaken. Such an analysis is only possible through the isolation of certain units capable of chemical investigation and to whose activity the individual functions of living substance may be referred. I wish to speak of these units, which I shall refer to as the 'Bausteine' or building-stones of protoplasm.
>
> The word 'Baustein' indicates that these units may be united to form larger structures and that their union takes place according to a determined plan or architectural idea. Through the union of these Bausteine larger aggregates are formed which we call either proteins, fats, nucleic acids, phosphatides, or polysaccharides, as the case may be.[43]

Polysaccharides such as cellulose and glycogen were composed of only one type of *Baustein*, the monosaccharide glucose, but Kossel emphasized that this was not normally the case. In the case of the proteins, the *Bausteine* were the amino acids, of which Kossel listed seventeen different types: 'glycollol', alanine, valine, leucine, proline, phenylalanine, 'glutaminic acid', aspartic acid, cystine, serine, tyrosine, 'oxyproline', histidine, arginine, lysine, 'tryptophane' and 'diaminotrioxydodecoic acid'. Fats were composed of one molecule of glycerine (glycerol) and three of fatty acid; nucleic acids of purine and pyrimidine bases, carbohydrate and phosphate. The complex cellular molecules were therefore more akin to a mosaic, in which different colored stones could form different patterns, than to a house composed of identical bricks.

The physiological conclusions Kossel drew from the *Baustein* hypothesis were not always confirmed by later studies. For example, there appears to be no particular significance to the fact that both the nucleic acid bases and the predominant amino acids of nuclear proteins, arginine and histidine, contain alternating carbon and nitrogen atoms. Likewise, Kossel's observation that proteins serving mechanical functions have few basic amino acids (lysine, arginine and histidine) did not lead to any insights into the relationship between structure and function.

Rather, the historical significance of the *Baustein* hypothesis lies in its explicit recognition that the same simple molecules could be assembled into a number of polymeric structures with different chemical or biological properties. Studies in Kossel's laboratory had convinced him that proteins in the diet of animals were broken down to single amino acids by 'ferments', and then reassembled by other ferments into different proteins: 'I should like to compare this rearrangement which the proteins undergo in the animal or vegetable organism to the making up of a railroad train'.

As the passage quoted earlier makes clear, Kossel included the nucleic acids among the 'protoplasmic' molecules composed of *Bausteine*. He had proposed in 1893 that the nucleic acids were polyphosphates; that is, the backbone of the molecule was composed of linked phosphate groups, and the bases and sugars were somehow attached to this backbone. Kossel appears to have believed that the nucleic acids were polymers. In his 1911 Harvey lecture, he stated: 'the nucleic acids . . . have acid properties, and long since I have put forward the view that they are to be considered as *poly*-metaphosphoric acids containing, therefore, a *chain* of phosphorus and oxygen atoms to which the previously mentioned pyrimidine and purine derivatives as well as carbohydrates are attached' [author's emphases]. Likewise, in his Nobel lecture of 1910, Kossel stated that 'nucleic acid appears as a complex of at least twelve building blocks [i.e., a tetranucleotide], but in the living cell the structure is probably larger, because some observations suggest that in the organs several of these complexes are combined with each other'.

However, there is no suggestion in Kossel's Harvey and Herter lectures that rearrangements of *Bausteine* could occur in the nucleic acids. Unlike the proteins, which clearly differed in amino acid composition, the nucleic acids appeared to occur in only two different forms, plant and animal. Kossel's idea of what we might now call heteropolymers was therefore applied only to proteins. Even in this limited domain, however, the *Baustein* hypothesis was to influence not only the biochemists, who quickly adopted the concept of building blocks, but also geneticists such as Hermann Muller and J. B. S. Haldane. The *Baustein* hypothesis therefore represents a major step towards a concept that was to become of central importance in the new science of biochemistry – the aperiodic polymer.

Chapter 6
The Chemical and Geometrical Phenomena of Heredity

'Studies on Plant Hybrids'

Reproduction had long been recognized as one of the characteristically 'vital' properties. The ability of living things to produce other living things had been introduced into the vitalist–materialist debate of the nineteenth century by both sides: Rudolf Virchow noted that 'an animal can spring only from an animal . . . ' (see Chapter 3), Lionel Beale that 'everything that lives . . . grows of itself, builds itself up and multiplies . . . ' (see Chapter 4). Biological reproduction was not mere copying, though – a child was never identical to its father or mother, and might more closely resemble a grandparent. In order to incorporate reproduction into a materialistic biology, a theoretical explanation of the phenomena of heredity was required. Although such a theory was provided by Mendel as early as 1866, it was 'lost' until the end of the century. The science of heredity therefore entered biology around the same time as the enzyme theory, and it was these two conceptual innovations that were to differentiate twentieth-century biochemistry from nineteenth-century physiological chemistry.

Johann Mendel was born in 1822 in Heinzendorf in Austrian Silesia. He attended the Gymnasium in Troppau, graduating in 1840. Subsequent studies at the Philosophical Institute in Olmütz were ended by a combination of financial and health problems. Mendel then entered the Augustinian monastery of St Thomas in Brünn (now Brno) as a novice and was given the name 'Gregor'.

According to Robert Olby, Brünn was then 'at the centre of Moravian culture and scientific study', and the monastery was by no means isolated from the latter. It had been directed by imperial decree in 1802 'to devote itself to the teaching of science in schools of higher learning'; many of its monks taught at the Brünn Gymnasium or Philosophical Institute, and several went on to take university positions. Scientific research, particularly in plant breeding, was also a common activity at the monastery.

The initial indications were that Mendel was not cut out for monastic life. He confessed in his autobiography that he felt no particular calling to the church, and his pastoral duties made him ill. In 1849, he was appointed as an occasional teacher to the Gymnasium at Znaim. In order to achieve a full-time appointment, however, he had to pass a teaching examination. He failed this examination rather miserably, and was sent to the University of Vienna in order to learn more science. The four terms he spent in Vienna were mainly occupied studying physics and botany, and his performance was once again unspectacular. However, he became acquainted with the extant controversy over whether species were fixed or variable, and also learned the statistics that he would later apply to his own breeding experiments.

Back at Brünn in 1854, Mendel became an occasional teacher of physics and natural history at the *Oberrealschule* (Royal High School). He sat the teaching examination again in 1856, but was unable to complete it because of a recurrence of his chronic nervous condition. His life to date showed no promise of scientific genius, but Mendel had already begun the great program of plant breeding experiments that would bring him immortal fame as the founder of genetics.

Mendel wanted to study the fixity of species by studying the inheritance of characteristics by the hybrid progeny of variant plant strains. For several reasons, he used the genus *Pisum* (pea): species of this genus produce fertile hybrids, self-pollinate, are easily cultivated and have a relatively short generation time. He studied thirty-four varieties of three species (*P. sativum*, *P. quadratum* and *P. umbellatum*) over eight years. These varieties differed in seven characteristics: seed shape, endosperm color, seed-coat color, pod shape, unripe pod color, flower position and stem length. Mendel performed cross-pollinations between varieties differing in each of these characteristics. In addition, he performed two series of crosses involving varieties differing in two characteristics and one series involving varieties differing in three.

When Mendel crossed two different varieties, he found that the hybrids (F_1 generation) all resembled one of the parental strains. For example, crosses between a strain with a round seed and one with a wrinkled seed always produced hybrids with round seeds. Likewise, hybrids of strains with yellow and green endosperm always had yellow endosperm. Mendel therefore described the traits round and yellow as 'dominant' and the traits wrinkled and green as 'recessive'.

When Mendel performed self-crosses on the hybrids – fertilizing each hybrid plant with its own pollen – he obtained a generation of plants (F_2) in which the recessive characters reappeared. The F_2 generation was a mixture of the dominant and recessive traits in a proportion of approximately 3:1. In the case of seed shape, for example, Mendel obtained 5474 F_2 plants with round seeds (previously established as the dominant trait) and 1850 with wrinkled seeds (the recessive trait), for a ratio of 2.96:1. In the case of endosperm color, he obtained 6022 yellow and 2001 green, for a ratio of 3.01:1. For all seven characteristics studied, the average ratio of dominant to recessive was 2.98:1.

The next step was to perform self-crosses of the F_2 plants to produce an F_3 generation. All the F_2 recessives bred true – that is, produced F_3 progeny with the same appearance as the parent – and therefore contained no trace of the dominant characteristics. Of the F_2 dominants, however, only one-third bred true. The remaining two-thirds of the F_2 plants exhibiting the dominant trait gave rise to a mixture of dominant and recessive F_3 plants in the same 3:1 proportion seen in the entire F_2 generation.

The fact that Mendel had studied seven different characteristics in a large number of plants gave him confidence that the 3:1 ratio of dominant/recessive observed in the F_2 generation was genuine. Mendel's great stroke of genius, however, was in realizing the meaning of these ratios – that each plant has two sets of hereditary determinants, one inherited from each parent.

Mendel represented a pure-breeding strain of plant exhibiting a dominant characteristic as AA, and one exhibiting a recessive characteristic as aa.[a] Crossing these two strains would produce a hybrid (F_1) generation of plants that contained one hereditary determinant from each parent, Aa. The offspring of a cross between two Aa hybrids could be either AA, Aa, aA or aa, depending on whether they inherited the A or a trait from either or both parents. The first three types (AA, Aa and aA) would all exhibit the dominant characteristic, and the fourth (aa) would exhibit the recessive characteristic, thus explaining the 3:1 ratio in the F_2 generation. The offspring AA and aa were pure strains and would therefore breed true. However, Aa and aA were hybrids, and would produce F_3 progeny exhibiting a 3:1 ratio of the dominant and recessive characteristics.

---

a   Mendel actually represented the pure-breeding dominant and recessive strains as A and a, respectively. However, the modern symbols AA and aa are more logical and entirely consistent with Mendel's interpretation.

Crossing *Pisum* varieties that differed in two characters provided a rigorous test of Mendel's interpretation. The hybrids from such crosses, as expected, all exhibited the characteristics that Mendel had previously found to be dominant. Crossing plants having round, yellow seeds with plants having wrinkled, green seeds gave a hybrid generation in which all the plants had round, yellow seeds. When these hybrids were then self-crossed, the F_2 progeny fell into four groups: 315 round/yellow, 101 wrinkled/yellow, 108 round/green, and 32 wrinkled/green – a ratio of 9:3:3:1. Taking these plants to an F_3 generation revealed which of the F_2 plants were hybrids. The results obtained were exactly those expected if the hereditary units governing seed shape and endosperm color distributed randomly among progeny plants. Assuming that the parental strains were *AABB* (round/yellow) and *aabb* (wrinkled/green), sixteen permutations of the hereditary determinants could result:

$$AABB + AABb + AAbB + AAbb + AaBB + AaBb + AabB + Aabb + aABB + aABb + aAbB + aAbb + aaBB + aaBb + aabB + aabb$$

Of these sixteen permutations, nine would have round, yellow seeds (*AABB*, *AABb*, *AAbB*, *AaBB*, *AaBb*, *AabB*, *aABB*, *aABb* and *aAbB*), three would have round, green seeds (*AAbb*, *Aabb* and *aAbb*), three would have wrinkled, yellow seeds (*aaBB*, *aaBb* and *aabB*) and one would have wrinkled, green seeds (*aabb*).

As well as confirming the existence of two sets of hereditary determinants, these bifactorial crosses showed that the hereditary determinants segregated independently, so the progeny had all possible combinations of the two characteristics. That is, although *A* and *B* had entered the cross together, as had *a* and *b*, in the hybrids *A* was as likely to end up with *b* as it was with *B*.

Mendel reproduced his main findings with other plant species before presenting his data on *Pisum* to the Natural History Society of Brünn in two lectures in 1865. This work was published the following year in the *Proceedings of the Natural History Society of Brünn* as *Versuche über Pflanzen-Hybriden* (Studies on plant hybrids) – a rather bland title for one of the most revolutionary works in the history of biology. This publication was taken by 115 libraries and scientific institutions in Europe, including the Royal Society of London. Mendel sent out a number of reprints of his article, perhaps as many as forty, and initiated a long correspondence with Carl von Naegeli, professor of botany at the University of Munich. Nonetheless, Mendel's mechanism of inheritance disappeared without a trace. His work did not even rate a mention in Naegeli's 1884 book *Mechanisch-physiologische*

Theorie der Abstammungslehre (*Mechanical-Physiological Theory of Inheritance*). It would be thirty-five years before the scientific world would suddenly realize the significance of Mendel's work and the science of genetics would begin.

By that time, Mendel was dead. He became prelate of the Brünn monastery in 1868 and published a study on hybridization of *Hieracium* (hawkweed) the following year. In 1871, his administrative responsibilities forced him to give up his experiments on plant breeding. Gregor Mendel died in 1884.

The Physiological Function of Nuclein

Mendel's paper may have lain unappreciated from 1866 to 1900, but others were interested in the mechanism of heredity. Experimentally, most progress came from embryologists studying fertilization and cytologists studying cell division. Unprecedented insights into these processes came from the use of light microscopy, which came close to its technical limits in the 1870s. In terms of optical hardware, the only significant advance since the 1830s was the introduction of the apochromatic objective, containing boron/phosphorus glass, which eliminated the residual chromatic aberration. More significant was the introduction of new fixation, sectioning and staining techniques, particularly the aniline dyes.

Using these methods, intracellular structures could now be resolved. Of particular interest was the nucleus, which, as noted in Chapter 3, had been implicated as the site of new cell formation by Mathias Schleiden. A role for this organelle in heredity was now suggested by the studies of Oskar Hertwig, who in 1876 showed that fertilization in the sea urchin and starfish involved penetration of the egg by the sperm followed by fusion of the two nuclei. Because sperm and egg had equal-sized nuclei but very different amounts of cytoplasm, it was inferred that it must be the nucleus that specified hereditary characteristics: 'The union of the egg nucleus with the sperm nucleus is necessary to produce a nucleus endowed with the living forces adequate effectively to stimulate the later developmental processes in the yolk, and to control them in many respects.'

In 1879, Walther Flemming, professor of anatomy at Kiel, gave the name 'chromatin' to the nuclear component that was stained by basophilic dyes. Three years later, he suggested that chromatin was the cytological equivalent of nucleic acid. In 1883, Eduoard van Beneden reported the existence of rod-like structures in the nucleus of the worm *Ascaris megalocephala* – these bodies were later named chromosomes. Van Beneden's studies showed that sperm and egg had half the number of chromosomes of somatic cells, and that the fusion of the germ cells restored the normal complement of chromosomes: 'each daughter nucleus [of the fertilized egg] receives half of its chromatic substance from the spermatozoon, the other from the egg'.

As Edmund Beecher Wilson, professor of zoology at Columbia University, wrote in the first (1896) edition of his book *The Cell in Development and Inheritance*, the fact that the nuclei of sperm and egg contributed equally to the embryo, whereas the contribution of cytoplasm was unequal: 'points unmistakably to the conclusion that the most essential material handed on by the mother-cell to its progeny is the chromatin, and that this substance therefore has a special significance in inheritance'. Based on his studies of the changes in chromosome structure during cell division (mitosis), Wilhelm Roux also proposed that these bodies carried the hereditary material. August Weismann agreed, writing in 1892: 'For the present it is sufficient to show that the complex mechanism for cell-division exists practically for the sole purpose of dividing the chromatin, and that thus the latter is without doubt the most important portion of the nucleus. Since, therefore, the hereditary substance is contained within the nucleus, *the chromatin must be the heredi-tary substance*' [emphasis in original].[44]

'Morphologico-phantastical Speculations'

The discovery of chromosomes and the realization of the importance of the nucleus in inheritance prompted a wave of speculation about the mechanism of heredity. As many of the proponents were embryologists, such theories were concerned not just with the transmission of the 'germ-plasm' between generations, but also how it controlled the development of the embryo. There were thus two general phenomena to be explained: how all the specialized tissues of the body could arise from a single fertilized cell; and, conversely, how specialized tissues of the adult organism could produce cells capable of generating a whole new organism.

The first serious attempt to address these issues came in Charles Darwin's 1868 book *The Variation of Animals and Plants under Domestication*. Darwin's 'provisional hypothesis of pangenesis' proposed that the formation of germ cells involved recruitment of tiny bodies that he called 'gemmules' from all the specialized tissues of the body. In the development of the embryo, the gemmules gave rise to the tissues from which they were originally derived. However, the pangenesis hypothesis appeared inconsistent with the well-known features of plant grafting, in which the root stock did not contribute to the properties of the fruit. Nor was it supported by experiments conducted by Darwin's cousin, Francis Galton, on the transfusion of blood between different strains of rabbit. If Darwin were correct, the gemmules present in the blood should induce 'mongrelism' in the offspring of the transfused rabbits; however, this was not observed.

An alternative to the hypothesis of pangenesis was proposed by August Weismann of the University of Freiburg. Instead of the germ cells being composed of contributions from all the specialized somatic tissues, he suggested that they represented a pool of cells that remained undifferentiated from generation to generation. Weismann wrote in 1889: 'the germ-cells are not derived at all, as far as their essential and characteristic substance is concerned, from the body of the individual, but they are derived directly from the parent germ-cell'. This theory obviated the necessity of the material determinants of hereditary properties having to find their way from all parts of the body to the germ cells, and was consistent with the results of plant grafting and blood transfusion.

Weismann attributed the different characteristics of germ cells and somatic cells to their containing different materials, which he called 'germ-plasm' and 'somatoplasm', respectively. These can be viewed as being analogous to Beale's 'germinal matter' and 'formed matter' and as the mediators of Bernard's 'legislative force' and 'executive force' (see Chapter 3). However, a later (1892) version of Weismann's theory proposed that the formation of specialized adult tissues involved a progressive loss of the hereditary material. This blurred the distinction between the generative and generated by postulating that the same substance could constitute the hereditary material of the nucleus and the structural material of the cytoplasm.

According to Weismann's 1892 theory, the hereditary material, now called 'idioplasm', consisted of a set of 'idants', which corresponded to individual chromosomes. The idants could be further subdivided into 'ids', which corresponded to microscopically visible chromosomal 'granules'. Ids, in turn, consisted of thousands or hundreds of thousands of units called 'determinants'. The terminal differentiation of a cell left it with a single determinant, which then disintegrated into the fundamental units of heredity, 'biophors'. Complexes of about a thousand molecules, the biophors were small enough to migrate through the nuclear membrane to the cytoplasm, where they formed the specialized structures typical of that cell, and also performed essential life functions such as assimilation and growth.

Oskar Hertwig, among others, thought that Weismann's unequal cell division was inconsistent with the cytological evidence. The fact that all somatic cells contained the same number of chromosomes seemed to rule out the progressive loss of hereditary elements as cells became more specialized. Hertwig therefore suggested in 1894 that mitosis divided the hereditary material equally, and that cells differentiated because of differences in their environments. His hereditary determinants were called 'idioblasts', which were 'the smallest particles of material into which the hereditary mass or idioplasm can be divided . . . Metaphorically they can be compared to the letters of the alphabet.'

Another weakness of Weismann's theory of heredity was that it appeared inconsistent with regeneration and healing phenomena. How could an amphibian regenerate a severed limb, for example, or a fracture callus produce new bone, if somatic cells contain only single determinants? To deal with this problem, and also with Hertwig's objection, Hugo Marie de Vries, professor of botany at the University of Amsterdam, proposed in 1899 a 'theory of intracellular pangenesis'.

De Vries, like Darwin and Weismann, assumed the existence of material particles that corresponded to visible heritable characters. However, he rejected Darwin's mechanism of transport of these particles from differentiated tissues to germ cells, and Weismann's loss of the hereditary particles during cell differentiation. All cells of an individual, according to de Vries, contained the same set of hereditary particles, or pangens, but in different tissues different sets of pangens were active: 'in higher organisms, not all the pangens of any given cell probably ever become active, but in every cell one or more of the groups of pangens dominates and impresses its character on the cell'. Pangens constituted not only the chromosomes of the nucleus, but also the protoplasm of the cytoplasm. In any specialized cell type, however, only some pangens migrated from nucleus to cytoplasm, and thereby became active.

The theory of intracellular pangenesis represents the zenith of the 'factor theories' of inheritance and development that flourished in the period prior to the rediscovery of Mendel's work. Like its predecessors, however, it was based almost entirely on theoretical considerations. To some biologists, this was a fatal weakness. Edmund Wilson's 1900 criticism of the 'metaphysical character' of Weismann's unequal division of the germ-plasm could be directed at any of the factor theories: 'The fundamental hypothesis is thus of a purely *a priori* character; and every fact opposed to it has been met by subsidiary hypotheses, which, like their principal, relate to matters beyond the realm of observation.' Similarly, Wilhelm Johannsen decried the 'morphologico-phantastical speculations of the Weismann school'.

All the factor theories of inheritance assumed the existence of subcellular particles that determined the specialization of tissues. According to Darwin and Weismann, only a liver cell contained liver-specifying particles; according to Hertwig and de Vries, all cells contained liver-specifying particles, but only in liver were these active. The physical nature of these tissue-specifying particles was not a central concern of the factor theorists, but certain inferences about this could logically be drawn from their mechanisms. Weismann suggested that biophors were composed of approximately a thousand molecules, and Darwin compared the size of his gemmules to the infectious particles now known as viruses. De Vries's pangens were much larger than the 'molecules of chemistry' but still 'invisibly small'. The hereditary particles were presumably related to one another chemically, but this basic chemical structure had to be capable of existing in as many variants as there were tissues of the body and species on the earth. The need to hypothesize the existence of many different, but related, substances therefore linked heredity and development with physiological chemistry.

As noted above, the evidence available at the end of the nineteenth century suggested that the material of heredity was likely to be the substance that cytologists called chromatin and physiological chemists called nuclein. Friedrich Miescher wrote in 1874: 'If one wants to assume that a single substance . . . is the specific cause of fertilization, then one should undoubtedly first of all think of nuclein'. At least at this time, however, Miescher did not believe that a single substance *was* the specific cause: 'the riddle of fertilization is not hidden in a particular substance'. Another great nucleic acid chemist, Albrecht Kossel, imputed an important role to nuclein in development when in 1882 he rejected the notion that this substance was simply an intracellular storage material: 'morphological observations make it likely that we have to look in another direction for the physiological function of nuclein, namely in relation to the formation of new tissue.' Oskar Hertwig wrote in 1884: 'I believe that I have at least made it highly probable that nuclein is the substance that is responsible not only for fertilization but also for the transmission of hereditary characteristics.'

Nuclein, as Richard Altmann showed in 1889, consisted of both nucleic acid and protein. In his 1896 book, Edmund Wilson specifically implicated the former in heredity and development: 'When this is correlated to the fact that the sperm nucleus, which brings the paternal lineage, likewise consists of nearly pure nucleic acid, the possibility is opened that this substance may be in a chemical sense not only the formative centre of the nucleus but also a primary factor in the constructive processes of the cytoplasm.'[45]

In this statement, Wilson attributed to nucleic acid two distinct functions – the transmission of the paternal lineage and the synthesis of cellular molecules. The former was an aspect of Claude Bernard's legislative force, the latter an aspect of his executive force. Other biologists also tended not to distinguish between the legislative and executive functions of the cell. Weismann and de Vries agreed that the hereditary particles determined tissue specificity by physically migrating from nucleus to cytoplasm and forming the differentiated structures characteristic of that cell type. An exception was Carl von Naegeli, who in 1884 proposed that cells consisted of two types of matter: idioplasm, which contained the heritable properties of the organism, and trophoplasm, which played a nutritive role. The idioplasm consisted of 'micellae', dynamic molecular complexes which catalyzed the reactions of development. Naegeli's distinction between idioplasm and trophoplasm bore some similarity to Lionel Beale's distinction between 'germinal matter' and 'formed matter' (see Chapter 3). However, Naegeli had, in an important refinement, relocated this concept to the subcellular level.

Although Weismann's biophors served both legislative and executive functions, a distinction between the legislative and executive underlay his theory of the continuity of the germ-plasm. Unlike Darwin, who proposed that somatic tissues and germ cells could be interchanged by the migration of gemmules, Weismann thought of the germ line as a continuous cell lineage separate from that of specialized somatic tissues: germ cells can give rise to somatic tissues, but somatic tissues cannot give rise to germ cells. The theory of the continuity of the germ-plasm is therefore a conceptual forerunner of Francis Crick's central dogma of molecular biology (see Chapter 13).

The Rediscoverers

Having escaped the notice of plant breeders for thirty-five years, Mendel's laws of inheritance were rediscovered in 1900 by three separate investigators. Each of these individuals claimed to have discovered the same character ratios as Mendel in the progeny of plant hybrids, and to have come to the same general interpretation as Mendel, before encountering his 1866 paper. This remarkable coincidence has understandably attracted much skepticism from historians of science.

Hugo de Vries was the first of the 'rediscoverers'. He was studying the transmission of inherited characters in a variety of plant species in an attempt to prove that species are generated by discontinuous change (a process for which he coined the term 'mutation') rather than by gradual variation. In studies apparently dating back to 1892, de Vries found 3:1 ratios of characters in crosses involving *Oenothera* (evening primrose), *Zea* (maize), *Lychnis* (campion) and *Papaver* (poppy). He also found the 9:3:3:1 ratio (actually 9.3:3.2:3.2:1) in two-character crosses of *Trifolium* (clover).

Carl Correns was working, as Mendel had, with *Pisum*, as well as with maize. He had noted in 1899 'very complicated but interesting relationships' between characters in his crosses. In April 1900, Correns received a copy of a French language version of de Vries's paper, which did not include the reference to Mendel's work. By the next day, he had written up his own studies and sent off the manuscript for publication. In crosses between varieties of peas with yellow and green embryos, Correns showed a 3:1 ratio of green/yellow in the F_2 generation, and that only one-third of the yellow F_2 plants subsequently bred true. He also referred to two-character crosses in maize in which he obtained ratios of 308:104:96:37 (8.3:2.8:2.6:1).

The third of the purported rediscoverers, Erich Tschermak von Seysenegg, presented breeding data on peas and wallflowers as part of his *Habilitationsschrift* (doctoral thesis) at the Institute for Agriculture of the University of Vienna in 1900. He obtained 3:1 and 9:3:3:1 ratios and quoted Mendel's 1866 paper. However, Tschermak attributed the prevalence of certain characters to greater 'hereditary potency' rather than to the presence of dominant and recessive determinants, and failed to realize that some of the dominant F_2 plants were in fact hybrids.

Whether de Vries, Correns and Tschermak rediscovered Mendel's laws in the laboratory or Mendel's paper in the library, there is no reason to doubt that an independent corroboration of Mendelian ratios was only a matter of time. As Iris and Laurence Sandler observed, studies on corn performed around 1890 had reported not only 3:1 ratios of heritable characteristics but also independent segregation of two such characteristics.

The thirty-five-year gap between the propounding of Mendel's laws of inheritance and their discovery/rediscovery represents one of the most curious episodes in the history of biology, and the most prominent example of what Gunther Stent described as 'prematurity of discovery'. How could Mendel have realized the significance of certain ratios when it eluded others for so long? One answer is that the thoroughness of Mendel's experimentation put the professional scientists to shame. As noted by the Sandlers, Mendel was unusual in insisting upon the purity of parental strains, deriving many generations from hybrid progeny, and keeping extensive records of individual plants. According to Robert Olby, his study on *Pisum* was 'a superbly planned series of experiments, much better than those of the rediscoverers'. Mendel's application of statistical techniques to biological data was also unusual in 1865. By 1900, however, thanks to the efforts of 'biometricians' such as Karl Pearson, this had become common, and Mendel's algebraic analysis of segregation did not appear as alien.

However, probably the most important difference between 1866 and 1900 was the progress in the cytology of the nucleus, particularly the discovery of chromosomes and their behavior during germ-cell formation (meiosis) and fertilization. Mendel posited the existence of pairs of hereditary determinants, one inherited from each parent, and randomly distributed among their progeny. By 1900, it was known that the nucleus contains pairs of chromosomes, half of which were contributed by each parent.

The person who first realized that Mendel's hereditary determinants must be carried on the chromosomes was Walter Stanborough Sutton, a graduate student in the laboratory of Edmund Wilson at Columbia. If the chromosomes carried the hereditary units, Sutton reasoned, they must be distributed randomly in the germ cells. That is, in order to account for Mendel's second law – independent assortment of different heritable characteristics – the process of meiosis must produce germ cells containing all possible combinations of chromosomes inherited from both parents, rather than all the 'maternal' and all the 'paternal' chromosomes segregating together into the germ cells. By carefully following individual chromosomes throughout meiosis in the grasshopper *Brachystola magna*, Sutton was able to show that such random segregation of chromosomes did occur. In his 1903 paper, he wrote:

> We have seen reason, in the foregoing considerations, to believe that there is a definite relation between chromosomes and allelomorphs or unit characters but we have not before inquired whether an entire chromosome or only a part of one is to be regarded as the basis of a single allelomorph. The answer must unquestionably be in favour of the latter possibility, for otherwise the number of distinct characters possessed by an individual could not exceed the number of chromosomes in the germ-products; which is undoubtedly contrary to fact. We must, therefore, assume that some chromosomes at least are related to a number of different allelomorphs. If then, the chromosomes permanently retain their individuality, it follows that all the allelomorphs represented by any one chromosome must be inherited together.[46]

The logic was remorseless. Hereditary determinants are carried on chromosomes; there are many more such determinants than there are chromosomes;

each chromosome must therefore carry many hereditary determinants; characteristics specified by determinants carried on the same chromosome should be inherited together. In 1903, however, instances of such co-inheritance were not known. Why had not Mendel, de Vries, Correns, Tschermak and all the other plant breeders observed this phenomenon? Mendel, in all probability, had selected his *Pisum* varieties and characteristics to give independent segregation. The 'rediscoverers', being ignorant of Mendel's work, were working under no such constraints. The rediscovery would have much more credibility if these workers had found instances of exceptions to Mendel's second law as well as instances of agreement with it.

Mendel's Bulldog

The rediscovery of Mendel's paper sparked an intense debate on the validity of Mendel's interpretation and the extent of its applicability. The most important proponent of Mendelism, at least in the English-speaking world, was the Cambridge zoologist William Bateson. In 1901, he published the first English translation of Mendel's 1866 paper, which he claimed was 'worthy to rank with those that laid the foundation of the Atomic laws of Chemistry'. The following year, Bateson published *Mendel's Principles of Heredity – a Defence*, written in response to criticisms of those principles and including translations of the papers on *Pisum* and *Hieracium*. His 1909 book *Mendel's Principles of Heredity* was a much more developed treatment of the same theme.

But Bateson was more than just 'Mendel's bulldog' – he made major contributions to the new science of heredity on both the theoretical and experimental sides, including most of its nomenclature. In 1902, he wrote:

> This purity of the germ cells, and their inability to transmit both of the antagonistic characters, is the central fact proved by Mendel's work. We thus reach the conception of unit-characters existing in antagonistic pairs. Such characters we propose to call *allelomorphs*, and the zygote formed by the union of a pair of opposite allelomorphic gametes we shall call a *heterozygote*. Similarly, the zygote formed by the union of gametes having similar allelomorphs, may be spoken of as a *homozygote*.[47]

The term 'allelomorph' was subsequently shortened to 'allele'. Three years later, Bateson suggested the term 'genetics' for the study of heredity.

Experimentally, Bateson's most important contribution also came in 1905, when he and his co-workers discovered that certain pairs of characters in

Lathyrus odoratus (sweet pea) did not segregate independently. In these studies, blue flowers (*B*) were found to be dominant over red (*b*), and long pollen (*L*) dominant over round (*l*). Hybrids between blue/long and red/round plants should, according to Mendel, give equal numbers of the progeny types *BL*, *Bl*, *bL* and *bl*. What was found, however, was a seven-fold excess of the parental types *BL* and *bl* over the 'hybrid' types *Bl* and *bL*. Bateson described this phenomenon as 'gametic coupling' and thought that it was due to the preferential formation of germ cells containing certain combinations of factors.

Other pairs of traits exhibited the opposite phenomenon. Instead of being inherited together more frequently than Mendel's law of independent assortment would suggest, these traits were *never* inherited together. This Bateson referred to as 'repulsion'.

The correct interpretation of gametic coupling and repulsion came from the studies on *Drosophila* of Thomas Hunt Morgan. From this work, it was clear that gametic coupling was an example of the phenomenon predicted by Sutton in 1903: 'all the allelomorphs represented by any one chromosome must be inherited together'. Genes exhibiting repulsion are on the sex chromosomes (see below).

Inborn Errors of Metabolism

To Bateson also goes the credit for the realization that Mendel's laws of inheritance applied in humans. In his 1902 book, Bateson described studies showing that the condition of alkaptonuria was 'an individual variation from normal metabolism with a familial distribution', and noted that this was suggestive of a Mendelian mechanism:

> Now there may be other accounts possible, but we note that the mating of first cousins gives exactly the conditions most likely to enable a rare and usually recessive character to show itself. If the bearer of such a gamete mate with individuals not bearing it, the character would hardly ever be seen; but first cousins will frequently be bearers of *similar* gametes, which may in such unions meet each other, and thus lead to the manifestation of the peculiar recessive character in the zygote.[47]

The studies to which Bateson referred were performed by Archibald Edward Garrod. Born in London in 1857, Garrod was the son of Sir Alfred Garrod, a distinguished physician who discovered that uric acid deposits were the cause of gout. The younger Garrod was educated at Oxford, graduating in

1880. Apart from a short time in Vienna, he spent the entire period between 1880 and World War I at St Bartholomew's Hospital in London.

One of Garrod's colleagues described him as follows: 'An able, practical physician when the need arose, patients, as he said himself did not really interest him, and the complex problems presented by an individual who is ill did not really appeal to him.' What did appeal to Garrod were the causes of disease. In the 1890s he became interested in cases of pigmented urine, particularly alkaptonuria. This was shown in 1899 to result from the excretion of homogentisic acid, a breakdown product of the amino acids tyrosine and phenylalanine, which oxidizes and blackens upon exposure to the air. At this time almost all diseases were thought to be due to infections, and alkaptonuria was conventionally attributed to an intestinal infection that disrupted the metabolism of tyrosine. Garrod, however, noticed that this condition appeared to run in families. His hypothesis that alkaptonuria was a congenital disease proved easy to test, as the mother of one of his patients, a four-year-old boy, was pregnant again. When the baby was born, Garrod had its urine examined, diaper by diaper, until, after 57 hours of life, it proved to be alkaptonuric. The parents were first cousins.

In a paper published in 1902, Garrod noted that alkaptonuria preponderantly occurred in males (29 of 40 reported cases) and that three of the four British families known to carry this trait involved first-cousin marriages. Further investigation revealed that approximately 60% of all known cases of alkaptonuria involved such marriages. He concluded: 'There are good reasons for thinking that alkaptonuria is not the manifestation of a disease but is rather of the nature of an alternative course of metabolism, harmless and usually congenital and lifelong.' He also identified albinism (absence of melanin pigments) and cystinuria (excretion of cystine in the urine) as other possible examples of such 'metabolic sports'.

By 1908, when he delivered the Croonian lectures to the Royal College of Physicians, Garrod had identified pentosuria (urinary excretion of the five-carbon sugar arabinose) as a fourth example of what he now called 'inborn errors of metabolism'. In these lectures, he made two main points. First, these conditions were inherited in Mendelian fashion. This could not be shown directly in humans, of course, but albinism was a Mendelian recessive in animals. Second, these conditions occurred because specific steps in metabolic processes were missing, with the result that 'intermediate products are excreted incompletely burnt'. In a series of reactions, the failure of one particular step would result in the accumulation of the product of the preceding step: 'If any one step in the process fail [sic] the intermediate product in being at the point of arrest will escape further change, just as when the film of a biograph is brought to a standstill the moving figures are left foot in air'.

This latter insight associated the two great biological movements of the turn of the century: Mendelism and the enzyme theory of life. The discovery of the inborn errors of metabolism could logically be taken to mean that genes acted through the production of enzymes. However, Garrod was as far ahead of his time as Mendel had been – it would be another forty years before the relationship between genes and enzymes was made explicit.

There appear to be several reasons why the relationship between genes and enzymes took so long to establish. First, the concept of genes as autocatalytic molecules (see below) blurred the distinction between genes and gene products. Second, Garrod described the effect of inborn errors as involving metabolic steps, not enzyme-catalyzed reactions. Third, as Garrod himself pointed out in 1902, heritable genetic defects that produced 'no obvious peculiarities' would be as hard to find as 'a needle in a haystack'. Indeed, it was only when George Beadle and Edward Tatum found a means of detecting such mutants that the relationship between genes and enzymes was established (see Chapter 9).

Nonetheless, the science of heredity had started with a bang. Within only three years of the rediscovery, it had been shown that the Mendelian factors were carried on the chromosomes, and that human metabolic abnormalities were inherited in Mendelian fashion. These events marked the end of the age of factor theories, of abstract entities such as idioplasm, micellae and bioblasts; and the beginning of the age of genetics, when heredity and development were thought of in terms of concrete genes that resided on chromosomes and controlled metabolic reactions.

Genotype and Phenotype

Garrod's theory of inborn errors of metabolism not only provided a link between heredity and biochemistry, but also between Bernard's legislative and executive forces. The Mendelian inheritance of the inborn errors of metabolism demonstrated that the defect resided in the legislative apparatus of the organism; the accumulation of specific metabolites demonstrated that this defect was manifested through the executive apparatus.

The distinction between legislative and executive was incorporated into genetics by Wilhelm Ludvik Johannsen, professor of plant physiology at the

University of Copenhagen. In his 1909 book *Elemente der exacten Erblichkeit-slehre* (*Elements of Heredity*), Johannsen proposed the terms 'genotype' and 'phenotype' as the genetic counterparts of the legislative and executive forces. He also introduced the term 'gene' (a contraction of 'pangene') for Mendel's 'unit-characters'. In a 1911 paper, Johannsen defined genotype as: 'the sum total of all the "genes" in a gamete or in a zygote'. Phenotypes were described as: 'All "types" of organisms, distinguishable by direct inspection or only by finer methods of measuring or description'. The genotype was thus a given combination of genetic determinants; the phenotype was the combination of physiological characteristics that resulted from a particular genotype.

Johannsen's genotype and phenotype formalized something that had been part of human experience since antiquity – that the hereditary potential of an organism was different from its tangible characteristics. This point was clearly made by Galton in his 1872 paper 'On blood-relationship':

> From the well-known circumstance that an individual may transmit to his descendants ancestral qualities which he does not himself possess, we are assured that they could not have been altogether destroyed in him, but must have maintained their existence in a latent form. Therefore each individual may properly be conceived as consisting of two parts, one of which is latent and only known to us by its effects on his posterity, while the other is patent, and constitutes the person manifest to our senses.[48]

Galton's distinction between latent and patent was also inherent in the work of Mendel. As noted above, only one-third of Mendel's F_2 dominant plants bred true, the remaining F_2 dominants producing a 3:1 mixture of dominant and recessive progeny. The two groups of F_2 dominants were identical in physical appearance – phenotype – but different in hereditary potential – genotype. As noted by Frederick Churchill, Johannsen had found similar differences between appearance and hereditary potential in his own studies on *Phaseolus vulgaris* (princess bean).

In proposing his distinction between genotype and phenotype, Johannsen may also have been influenced by the nineteenth-century factor theorists: Weismann had proposed in 1883 that organisms consisted of 'germ-plasm' and 'somatoplasm', only the former of which had the ability to replicate the species. Gustav Jäger had made a similar distinction between 'phylogentic' and 'ontogentic' and Naegeli between 'idioplasm' and 'trophoplasm'. Lionel Beale's 'germinal matter' and 'formed matter' (see Chapter 3) may well have been seen as histological analogs of these conceptual entities.

Finally, the concepts of genotype and phenotype may have been derived in part from Claude Bernard's legislative and executive forces. According to Öjvind Winge, Johannsen avidly read Bernard's *Phenomena of Life* during the period 1890–4.

By 1909, therefore, the idea was firmly established in biology that inheritance represented one thing – the genotype – while the production of the specialized somatic tissues represented another – the phenotype. At least in the minds of Bateson and Garrod, the phenotype was produced by the action of enzymes. Kossel's contemporaneous proposal that complex biological molecules were 'mosaics' of simple *Bausteine* (see Chapter 5) meant that both the major themes in the molecular analysis of life – the aperiodic polymer and the distinction between legislative and executive – were available to biologists as early as 1911. However, it would take another fifty years before these concepts could be sorted out from the crowd of competing ideas and speculations and used to construct a coherent molecular description of living systems.

The Protein Theory of the Gene

As described above, the concept of a genetic role for nucleic acid was popular in the late nineteenth century. By the first decade of the twentieth century, however, this idea was falling out of favor. The reasons for this change provide important insights into the way heredity and development were viewed at the time.

In the first two editions (1896 and 1900) of his highly influential textbook of cell biology, Edmund Wilson had been supportive of the idea that chromatin, specifically nucleic acid, was the material basis of inheritance. This was based on the fact that sperm, which contributed the paternal lineage to the embryo, contained very little cytoplasm. Further support for a genetic role of chromatin came from the 1903 observation of Wilson's student, Sutton, that chromosomes, like the Mendelian factors, segregated independently. Strictly speaking, however, these observations only suggested that the genes resided on chromosomes,[b] which were known to be composed of both proteins and nucleic acid. By the time the third edition of his book, now retitled *The Cell in Development and Heredity*, appeared in 1925, Wilson believed that it was the protein of chromosomes, not the nucleic acid, that formed the genetic material.

---

b   Johannsen did not even accept that genes were present in the nucleus, writing in 1911: 'The question
    of *chromosomes* as the presumed "bearers of hereditary qualities" seems to be an idle one. I am not
    able to see any reason for localizing "the factors of heredity" (i.e. the genotypical constitution) in
    the nuclei' [emphasis in original].

Two reasons were given for this change of mind. The first was that cytological staining techniques showed that the nucleic acid component of chromosomes seemed to disappear at certain stages of the cell cycle, while the protein component was always present. The disappearance of the chromosomes also influenced Eduard Strasburger, who wrote in 1910: 'The chromatin cannot itself be the hereditary substance, as it afterwards leaves the chromosomes, and the amount of it is subject to considerable variation in the nucleus, according to its stage of development.'

The second reason why it seemed to Wilson that chromatin could not function as the material of heredity was that it appeared to be chemically and microscopically undifferentiated. Apart from the well-established difference between plants and animals, nucleic acids from different cell types were 'remarkably uniform'. This chemical homogeneity was in stark contrast to the 'inexhaustible variety' of the proteins. Bateson agreed, writing in 1916: 'The supposition that particles of chromatin, indistinguishable from each other and indeed homogeneous under any known test, can by their material nature confer all the properties of life surpasses the range of even the most convinced materialism.'

Even those chemists who specialized in the study of the nucleic acids were convinced that proteins were more likely to be responsible for species- and tissue-specific differences. In his Harvey lecture of 1911, Albrecht Kossel wrote: 'The number of *Bausteine* which may take part in the formation of the proteins is about as large as the number of letters in the alphabet. When we consider that through the combination of letters an infinitely large number of thoughts may be expressed, we can understand how vast a number of the properties of the organism may be recorded in the small space which is occupied by the protein molecules'.[43] These properties included heredity, as Kossel made clear in his Herter lecture of the same year: 'We may readily understand how peculiarity of species may find expression in the chemical nature of the proteins constituting living matter, and how they may be transmitted through the material contained in the generative cells.' Thomas Osborne, who had discovered triticonucleic acid (see Chapter 5), wrote in 1911 that 'the morphological differences between species find their counterpart in the protein constituents of their tissues'. Phoebus Levene wrote in 1917 that the nucleic acids 'are indispensable for life, but carry no individuality, no specificity, and it may be just to accept the conclusion of the biologist that they do not determine species specificity, nor are they carriers of the Mendelian characters'. Levene may have believed that this conclusion derived from biology, but Albert Mathews, who had trained with Kossel, in 1921 cited chemical evidence for his view that nucleic acid was likely to be a 'skeletal substance' of chromosomes. Mathews was prepared to attribute the variety of species to lipids and carbohydrates as well as proteins – but not to nucleic acids.

The material of heredity, if heredity did indeed have a material basis, need not differ between tissues; the developmental theories of Hertwig and de Vries both postulated that all cells contained an identical set of hereditary particles. However, the material of heredity must, by definition, differ between species. If a single substance were responsible for the transmission of hereditary characteristics, this substance must be capable of existing in at least as many forms as there were species of organism, and it must also be present in different forms in different organisms. In the early twentieth century, proteins appeared to fulfill both these criteria much more fully than did nucleic acid.

Proteins were known to consist of a number of different amino acids – more than twenty had been identified – attached head to tail into a polypeptide. Even if proteins had a maximum size of about forty amino acids, as Emil Fischer believed, the number of possible structures was 20^{40}. As Garrod wrote in 1914: 'The number of distinct fractions (amino acids) is by no means large, but every change of grouping, however slight, brings into existence a new protein.' Proteins were also known to possess species-specific differences. By the turn of the century, the hemoglobins of different vertebrate species could be distinguished on the basis of their crystal shapes. Species-specific differences in amino acid composition were known for several proteins, and it had also been shown that the proteins of different species differed in serological activity.

Nucleic acids were known to consist of sugar, phosphate and four bases. The bases appeared to be present in equimolar amounts (see Chapter 5). Nucleic acids did not appear to differ between species; only two types were known, thymus nucleic acid, which was also found in other animal tissues, and plant nucleic acid, which was also found in yeast. Other types of nucleic acids, such as 'paranucleic acid' (a phosphoprotein) had been discredited. If the same nucleic acid was present in yeast and wheat, how could this be the material of heredity?

Proteins had been discovered much earlier than nucleic acids, but the same techniques were available to analyze both classes of molecules in the early twentieth century. Why then was it recognized that proteins were of 'inexhaustible variety' while nucleic acids were thought to be 'remarkably uniform'? There are two main reasons for this. One is that the amino acids that make up proteins are chemically diverse – only two, for example, contain sulfur. Differences between proteins from different sources were therefore apparent on the basis of a simple elemental analysis. This had allowed Liebig's laboratory to disprove the 'protein radical' theory of Mulder as early as the 1840s (see Chapter 2). The nucleic acid bases, in contrast, are chemically very similar. The second reason why the capacity of nucleic acids for structural diversity was underestimated was that, until these molecules were recognized to be macromolecules, the potential for arranging four different nucleic acid bases into many different sequences was not realized.

Autocatalysis

There was another powerful argument in favor of genes being proteins – the important biological role of enzymes. The early twentieth century was, as Robert Olby put it, 'the golden age of enzymology', and it appeared that the mechanism of hereditary transmission would, like other vital phenomena, eventually be explained by the action of enzymes. And enzymes appeared to be proteins.

In the 1909 edition of his book *Mendel's Principles of Heredity*, Bateson proposed that: 'a dominant character is the condition due to the *presence* of a definite factor, while the corresponding recessive owes its condition to the *absence* of the same factor.' Mendel's recessive green peas, for example, lacked the factor that made yellow peas yellow. From the work of Garrod, it was clear what such factors must be: 'What the physical nature of the [hereditary] units may be we cannot yet tell, but the consequences of their presence is in so many instances comparable with the effects produced by ferments that with some confidence we suspect that the operations of some units are in an essential way carried out by the formation of definite substances acting as ferments'.[49] Thus, wrinkled seeds were caused by the absence of an enzyme that turned sugar into starch, and albinos resulted from the absence of an enzyme that produced the skin pigment.

In this work, Bateson also wrote: 'It is scarcely necessary to emphasize the fact that the ferment itself must not be declared to be the factor or thing transmitted, but rather, the power to produce that ferment, or ferment-like body.' Here Bateson was distinguishing between the legislative function of producing an enzyme and the executive function of catalyzing a metabolic reaction. However, it certainly *was* necessary to emphasize this – the distinction between genes and enzymes was not only ignored but explicitly denied.

The main exponent of the correlation between genes and enzymes was a virtual outsider in the biological establishment, Leonard Thompson Troland. After studying biochemistry at the Massachussetts Institute of Technology, Troland obtained a doctorate in psychology at Harvard in 1915. He was thus still a graduate student in 1914 when he published a paper entitled 'The chemical origin and regulation of life'. In this, Troland identified 'five fundamental mysteries of vital behaviour': '(1) the origin of living matter, (2) the origin of organic variations, (3) the ground of heredity, (4) the mechanism of individual development, and (5) the basis of physiological regulation in the mature organism.' One class of substances was thought to be responsible for all these phenomena:

> We have said that enzymes and catalysts in general have the power to assist in the production of specific chemical substances. Now there is no reason why the same enzyme should not aid in the formation of more than one substance and also why one of these substances should not be identical with the enzyme itself. A process of the last mentioned variety, in which the presence of a catalyzer in a chemical mixture favours the production of the catalyzer itself is known as *autocatalysis* [emphasis in original].[50]

Autocatalysis could explain the origin of life because a substance with this property, no matter how it arose, would automatically be capable of producing many copies of itself. It would explain organic variations – mutations – because any change in the enzyme would be reflected in its copies. To explain the mechanism of heredity, one merely had to suppose that the genetic enzymes were heterocatalytic – capable of catalyzing metabolic reactions – as well as autocatalytic.

This left only the 'mechanism of individual development'. Here Troland proposed that 'differentiation may be an expression of processes of chemical suppression or interference, rather than the actual loss or exhaustion of the original determinants of the germ-cell.' This was one of Troland's most profound ideas, presaging the mechanism of gene inhibition worked out by François Jacob and Jacques Monod in the 1950s (see Chapter 13). A system of cellular differentiation based on suppression of gene expression rather than the physical loss of genes obviated the problems that earlier theories, such as that of Weismann, had in explaining regeneration of differentiated tissues. If genes could be switched on and off, regenerative phenomena such as wound healing and fracture repair would only require that certain genes suppressed in the differentiated tissue be turned on again to recapitulate the formation of new skin or bone. However, Troland does not appear to have fully realized this, proposing instead that these regenerative phenomena were mediated by undifferentiated cells with which adult tissues were seeded.

Troland was not the first person to propose that genes were autocatalytic enzymes. Johannsen, for example, had written in 1911: 'we have to point out the fact that "living matter" – or, with a more precise definition, those substances or structures the reactions of which we call "manifestations of life" – is *inter alia* characterized by the property of *autocatalysis*' [emphasis in original]. However, Troland's 1914 paper, and another three years later, represent the fullest exposition of this idea. Despite his lowly status in the social organization of science, Troland was playing to a receptive audience. Studies on sexual dimorphism in the moth *Lymantria dispar* led the eminent geneticist Richard Goldschmidt to conclude in 1917: 'From these facts only one conclusion can at present be drawn: that the sex-factors are enzymes (or bodies with the properties of enzymes) which accelerate a reaction according to their concentration.'

Remarkably, Troland was willing to accommodate within his theory of autocatalysis the then-unfashionable view that the genetic material was nucleic acid, even though in order to do so he had to ascribe to it catalytic properties that had never been observed:

> If, as now seems probable, the genetic enzymes must be identified with the nucleic acids, we shall be forced to suppose that these substances, although homogeneous – in animal or plant – from the point of view of ordinary chemical analysis, are actually built up in the living chromatin, into highly differentiated colloidal, or colloidal-molar, structures. The apparent homogeneity results from the fact that ordinary chemical analysis provides us only with the *statistics of the fundamental radicles* which are involved [emphasis in original].[51]

This was the first suggestion that the apparent chemical blandness of nucleic acids could obscure a structural diversity at a different level of organization. Thirty years later this concept would be revived in John Gulland's idea of the 'statistical tetranucleotide' (see Chapter 12). At the time, however, the suggestion that the 'genetic enzymes' were nucleic acids seems to have been ignored. What was taken from Troland's thesis was the idea that autocatalysis – self-replication – was a necessary property of genes, and that, in biological systems, only enzymes had catalytic properties. The concept of the gene as an autocatalytic protein was to dominate thinking in genetics for a generation.

The Fly Room

In 1909, Bateson wrote of the fledgling science of genetics: 'Perhaps the greatest advance that can be foreseen in this department of physiology will be made when the nature of the interaction between the chemical and the geometrical phenomena of heredity is ascertained.' This interaction would not be fully ascertained until 1953, but a major step forward in understanding the 'geometrical phenomena of heredity' was, in 1909, just around the corner.

In 1866, the year Mendel's *Pisum* paper was published, Thomas Hunt Morgan was born in Lexington, Kentucky. He graduated with a BS from the State College of Kentucky in 1886 and with a PhD in zoology from Johns Hopkins University in 1891. After a period at Bryn Mawr College, Morgan became professor of experimental zoology at Columbia University in 1904.

Morgan worked on a variety of research projects, leading the Harvard zoologist Frank Lutz to joke that Morgan had 'more irons in the fire than an ordinary man has coals'. In 1907, however, Lutz added another iron to the fire when he suggested that Morgan work on the fruit fly *Drosophila melanogaster*. Fruit flies had been used as experimental animals since around the turn of the century. However, they were not used in genetic studies. The new Mendelists preferred to study inbred populations that came in well-defined varieties – what Robert Kohler termed 'the inhabitants of second nature' – such as the sweet peas of Mendel and the pigeons of Darwin. Instead, *Drosophila* was used in evolutionary studies. Lutz, for example, used fruit flies to determine which traits were inherited and which were environmental. The traits he was interested in, however, were those that contributed to the fly's survival in the wild. Studying traits that did not have evolutionary significance was, to Lutz, 'mere breeding'.

Morgan had visited Hugo de Vries in 1893, and was interested in his idea that new species arose as a result of periods of rapid mutation. Unlike de Vries, however, Morgan believed that new species arose by accumulation of discontinuous mutations that were within the normal range of natural variation. *Drosophila*, which had large families and a short generation time, seemed well suited to such studies. Morgan subjected flies to heat, cold, different diets and even X-rays in an attempt to induce mutation, but obtained only inconsistent results. In the fall of 1909, he decided to out-breed his laboratory populations with wild flies and select for those having a prominent trident-shaped pattern of wing veins, a variable trait in natural populations. Ross Harrison, who had been a graduate student in the same laboratory as Morgan, visited Columbia at the beginning of 1910. Morgan, pointing to shelves of bottles, said: 'There's two years' work wasted. I've been breeding these flies for all that time and have got nothing out of it.'

Within days, a decisive breakthrough came when Morgan found a mutant fly with a prominent trident pattern. Other mutants quickly followed, including an eye color mutant, *white*. The more crosses Morgan had to do to characterize his mutants, the more new mutations arose. Morgan had invented an autocatalytic process for generation of mutants, and was soon overwhelmed. Ten new mutants were discovered in 1911, and by the beginning of 1912 Morgan wrote that he was 'head over ears with my flies'.

To assist him in maintaining the *Drosophila* stocks and counting the mutants among newly hatched flies, Morgan used the services of undergraduate students. In the fall of 1910, two students who had attended his introductory biology course asked Morgan if they could work in his laboratory. Alfred Sturtevant was taken on as a research assistant, while Calvin Bridges became the bottle-washer. Following their graduation in 1912, both men remained with Morgan as graduate students. Bridges would work with Morgan until his death in 1938; Sturtevant until Morgan's retirement in 1941.

A third graduate student, Hermann Muller, joined Morgan in 1912. As an undergraduate at Columbia, Muller had formed a biology club attended by Sturtevant and Bridges. When he graduated in 1909, Muller asked to work with Morgan, but was turned down. Instead, he performed an MSc in physiology at Columbia and worked for a year as a teaching assistant at Cornell University. In 1912, Muller obtained a teaching assistantship in zoology at Columbia that allowed him to work part-time with Morgan. However, he was not given a desk in Morgan's laboratory, apparently because the latter felt that Muller's strong personality would dominate the other students. Despite his undeniable brilliance, Muller became the Cinderella of the *Drosophila* group. He resented the fact that he, unlike Sturtevant and Bridges, did not receive financial support from Morgan. Muller was also upset that his close friend Edgar Altenburg was denied a place in the fly group. For his part, Morgan disliked Muller's lack of humor and habit of publicly correcting the 'Boss'. Nonetheless, it was Muller who was to be Morgan's most distinguished protégé.

Morgan's laboratory at Columbia, soon known simply as the 'fly room', measured 16 feet (about 5 m) by 23 (or 26) feet (7 or 8 m), with a small adjoining office. Initially there were no incubators, so the racks that held the bottles were heated by a row of light bulbs. Bananas were used as food, and housing was supplied by half-pint milk bottles that Morgan's students 'foraged' from doorsteps on their way to the laboratory. Visitors were typically struck by the squalor of the surroundings; Kohler wrote that: 'Morgan's place was especially disgusting, owing to his habit of squashing flies on his counting plate (he never could get used to Bridges' tidy 'fly morgue'), and legend had it that a rare cleaning of Sturtevant's desk disinterred a mummified mouse.'

The *Drosophila* group was a new kind of research unit. On one hand it was highly egalitarian, with all members working side by side at the bench and with all ideas being openly discussed and freely shared. On the other hand, it had distinctly feudal aspects. Until 1928, when the fly group relocated to the California Institute of Technology (Cal Tech), Morgan was the only member who held a university appointment. From 1915 onwards, the group was wholly supported by grants from the Carnegie Institution of Washington; Morgan was personally responsible for the money and kept his purse-strings extremely tight.

Despite the lack of pure-bred strains, *Drosophila* proved to be a reagent of unprecedented power for genetic analysis. It had only eight chromosomes, produced a new generation every two weeks, and as a bisexual species mutants could easily be characterized by back-crossing. The mutations located on the X chromosome proved particularly valuable. Studies on the inheritance of the white-eye trait produced the first major insight from *Drosophila* genetics. Crosses between *white* males and wild-type (red-eyed) females produced an F_2 generation with a Mendelian ratio of 3 red-eyed to 1 white-eyed progeny; however, all the latter were males. By the summer of 1910, Morgan had concluded that this pattern of inheritance was because of a physical association between the red-eye (R) gene and the female sex-factor (X): 'The fact is that this R and X are combined, and have never existed apart'. This 'combination' provided a physical basis for Bateson's 'coupling' of heritable traits; such traits were, as had been predicted by Sutton, carried on the same chromosome.

Another breakthrough came in 1911 when it was found that females carrying different mutations on both X chromosomes sometimes produced male off-spring that had inherited both. This showed that genes that were normally inherited together, presumably because they occurred on the same chromosome, could sometimes become separated during the formation of the germ cells. To explain this, Morgan invoked the idea of 'crossing-over'. It was known that homologous chromosomes paired during the process of meiosis (germ-cell formation). In females, for example, the X chromosome inherited from the father would pair with the X chromosome inherited from the mother. It appeared that the two paired homologous chromosomes were intertwined during this process of 'chiasmatype'. This observation led Morgan to propose that the breakage and rejoining of chromosomes during chiasmatype could result in the exchange of genetic material.

The spring of 1912 was a watershed for the *Drosophila* group. By this time it was becoming clear that the large number of mutations could not be accommodated within a traditional Mendelian scheme, which had been developed for small numbers of paired variants that segregated independently of one another. In *Pisum*, Mendel had studied seven characters, each of which occurred in two forms and segregated independently. In *Drosophila*, some characters occurred in a large number of mutant forms and exhibited various degrees of linkage. Morgan had already revised his Mendelian analysis of eye color three times when the *purple* and *maroon* mutants appeared, overthrowing the scheme again. A new method of categorizing genetic data was needed.

Studies on the phenomenon of crossing-over had revealed that some pairs of mutations were more likely to be subject to cross-over events than others. Sturtevant and Bridges interpreted this as indicating that these genes were far apart on the chromosome. Genes further apart were more likely to be separated during meiosis by a cross-over event between them than genes that were close together. Therefore, the probability of two genes being inherited together should be inversely proportional to the distance between them on the chromosome. Careful measurement of the number of progeny flies that inherited different genes could therefore be used to determine the relative positions of these genes.

By early 1913, Sturtevant was able to publish the first genetic map, containing the locations of six genes (Figure 6.1). According to the Morgan biographer Garland Allen, 'Despite the refinements in mapping procedures developed since that time, Sturtevant's original map, made when he was a college senior, is astonishingly accurate, being very close to newer maps constructed some twenty-five years later.'

The abandonment of Mendel's 'organ system' method of grouping genes according to the characteristic to which they contributed was accompanied by the abandonment of Morgan's hopes of using *Drosophila* to study evolution, and by the end of his work on other species. From now on, Morgan's group would be devoted solely to genetic analysis. As soon as they had identified linkage between autosomal genes, Sturtevant and Bridges commenced the mapping of chromosomes 2 and 3.

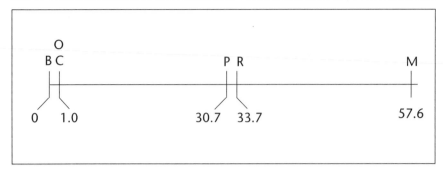

Figure 6.1: Alfred Sturtevant's 1913 map of six genetic elements in *Drosophila*; B is a body color trait; C, O and P are eye colors; R (rudimentary) and M are wing shapes. The 'scale' of the map represents the percentage probability of a cross-over event occurring between the two genes. Thus, the distance of 30.7 units between B and P means that 30.7% of the progeny of a fly carrying these markers will inherit only one of them. Adapted from Sturtevant, A. H. (1913) *Journal of Experimental Zoology* **14**, 49

Gene mapping was not without its problems. Linkage data had to be corrected for variable rates of crossing-over and for double cross-over events. Sturtevant and Bridges also had to eliminate sources of 'genetic noise' – lethal mutations, and genes that modified or suppressed crossing-over. The purity of the precious mutant stocks was constantly under threat from failure of the incubators and coolers used to maintain consistent year-round temperatures, genetic contamination by wild flies and attacks by marauding mice. The worst problem was periodic infestations of fly mites.

By 1915, when Morgan, Sturtevant, Muller and Bridges published their landmark book *The Mechanism of Mendelian Inheritance*, chromosomes 1 (X), 2 and 3 had been largely filled in. Sturtevant received his PhD in 1914, Muller in 1915 and Bridges in 1916. These three doctoral theses were, according to a later Morgan student, Jack Schultz, 'the foundation stones for a new science'.

Drosophila allowed genes to be treated as actual physical entities rather than abstract Mendelian units. The studies of Morgan's group showed that genes existed in linear arrays on chromosomes, and their discovery of crossing-over allowed the construction of physical maps of genes. As Muller later wrote: 'Morgan's evidence for crossing over and his suggestion that genes further apart cross over more frequently was a thunderclap, hardly second to the discovery of Mendelism, which ushered in that storm that has given nourishment to all of our modern genetics.' Morgan received the Nobel prize for physiology or medicine in 1933; he divided the prize money between his own children and those of Sturtevant and Bridges.

Not everyone was convinced by the chromosome theory, however. Among the skeptics were such prominent geneticists as Bateson, Johannsen and Goldschmidt. The problem was that genes were purely theoretical entities, the existence of which could be detected only by indirect means and in specific circumstances. Even in the 1930s and 1940s, genes would remain mysterious and controversial (see Chapter 9). Until the gene was shown to be a material particle of defined size and composition, acceptance of the chromosome theory would require some suspension of disbelief.

'To Grind Genes in a Mortar and Cook Them in a Beaker'

Morgan appears to have had little interest in the physical nature of genes. In his Nobel lecture of 1933, he stated: 'at the level at which the genetic experiments lie, it does not make the slightest difference whether the gene is a hypothetical unit, or whether the gene is a material particle.' Muller, in contrast, thought that the properties of genes could be profitably used to infer their structure. An uncanny ability to see straight to the heart of the matter made Muller the outstanding theoretical geneticist of his time. As Joshua Lederberg wrote: 'It is not easy to find an original thought in biological theory that has not, in some way, been anticipated' [in Muller's papers].

On graduating from Columbia in 1915, Muller became instructor in biology at the Rice Institute in Houston. Three years later, he obtained a similar position at Columbia. His initial contract was not renewed, however, for which he blamed Morgan (although it appears to have been Edmund Wilson's decision). By 1920, Muller was back in Texas as associate professor of zoology at the University of Texas, Austin. There he formed a strong *Drosophila* group with Theophilus Painter and John Patterson.

Muller's first great contribution to gene theory came as part of a symposium jointly organized by the American Society of Naturalists and the American Association for the Advancement of Sciences in Toronto in 1921. In his talk, published the following year as 'Variation due to change in the individual gene', Muller tried to deduce the chemical nature of the gene:

> The most distinctive characteristic of each of these ultra-microscopic particles
> – that characteristic whereby we identify it as a gene – is its property of self-
> propagation: the fact that, within the complicated environment of the cell
> protoplasm, it reacts in such a way as to convert some of the common
> surrounding material into an end-product identical in kind with the original
> gene itself. This action fulfils the chemist's definition of 'autocatalysis'; it is what

the physiologist would call 'growth', and when it passes through more than one generation it becomes 'heredity' . . . But the most remarkable feature of the situation is not this oft-noted autocatalytic action in itself – it is the fact that, when the structure of the gene becomes changed, through some 'chance variation', the catalytic property of the gene may become correspondingly changed, in such a way as to leave it still *auto*catalytic [emphasis in original].[52]

It was this seemingly paradoxical ability of the gene to change and yet be able to produce copies of itself that was the basis of evolutionary change:

Inheritance by itself leads to no change, and variation leads to no permanent change, unless the variations themselves are heritable. Thus it is not inheritance *and* variation which bring about evolution, but the inheritance *of* variation, and this is in turn due to the general principle of gene construction which causes the persistence of autocatalysis despite the alteration in structure of the gene itself [emphases in original].[52]

Troland had proposed that the autocatalytic property of genes was due to the 'similarity of their force-field patterns'. Muller pointed out, however, that similarity of 'force-field patterns' cannot be the basis of autoattraction, as in that case all substances would be autocatalytic, and no compound would be soluble. A further problem with autoattraction as the basis of gene replication was that only external parts of the gene would be capable of this mechanism. Muller preferred to believe that some 'special manner of construction' was responsible for autocatalysis.

Heritable changes in genes – mutations – may, Muller suggested, provide a way to analyze the 'manner of construction' of genes just as they provided a means of mapping them. The major limitation of this approach, as Garrod had earlier noted, was the scarcity of mutations that caused a detectable change in phenotype. Garrod had compared mutant-hunting to looking for needles in a haystack; Muller's analogy was finding dollar bills on the sidewalk. Within a few years, however, Muller himself would discover a means of inducing mutations with such frequency that the 'sidewalks' would be littered with 'dollar bills'.

There was, however, another possible approach. Agents that caused the lysis (dissolving) of bacteria had been discovered by Frederick Twort in 1915 and independently by Felix d'Hérelle two years later. Like genes, these d'Hérelle bodies could be indefinitely propagated and appeared to undergo mutation:

if these d'Hérelle bodies are really genes, fundamentally like our chromosome genes, they would give us an utterly new angle from which to attack the gene problem . . . It would be very rash to call these bodies genes, and yet at present we must confess that there is no distinction known between the genes and them. Hence we cannot categorically deny that perhaps we may be able to grind genes in a mortar and cook them in a beaker after all. Must we geneticists become bacteriologists, physiological chemists and physicists, simultaneously with being zoologists and botanists? Let us hope so.[52]

Twenty-five years later, Max Delbrück would come to the same conclusion and use d'Hérelle bodies, by then renamed bacteriophage, in his own attack on the structure of the gene.

Chapter 7
The Megachemistry of the Future

'The Dark Age of Biocolloidology'

Although vitalism had suffered a number of reverses during the nineteenth century, it had by no means been extinguished. The main issue over which vitalists and materialists had fought, the nature of alcoholic fermentation, had been resolved largely in favor of the latter by Eduard Buchner's demonstration of cell-free fermentation. It was now clear that the conversion of sugar to alcohol, and other reactions of 'physiological chemistry', were the properties of specific enzymes rather than of living cells as a whole. However, this merely shifted the materialist–vitalist debate from the cellular to the subcellular level. According to the neo-vitalist view, the agents of the vital properties were proteins: enzymes, which catalyzed the breakdown and, it now appeared, the synthesis, of organic molecules, and thereby accomplished the processes of growth and assimilation; protoplasm, a poorly characterized but clearly proteinaceous substance which mediated the quintessentially vital property of movement; and 'living protein', which was responsible for intracellular respiration and therefore for the phenomenon of 'animal heat'.

What was it about proteins that allowed them to perform so many functions never seen in the inanimate world? According to Thomas Graham, it was because proteins were colloids (see Chapter 5). Colloidal substances such as proteins and polysaccharides were 'dynamical', whereas crystalloids such as inorganic salts were 'statical'. For Graham, the colloid 'may be looked upon as the probable primary source of the force appearing in the phenomena of vitality'.

When the new 'physical' chemists began to study colloids in the late nineteenth century, they found that Graham's neat distinction between colloidal/living and crystalline/non-living did not hold up. It turned out that many inorganic substances, such as gold particles, formed stable suspensions in liquids. The discovery of such inorganic colloids led to a new definition: colloids were a *state* of matter, not a *type* of matter. In principle, any substance could exist as a colloid, just as any element could exist as a solid, a liquid or a gas. By the physicochemical definition, colloids were stable mixed phases in which one substance was non-homogeneously dispersed within another. This could be a solid dispersed in a liquid, as in the case of the colloids Graham had studied. However, it could also be a liquid dispersed in an immiscible liquid (an emulsion) or a gas dispersed in a liquid (a foam).

The discovery of inorganic colloids was inconsistent with Graham's view that colloids were responsible for the 'phenomena of vitality'. On the other hand, the finding that colloids had novel chemical properties seemed to vindicate his idea that these substances were 'dynamical'. Indeed, the behaviors of dispersed systems were often quite different from those of solutions. Phenomena such as adsorption from one phase to another depended not on the amount of the dispersed phase but on its surface area, in violation of the normal chemical rules of proportionality. Even more intriguingly, some reactions were catalyzed by colloids because of concentration of reactants at the boundary between the two phases.

One of the principal advocates for the importance of colloids was Wolfgang Ostwald, son of the eminent physical chemist Wilhelm Ostwald. In his influential 1917 book, *An Introduction to Theoretical and Applied Colloid Chemistry*, Ostwald defined colloids as particles between 1 nm and 100 nm in diameter[a] – a size range that he called 'the world of neglected dimensions'. As Ostwald pointed out, colloids often had catalytic activities and 'ferments' appeared to be colloidal. Not unreasonably, he drew from this the conclusion that enzymes had catalytic activity *because* they were colloids. He further proposed that a variety of biological processes, including fertilization, muscle contraction and secretion, were fundamentally 'colloid-chemical' in nature.

Unlike his father, Wolfgang Ostwald was not an influential scientist – he was described by the historian Pierre Laszlo as 'mostly an operator, self-aggrandizer, and propagandist'. However, many mainstream scientists shared his view that 'all life processes take place in a colloid system'. The zoologist Edmund Wilson wrote in 1923: 'No conception of modern biology offers greater promise of future progress than that the cell regarded as a whole is a colloidal system'. The biochemist Joseph Needham, who regarded the concept of protoplasm as 'a mere repository for ignorance', was far more charitable to colloids, even going so far as to suggest that consciousness was a colloidal phenomenon. The scope of the colloid theory of life is shown by

---

a   1 nm (nanometer) = 10^{-9} meters.

Volume 2 of *Colloid Chemistry: Theoretical and Applied*, which appeared in 1928. It included chapters on vitamins, bacteria, botany, serology, internal medicine and tuberculosis. Among the authors were such eminent main-stream scientists as William Bragg, Calvin Bridges and Felix d'Hérelle. The colloid theory had become a kind of scientific tulip craze. It was, as the historian Marcel Florkin put it, 'the dark age of biocolloidology'.

'The So-called Colloid Chemistry of Proteins'

Despite the enthusiasm of neo-vitalists for the colloid theory of life, the fact remained that many proteins had properties more typical of crystalloids than of colloids. For Graham, the characteristic features of colloids were that they diffused slowly and formed amorphous solids; the archetype of this class was the protein gelatin. Unlike gelatin, however, most proteins were quite soluble, and many, like hemoglobin and the plant globulins, formed crystalline solids. Nor did proteins diffract light when examined with an ultramicroscope, a characteristic property of inorganic colloids. By the late nineteenth century, there was good reason to believe that the crystalline proteins were in fact polypeptide chains of large, but definite, molecular weight.

In 1902, however, the same year that the polypeptide theory was proposed, it was suggested that certain molecules contained a weak type of chemical bond known as a 'secondary valence'. This hypothesis led to a view of protein structure that was in some ways intermediate between the colloid and the crystalloid. Perhaps proteins did contain huge numbers of amino acids, as suggested by the elemental compositions of some proteins; perhaps they did contain peptide bonds. If the polypeptides were rather short, however, the protein 'molecule' could consist of such small polypeptides aggregating by means of secondary valences into a colloidal particle.

This view benefited from the weighty support of Emil Fischer, who believed that the maximum size of polypeptides was 4000–5000 daltons, corre-sponding to about forty amino acids. 'Natural proteins' were probably mixtures of these small polypeptides, which was why techniques like elemental analysis sometimes suggested much larger molecular weights.

Fischer's views were clearly influenced by wishful thinking. He was devoting tremendous efforts to the chemical synthesis of polypeptides. As he confided to his mentor, Baeyer, 'my entire yearning is directed toward [making] the first synthetic enzyme'. However, this was a very difficult task using the methods then available, and by 1906 the largest molecule Fischer's laboratory had produced was only a decapeptide (ten amino acids). If proteins were much larger than 5000 daltons, Fischer would never succeed in synthesizing an enzyme.[b]

Fischer also believed that proteins contained structures known as piperazine rings, which were formed by two amino acids reacting together in head-to-tail fashion. After his death, Fischer's former student Emil Abderhalden went even further, suggesting that proteins were colloidal aggregates of these cyclic dipeptides (diketopiperazines) (see Figure 5.4). Another student of Fischer, Max Bergmann, extended this theory to polysaccharides, proposing that substances such as cellulose consisted of colloid aggregates of simple sugars.

Despite the popularity of the colloid theory of life and the influential view of Fischer and his school that proteins were aggregates of small molecules, the idea that proteins were giant amino-acid polymers was still widely accepted among biochemists. Thomas Osborne, a leading authority on plant proteins, pointed out in a 1911 Harvey lecture that the hempseed globulin edestin had been fractionated by a large number of methods without exhibiting any change in elemental composition or optical rotation: 'It is impossible after my experience to believe, as Fischer suggests, that many of the seed proteins are mixtures of several substances of simpler constitution.' Søren Sørensen showed in 1917 that ovalbumin dissolved in water behaved like a true solution, not a suspension, and derived a molecular mass of 34000 for this protein. In the 1921 edition of his book *Physiological Chemistry*, Albert Mathews pointed out that the uptake of water by proteins when they were broken down to amino acids was consistent with the polypeptide structure. He also noted that measurements of the molecular mass of hemoglobin by the techniques of sulfur content, iron content, oxygen uptake and osmotic pressure measurements all agreed in giving a value of 16000–17000. Studies conducted by Jacques Loeb of the Rockefeller Institute for Medical Research suggested that proteins combined with ions in the manner of crystalloids rather than that of colloids. In his 1922 book *Proteins and the Theory of Colloid Behaviour*, Loeb concluded that: 'It is, therefore, obvious that the so-called colloid chemistry of proteins is a system of errors based on inadequate and antiquated methods of experimentation.'

The colloid theory of protein structure thus became the subject of an acrimonious debate in the 1920s. The waters of this debate were muddied by

b   The largest polypeptide Fischer synthesized contained only eighteen amino acids, all of which were either glycine or leucine. By 1910 the quest for the first synthetic enzyme had been abandoned.

confusion about the definition of colloids and the physical nature of proteins. By Graham's original definition, colloids were large and soluble (or sparingly soluble) molecules. According to the physical chemists, colloids were insoluble particles with a range of sizes (polydispersity). The 'physiological chemists' believed that proteins were large molecules of defined composition; the organic chemists believed that proteins were mixtures of small polypeptides (Fischer) or aggregates of dipeptides (Abderhalden). Therefore, the key points at issue were the size, polydispersity and solubility of proteins. To determine molecular size, physical chemists examined properties that depended upon the numbers of molecules per unit volume; such techniques were limited, in theory or in practice, to small, monodisperse, soluble molecules or ions. New methods of structural analysis would be needed to resolve the conflicting views of protein structure. One of these methods would arise from the study of crystals, another from the study of colloids.

'New Worlds Unfolding Before Us'

The year after Thomas Graham's paper on colloids appeared, William Henry Bragg was born near Wigton in the north-west of England. He studied mathematics at Trinity College, Cambridge, graduating early in 1885 with first-class honors, third wrangler (third in that subject of all the Cambridge colleges). Following the completion of his degree, Bragg stayed on at Cambridge, studying physics at the Cavendish Laboratory. In late 1885, the Cavendish professor, J. J. Thomson, was part of the selection committee for a chair in mathematics and physics at the University of Adelaide in South Australia. Bragg found out about this position only by accident and on the last day of competition, but managed to get his application submitted in time by telegraph. Among the other applicants was the first wrangler from a previous year. Thomson made amends for his failure to inform Bragg of the vacant position by informing a 'certain Adelaide man' that the senior candidate 'occasionally disappeared under the table after dinner'. The University of Adelaide took the hint and offered the position to the abstemious Bragg. The 'Adelaide man' who conveyed Thomson's message was the government astronomer and postmaster-general of South Australia, Sir Charles Todd. Earlier in his career, he had laid the telegraph line that carried Bragg's application. In 1889, Todd became William Bragg's father-in-law.

Teaching physics did not present a difficulty to Bragg, as the mathematical tripos he had taken at Cambridge included optics, gravitational theory, astronomy, electricity and magnetism. However, he was overwhelmed by the staggering amount of work. In his second year at Adelaide he had 672 contact hours with students, 168 of them in the evening, and single-handedly set 21 examinations. The only support he had was a part-time assistant to help with laboratory teaching. In 1888, the University of Adelaide found money for an assistant lecturer to share Bragg's teaching. However, he performed no original research until he was forty-one, an age at which most mathematicians, and many physicists, find their best work behind them.

Although unable to do any research himself, William Bragg followed new developments in physics with great interest. One such development came in 1895, when Wilhelm Röntgen of the University of Würzburg in Germany discovered a novel form of radiation. Like many such discoveries, it was the result of either sheer luck or brilliant observation, depending on one's point of view. Röntgen was experimenting with electrical discharge in evacuated glass tubes when he noticed that coated paper he had prepared for studies on ultraviolet radiation fluoresced brilliantly in the vicinity of the discharge tube. Further investigation revealed that the source of these emissions, which Röntgen referred to as X-rays, was the point on the tube where the cathode rays (electrons) struck the glass. However, X-rays had quite different properties from cathode rays. Cathode rays were deflected by magnetic fields; X-rays were not. Cathode rays were absorbed by matter; X-rays passed freely through everything except metals. In the body, only the mineralized tissues were opaque to X-rays: 'if a hand be held before the fluorescent screen, the shadow shows the bones clearly with only faint outlines of the surrounding tissues.' By analogy with cathode rays, and in keeping with the physics of the day, Röntgen proposed that X-rays were 'longitudinal waves in the ether'. For the discovery of these mysterious emissions, Wilhelm Röntgen was in 1901 awarded the first Nobel prize for physics.

Bragg's first contact with the form of radiation that would make him famous came, fittingly enough, through his son. The same year that X-rays were discovered, the five-year-old William Lawrence Bragg[c] fell off his tricycle and suffered a severe fracture of the left elbow. To assess the damage, William Bragg set up a replica of the Röntgen apparatus. The buzzing noise, electrical sparks and eerie green glow produced by this machine terrified young Lawrence. However, it also generated a clear image of the damaged joint – the first diagnostic X-ray in Australia. Another innovative relation, his uncle Charlie Todd, a physician, devised a therapy to prevent the child's elbow from fusing.

---

c   William Lawrence Bragg was known throughout his life as 'Willie'. He is referred to here as Lawrence
     Bragg to distinguish him from his father.

In 1904, William Bragg was invited to give a presidential address to the Australian Association for the Advancement of Science. Studies on radioactivity and the electron were producing exciting new findings at this time, and Bragg's presidential lecture inspired him to construct equipment to measure the absorption of alpha particles (helium nuclei) by air and by solids. He showed that the 'stopping power' of elements was proportional to the square root of their atomic weight, and also studied the 'secondary electrons' that were produced by the interaction of beta (electrons) and gamma (electromagnetic) radiation with matter. These studies were well received, and Bragg was elected as a Fellow of the Royal Society in 1907, just three years after the publication of his first original paper. This honor led to the opportunity to move back to Britain, and in 1909 he became professor of physics at Leeds University.

Lawrence Bragg had graduated with first-class honors in mathematics from the University of Adelaide the previous year. However, he took greater pleasure in winning the university prize for the best English essay 'from under the noses of the professionals'. When the family moved to England, Lawrence Bragg followed in his father's footsteps to Trinity College, Cambridge, graduating in 1911 with another 'first', this time in physics. He then took up a lectureship at Trinity and performed research at the Cavendish Laboratory.

Whether Röntgen's X-rays were particles, like the alpha particles produced by the radioactive decay of radium, or waves, like light, was the subject of an active debate in the first decade of the twentieth century. William Bragg's studies on alpha particles in Adelaide had made him a proponent of the particle theory of X-rays. Max von Laue, a theoretical physicist at the Institute for Theoretical Physics in Munich, favored the wave theory. In 1912, Laue thought of an experiment that could settle the issue. He realized that the wavelength of X-rays should be similar to the distance between atoms in a crystal; if X-rays were waves, they should be scattered by passage through the crystal in the same manner as light is scattered by an optical diffraction grating.

If X-rays could be said to be a phenomenon in search of a theory, the opposite could be said of crystals. During the nineteenth century, an elaborate mathematical theory of crystal structure had been constructed in the complete absence of any information about the internal arrangement of these substances. According to this theory, crystals were composed of one or more types of ions or molecules arranged in a three-dimensional lattice. Such lattices could be considered to be made up of blocks known as 'unit cells', which were the smallest structural units that repeated in space throughout the volume of the crystal. It had been shown that there were only seven different shapes of unit cell that could be packed together in three dimensions without leaving spaces. These types of unit cells, differing in the relative lengths of their three axes and the angles these axes formed with one another, were used to define seven 'crystal systems': cubic, tetragonal, orthorhombic, trigonal, hexagonal, monoclinic and triclinic. The cubic system was the simplest, as its unit cell axes were all the same length and were mutually perpendicular (that is, formed 90° angles).

Towards the end of the nineteenth century, it had been recognized that each crystal system could be further subdivided based on the symmetry elements present in the unit cell. One common symmetry element was an axis of rotational symmetry, an imaginary line around which rotation of an object by a certain angle brings it to an identical position. Cubic unit cells, for example, could have a large number of axes of rotational symmetry. A line running from one corner of a cube to the diagonally opposite corner was known as a three-fold rotational axis, because a 120° (360°/3) rotation of the cube around this line brought it back to the same orientation. Similarly, a line running from the center of one face of a cube to the centre of the opposite face was known as a four-fold rotational axis, as rotation of 90° (360°/4) resulted in an identical orientation of the cube. However, depending on how the lattice points of the crystal were distributed within the unit cell, some of the symmetry elements characteristic of that crystal class could be lost. In order to define all the possible types of crystal lattice, therefore, it was necessary to take into account not only the shape of the unit cell but also its symmetry elements. Consideration of all possible combinations of unit cells and symmetries allowed crystallographers to determine that there were exactly 230 types of three-dimensional lattice structures, or 'space groups'.

It should be emphasized that description of the seven crystal systems, identification of all the permissible three-dimensional symmetry operations, and derivation of the 230 space groups, had all been achieved by an abstract geometrical analysis inspired by the external symmetry of crystals, at a time when it was not possible to generate any information about their internal structure. For example, it was not known whether the 'lattice points' of space group theory corresponded in real crystals to atoms, molecules – or to nothing at all.

Laue's idea that X-rays could be diffracted by crystals therefore represented a test not only of the wave theory of X-rays but also of the lattice theory of crystal structure. As a theoretician, Laue felt unqualified to do the experiment himself. For that reason, he requested the assistance of Arnold Sommerfeld, the director of the Institute of Theoretical Physics. Sommerfeld thought the vibration of atoms in the crystal would prevent any diffraction effect, and refused to authorize any experiments on the subject. Turning his blind eye to the telescope, Laue persuaded Paul Knipping, a student of Röntgen, and Walther Friedrich, an assistant to Sommerfeld, to attempt the diffraction of X-rays by crystals.

Knipping and Friedrich's first attempt used a crystal of copper sulfate, which was assigned to the orthorhombic class. A photographic plate placed behind the crystal was blackened where it was struck by X-rays passing straight through the crystal, but also exhibited some smudgy dots arranged elliptically around the central spot, indicating that some X-rays had indeed been scattered. Friedrich, Knipping and Laue decided to try a crystal of the cubic class, zincblende (zinc sulfide). This second experiment was spectacularly successful. Directing the X-ray beam down a four-fold symmetry axis of the crystal produced a pattern of dots with four-fold symmetry; directing it down a three-fold axis likewise produced a pattern of corresponding symmetry. These experiments with zincblende not only provided clear proof of the wave nature of X-rays, but also validated the belief of the mathematicians that the external symmetry of crystals was a reflection of the internal arrangement of their atoms.

Laue now attempted to use the diffraction patterns obtained to deduce the characteristics of the crystal lattice. He succeeded in establishing a relationship between the atomic spacings in the crystal and the wavelength of the diffracted radiation. He also recognized that the relative lack of spots on the diffraction pattern resulted from destructive interference between the diffracted waves. However, Laue's analysis was hampered by two incorrect assumptions. The first was that the zincblende crystal had a cubic lattice of the 'primitive' type, in which the only lattice points – presumably corresponding to zinc and sulfide ions – were at the eight corners of the unit cell. The second was that crystals acted as three-dimensional diffraction gratings, and therefore all crystal atoms scattered X-rays. As a result, Laue's theory of X-ray diffraction could explain all the observed dots found in the zincblende pattern – but other spots predicted by his theory were missing. To account for the missing spots, Laue had to assume that the X-ray source used did not produce a continuous spectrum of radiation, but only certain wavelengths.

Lawrence Bragg heard of Laue's paper when he joined his parents for a vacation on the Yorkshire coast in the summer of 1912. A few weeks later,

during a walk along the Cambridge 'backs', the true meaning of the diffraction pattern came to him: 'I can remember the exact spot in the Backs
where the idea suddenly leapt into my mind that Laue's spots were due to
the reflection of X-ray pulses by sheets of atoms in the crystal.' The key to
this insight was Laue's observation that the spots produced on photographic
plates by diffracted X-rays became more elliptical as the plate was moved
further away from the crystal – in a similar way to a beam of light reflected
by a mirror. Bragg realized that X-rays are reflected by planes of atoms in
crystals in the same way that visible light is reflected by a polished surface
– except that in the former case the reflection is merely from the surface
layer of atoms, whereas in the latter it is from internal layers as well (Figure
7.1).

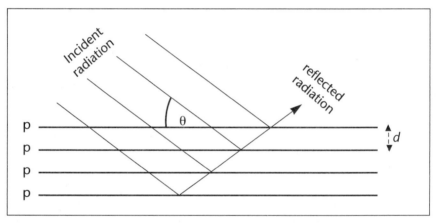

Figure 7.1: Lawrence Bragg's model of reflection of X-rays by atomic planes in a
crystal. Reinforcement of the reflected radiation will only occur at the angle (θ) at
which the distance (*d*) between planes (p) is a multiple (*n*) of the X-ray wavelength (λ):
$n\lambda = 2d \sin \theta$.

An even more important insight was that reflected X-rays would only be
detected when the angle of the X-ray beam to the surface of the crystal was
such that the wavelength of the radiation was related to the distance between
the planes of atoms in such a way that the X-rays reflected from different
layers of the crystal reinforced one another. Therefore, there was a simple
mathematical relationship, now known as the Bragg equation, between the
X-ray wavelength (λ), the angle between the X-ray beam and the crystal
face (θ) and the distance between the planes of atoms in the crystal (*d*):

$$n\lambda = 2d \sin \theta$$

where *n* is an integer.

Bragg confirmed this relationship by reflecting an X-ray beam from a sheet of mica. He took his first diffraction picture to the man who a quarter of a century earlier had helped the elder Bragg secure his first job: 'I remember so well taking my terribly crude picture of reflection to show to J. J. Thomson. He betrayed his excitement in a characteristically J. J. way by thrusting his spectacles up on his forehead, ruffling his hair violently, and making a peculiar mixture of grin and chuckle. It was a great moment.'[53] Even more convincing support for his reflection theory was contained within Friedrich's and Knipping's diffraction patterns for zincblende. Bragg realized that the dots obtained could all be assigned to specific planes of atoms within the crystal lattice if he made one assumption – that the zincblende lattice was not primitive cubic, but rather face-centered, in which there are additional lattice points at the center of each face of the unit cell.

With the help and encouragement of his Cambridge colleagues, Lawrence Bragg started to perform X-ray diffraction studies on alkaline halides such as potassium iodide and sodium chloride (Figure 7.2). Meanwhile, his father's laboratory at Leeds concentrated on equipment improvements. This

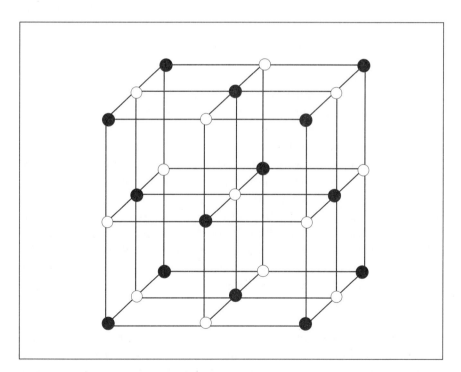

Figure 7.2: Lawrence Bragg's 1914 structure of the sodium chloride crystal. Sodium and chloride ions are arranged in a primitive cubic lattice. Adapted from Bragg, W. L. (1914) *Proceedings of the Royal Society of London A* **89**, 470, with permission

led to the development of monochromatic (single-wavelength) X-ray sources, which obviated the need to measure the wavelength of the diffracted radiation. William Bragg also developed the X-ray spectrometer, a device that allowed measurement of the angle and intensity of diffracted radiation. These innovations meant that the dimensions of the unit cell could easily be measured. From the symmetry of the Laue photograph, the space group of the crystal could also be determined. Putting together the space group and the unit cell dimensions, it was an easy matter to determine the three-dimensional atomic structure of simple salts and minerals. As William Bragg wrote in 1915: '. . . we find ourselves able to measure the actual distances from atom to atom and to draw a diagram as if we were making the plan of a building.'

In 1913, Lawrence Bragg moved to Leeds. Working together, the Braggs were quickly able to determine the structures of fluorspar (calcium fluoride), cuprite, iron pyrites (iron sulfide), sodium nitrate and calcite (calcium carbonate). They also solved the crystal structure of diamond, proving the tetrahedral arrangement of carbon valencies proposed forty years earlier by van't Hoff and Le Bel (see Chapter 4). As Lawrence Bragg later wrote: 'It was a glorious time, when we worked far into every night with new worlds unfolding before us in the silent laboratory.'

Within the short period of two years, the insights of Laue and Lawrence Bragg had proved the wave nature of X-rays and the lattice structure of crystals. Of even more significance, these insights also led to a revolution in the analysis of molecular structure. The old method of elemental analysis could produce the composition of a molecule, but give no information about how these components were put together; organic chemistry and spectroscopy could be used to infer structure, but were difficult and ambiguous. The use of X-ray diffraction offered the possibility of mapping the positions of individual atoms in a crystal by direct interpretation of a scattering pattern. Recognition followed quickly for the founders of X-ray crystallography. Max von Laue was awarded the Nobel prize for physics in 1914; the following year, this award was shared by William and Lawrence Bragg. At twenty-five years of age, Lawrence Bragg was – and remains – the youngest-ever Nobel laureate.

There were a couple of important limitations to the determination of structure by X-ray diffraction. First, the molecule had to be capable of forming crystals. Second, the simpler the molecule was, the easier its structure was to solve. On both grounds, X-ray crystallography seemed unlikely to resolve the debate over protein structure. If proteins were colloids, then – by

Graham's definition – they would not crystallize. If proteins were large molecules, their structures could not be solved by Bragg's method – even Abderhalden's diketopiperazines were much more complex than the simple salts analyzed by X-ray diffraction so far. The technique of X-ray crystallography seemed to offer little to the biochemists.

'Fibres Are What They Are Because Their Inner Molecular Structure is Also of a Fibrous Nature'

Scientific research in Europe was interrupted by the outbreak of war in August 1914. On the Allied side, Lawrence Bragg worked at the front, developing techniques for determining the position of enemy artillery from the pressure wave generated on firing. Archibald Garrod served in Malta with such distinction that he was knighted. The Central Powers also recruited scientists for the war effort. Emil Fischer promoted the use of nitric acid to replace Chilean saltpeter made unavailable by the British blockade. He also carried out research on food substitutes, resulting in the invention of margarine. Although he was fifty-four years old, Eduard Buchner volunteered for active service and died of shrapnel wounds in 1917.

Scientists were also used for propaganda purposes. In Germany, a group of ninety-three intellectuals, including the physicists Max Planck and Wilhelm Röntgen and the chemists Emil Fischer, Fritz Haber and Richard Willstätter, signed the *Aufruh an die Kulturwelt* (Appeal to the civilized world), a manifesto that defended the invasion of Belgium and denied the reported atrocities committed there. A British counter-manifesto entitled 'Reply to the German Professors' was signed, *inter alia*, by William Bragg and J. J. Thomson. Fischer soon recanted the views expressed in the *Aufruh*, and campaigned for another manifesto that would call for an end to the war without blaming either side. He also performed humanitarian work in locating British prisoners of war. For many of these men, patriotic antagonism to the enemy was reinforced by personal losses. Robert Bragg, son of William and brother to Lawrence, was killed in the Dardanelles. Two of Garrod's sons died in action, the third in the influenza epidemic of 1918. One of Fischer's sons committed suicide in 1916, another died of typhus in Romania the following year. Embittered by Germany's postwar condition, broken by his personal losses and suffering from a variety of chronic ailments, Emil Fischer took his own life in 1919.

The Treaty of Versailles that ended the war returned Alsace and Lorraine to France, costing Germany, among other things, their showpiece University of Strasbourg. The crippling reparations imposed by the victorious Allies and the chaos of the Weimar period effectively ended a German scientific hegemony that had lasted for almost a century. This was exacerbated by anti-German sentiment; as late as 1920, the American *Journal of General Physiology* refused to publish papers by Germans on the grounds that this would result in a boycott by subscribers.

As part of the reconstruction of Germany after World War I, a series of research institutes was set up to ensure new scientific advances for the use of industry. One of these was the Kaiser Wilhelm Institute for Fiber Chemistry, established in 1921 in Dahlem, a suburb of Berlin, under the direction of Reginald Herzog.

Herzog, who had worked under Albrecht Kossel in Heidelberg, had recently discovered that biological fibers such as the cellulose of plant cell walls could diffract X-rays. In this case, the diffraction patterns obtained indicated that a regular periodicity existed in only one dimension, along the axis of the fiber, as opposed to the periodicity in all three dimensions seen in crystals. One-dimensional fiber diffraction presumably meant that regularly spaced planes of atoms were present only along the long axis of the fiber, but not along axes at right angles to it. A crystal could be considered as a three-dimensional structure composed of unit cell blocks fitted together front to back, side to side and top to bottom; a fiber must consist of such blocks attached only top to bottom.

The study of biological fibers was one of the main aims of the new Institute of Fiber Chemistry. For this purpose, Herzog assembled a talented group of researchers that included Herman Mark, Rudolf Brill and Michael Polanyi.

It was Polanyi, a physician turned physical chemist – and later a distinguished philosopher – who first established a theory of fiber diffraction. Rather than discrete spots, the fiber diagram consisted of a series of short, parallel lines that lay perpendicular to the long axis of the molecule. Polanyi showed that these 'layer lines' were made up of reflections from all the atoms in a particular plane through the fiber. Each layer line represented atomic planes different integral numbers of repeat units apart.

The first fiber diffraction pattern analyzed by Polyani was that of cellulose. As hydrolysis of cellulose produced the simple sugar glucose, it was believed that cellulose was some kind of polymeric form of glucose. In 1921, however, Polanyi showed that the unit cell of cellulose contained just four glucose units, corresponding to two disaccharide (glucose–glucose) molecules arranged side by side.

A similar result was obtained for silk fibroin. It was known that this protein was very rich in the amino acids alanine and glycine, and alanine–glycine dipeptides had been identified among its hydrolysis products. In 1923, Brill reported that the unit cell of silk fibroin consisted of eight amino acids – four parallel glycine–alanine dipeptides.

In these early days of X-ray crystallography, the interpretation of the fiber diffraction patterns of cellulose and silk fibroin was heavily influenced by habits of thought derived from the diffraction of inorganic single crystals. From that point of view, the unit cell, defined as the repeating unit of the crystal, had to include one or more molecules. If a molecule was bigger than the unit cell, then the unit cell obviously could not repeat through the crystal lattice. This was apparently the view of William Bragg, who wrote in 1922: 'The crystal unit [cell] must contain the substance of an integral number of molecules.' Herzog, director of the Fiber Instititute, shared this opinion.

If the molecule could not be larger than its crystallographic unit cell, then the maximum size of the cellulose molecule must be two glucose units, and that of the silk fibroin molecule two amino acids. Since these substances behaved as if they were of much larger molecular weight, it seemed reasonable to suppose that large numbers of the disaccharide or dipeptide units were held together by secondary valences in some kind of colloidal aggregate. Ironically, the demonstration that some proteins and carbo-hydrates had paracrystalline properties was interpreted as evidence for their colloid nature.

Of all people, William Bragg should have realized that the unit cell of a crystal could be smaller than the molecules of which the crystal was composed. Bragg's structure of diamond had eight carbon atoms per unit cell; but as the diamond lattice was composed of carbon atoms linked by normal valence bonds, the whole crystal could be considered to be a single diamond molecule containing millions of carbon atoms. The same argument – that the unit cell was just the repeating subunit of a giant polymeric molecule – could be made with equal force for fibers. Just like true crystals, fibers produced not only an X-ray reflection that corresponded to the distance, d, between atomic planes, but also reflections that corresponded to multiples of d. This showed that the subunits of the fiber repeated in space along the fiber axis, rather than being randomly aggregated together as they would be in a colloidal particle. Quite reasonably, fiber crystallog-raphers took the view that the molecule was at least as large as the highest-order reflections indicated. Thus, if the unit cell of cellulose contained two glucose units, and reflections were observed on ten layer lines, then the cellulose molecule must be at least twenty glucose units in length. Improvements in crystallographic technique meant that such weak reflections could more easily be detected, and crystallographers gradually accustomed themselves to the idea of larger and larger fiber molecules.

In 1919, Lawrence Bragg was appointed to the Langworthy chair of physics at Manchester University, replacing Ernest Rutherford, who had become Cavendish professor at Cambridge. William Bragg had moved from Leeds in 1915 to become Quain professor of physics at University College, London. They divided their efforts, Bragg *père* studying organic crystals while Bragg *fils* studied inorganic ones.

In 1923, William Bragg, who had been knighted three years earlier, moved to the Royal Institution in London, where he acquired the comic-opera title of 'Professor of Chemistry in the Royal Institution, Director of the Laboratory of the Royal Institution, Superintendent of the House and Director of the Davy–Faraday Research Laboratory'. At the Royal Institution, Bragg supervised a group of researchers that would dominate British crystallography in the years ahead. This group included Kathleen Yardley (Lonsdale), one of the first two women to become Fellows of the Royal Society; Arthur Lindo Patterson, whose application of Fourier methods to X-ray diffraction was to make possible the solving of complex biological structures (see Chapter 8); John Desmond Bernal; and William Thomas Astbury.

Both Astbury and Bernal had attended Cambridge University and had been recommended to Bragg by Arthur Hutchinson, lecturer in mineralogy. However, their backgrounds were otherwise quite different. Astbury was born in 1898 in Longton, near Stoke-on-Trent, the son of a potter's turner. Scholarships supported all his education, which culminated in a first-class degree in physics from Jesus College, Cambridge, in 1921. Bernal was from a family of Anglo-Irish gentry, born in 1901 at the family estate in Brookwatson, Ireland. He attended Emmanuel College, Cambridge, between 1919 and 1923. In spite of (or perhaps because of) his brilliance, he was too unfocussed to be a good student. Much of his last year was spent on a private project on the mathematics of crystal lattices, which he discovered halfway through was a re-invention of the space group theory developed at the end of the nineteenth century. Largely as a result of this digression, Bernal achieved only a second-class degree.

In 1926, Hutchinson was appointed professor of mineralogy at Cambridge and a new lectureship in structural crystallography was created; the applicants for this position included Astbury and Bernal. Neither had a good interview technique. When asked his views on collaborative research, Astbury famously replied: 'I am not prepared to be anybody's lackey.' Bernal slumped in his chair and replied 'yes' or 'no' to every question he was asked. Finally, when one of the interviewers asked him what he would do with the crystallography laboratory, Bernal launched into an eloquent forty-five minute monologue. According to Hutchinson, 'There was nothing for it but to elect him.'

Astbury was three years Bernal's senior, and the appointment of the well-born younger man may have rankled. However, that same year a different kind of opportunity came his way. William Bragg was planning a lecture entitled 'The imperfect crystallization of common things', and asked Astbury to take X-ray photographs of biological fibers such as cotton and silk. The fuzzy diffraction images obtained from these fibers would have disillusioned most crystallographers, but Astbury took a liking to the odd patterns. He was therefore well qualified when a lectureship in textile physics at Leeds University became available in 1928. This time Astbury's application was successful and he stayed in Leeds for the rest of his career.

Astbury arrived just in time to take advantage of a breakthrough in fiber analysis: J. B. Speakman, the lecturer in textile chemistry, had shown that the X-ray diffraction pattern of the wool fiber, known to consist almost entirely of the protein keratin, was altered when the fiber was stretched. The patterns from unstretched and stretched keratin he called 'alpha' and 'beta', respectively. Astbury was able to claim that X-ray analysis was physics rather than chemistry, and thereby take over the wool project from Speakman.

By treating them with steam, Astbury was able to extend wool fibers to approximately twice their dry length. The major X-ray reflection obtained from unstretched wool fibers corresponded to an atomic spacing of 5.15 Å along the fiber axis.[d] On stretching the wool, this reflection disappeared and was replaced with one at 3.32 Å. Astbury realized that the properties of the macroscopic wool fiber reflected those of the submicroscopic polypeptide chain: 'fibres are what they are because their inner molecular structure is also of a fibrous nature.' To understand what that fibrous molecular structure was, however, he needed help from the German crystallographers.

Herman Mark had moved from the Institute for Fiber Chemistry to the I. G. Farben chemical company in Ludwigshafen in 1927. He and Kurt

---

d 1 Å (ångström unit) = 10^{-10} meters.

Meyer, director of the Ludwigshafen laboratory, studied the fiber structures of cellulose, silk fibroin and other natural and synthetic materials. Because fiber diffraction only provided information about atomic planes parallel to the fiber axis, it was not possible to arrive at unambiguous structures. Instead, models were built of the polymer, and adjusted until a conformation that agreed with the X-ray diffraction data was achieved.

Meyer's and Mark's 1928 model of cellulose introduced an important idea into the structural analysis of biological fibers. In this new structure, the glucose subunits of cellulose were not all oriented in the same way along the fiber. Instead, each glucose molecule was rotated 180° with respect to its neighbors. It was this, not the presence of disaccharides, that explained why the unit cell of cellulose was two glucose units long – because of the 180° rotation, the same atoms occurred along the fiber axis at every second glucose (Figure 7.3). In crystallographic terms, neighboring glucose units were related by a translation along the fiber axis followed by a rotation. This symmetry operation was known to occur in true crystals, where it was known

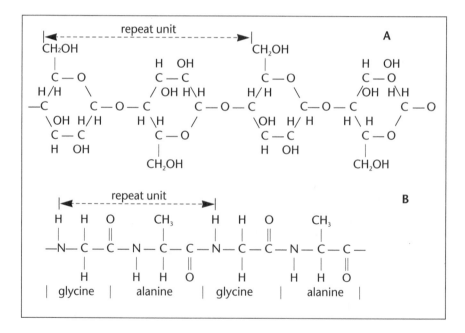

Figure 7.3: Crystallographic repeat structures of cellulose (A) and silk fibroin (B). In cellulose, each glucose unit is rotated 180° relative to its neighbors, resulting in a repeat unit of two glucose units. In silk fibroin, the repeat unit is a glycine-alanine dipeptide. As the side-chains of neighboring amino acids lie on opposite sides of the polypeptide, silk fibroin has a 'ribbon chain' structure similar to that of cellulose

as a screw axis.[e] In a fiber, the operation of a screw axis of symmetry would normally produce a helix. In the case of cellulose, however, as the rotation was 180°, all the subunits lay in the same plane, and this special case was referred to as a 'ribbon chain'.

Meyer and Mark also published a new structure for silk fibroin. The idea of glycine–alanine dipeptides held together by weak interactions was rejected in favor of polypeptide chains. Like cellulose, the unit cell of silk fibroin was two subunits long. In this case, the tetrahedral nature of the carbon atom dictated that the side-chains of adjacent amino acids would lie on opposite sides of the chain, resulting in planes of atoms repeating every two amino acids. Therefore, silk fibroin could also be considered to be a ribbon chain structure (Figure 7.3).

Cellulose and silk fibroin gave prominent X-ray reflections corresponding to periodicities along the fiber axis of 5.1 Å and 3.5 Å, respectively. Meyer and Mark interpreted these as being the distances between the same atoms in adjacent subunits of the two polymers. Astbury could not help but be struck by the resemblance between these values and those obtained from the two forms of keratin. If the 3.5 Å reflection of silk fibroin represented an extended polypeptide chain, as Meyer and Mark proposed, then it would make sense that a similar (3.32 Å) reflection was given by stretched β-keratin. Fibers of α-keratin were shorter, which must mean that the polypeptide chain was folded in some way. However, these unstretched fibers gave a diffraction pattern, so the folding of the polypeptide must produce some kind of ordered structure. The major reflection of α-keratin, at 5.15 Å, was almost exactly the same as that obtained from cellulose, which consisted of hexagonal glucose molecules.

From this, Astbury concluded that the alpha structure occurred by folding the polypeptide chain into a series of hexagons. These could be stabilized by 'side links' between the carbonyl (C=O) group of one amino acid and the amino (N–H) group of the adjacent amino acid (Figure 7.4). The spacing between the hexagons, corresponding to positions in the polypeptide chain three amino acids apart, was 5.15 Å. Stretching of the protein broke the 'side links' and straightened out the hexagons, placing amino acids three positions apart at a distance of 9.96 Å (3 × 3.32 Å), thus accounting for the approximate doubling in length of the stretched fiber. This model was the

e   The operation of a screw axis of symmetry gives a crystal 'handedness'. Such is the case for the right-handed and left-handed hemihedral forms of tartrates that allowed Louis Pasteur to discover optical isomerism (see Chapter 4).

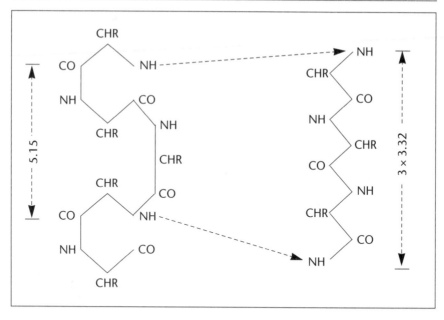

Figure 7.4: William Astbury's 1930 model of the α-β transition in keratin. Alpha-keratin (left) is organized into hexagonal folds; β-keratin (right) is a fully extended polypeptide chain. Adapted from Astbury, W. T., and Street, A. (1931) *Philosophical Transactions of the Royal Society of London A* **230**, 101, with permission

first attempt to understand how polypeptide chains could be folded in the native protein molecule. Unfortunately, and not for the last time, Astbury had constructed his theory around a numerical relationship that proved to be coincidental.

The German and British fiber crystallographers had discovered two general principles that necessitated a revised view of the structure of biological polymers. The first principle was that, contrary to what William Bragg and Reginald Herzog had thought, biological fibers could indeed consist of unit cells that were smaller than the molecule. This occurred when a polymeric molecule folded in such a way that its subunits repeated in space along the axis of the fiber. In the case of Meyer and Mark's cellulose structure, every second glucose unit was equivalent. In the α-keratin structure proposed by Astbury, every fourth amino acid was equivalent.

The second general principle was that subunits occupying equivalent positions in the fiber did not have to be identical. In the case of a fibrous protein, for example, a diffraction pattern would occur if the polypeptide chain folded into a repeating *structure*, such as Astbury's hexagons, but did not require that there be any regularity in the *sequence* of amino acids along the chain.

In other words, amino acids that were equivalent in space did not have to be equivalent in chemistry.

In coming to this reinterpretation of the fiber diffraction pattern in the late 1920s, crystallographers were explicitly accepting that biological fibers, including the fibrous proteins, were long molecular chains folded into a periodic structure, not identical small molecules interacting by secondary valences. Needless to say, this point of view was not compatible with the colloid theory of protein structure.

'I Have Chosen the Open Tundra'

Across the North Sea from Yorkshire was occurring a development of even more immediate importance for the analysis of protein structure – the invention of the ultracentrifuge by Theodor (The) Svedberg. Born in 1884, Svedberg graduated from the University of Uppsala in 1905 with a degree in chemistry. He remained at Uppsala for a PhD degree, performing his research on inorganic colloids. Following his graduation in 1907, Svedberg became a *docent* (tutor) in physical chemistry. Thanks to the support of Svante Arrhenius, who had won the Nobel prize for chemistry in 1903, a chair in physical chemistry was created for Svedberg at Uppsala in 1912.

Svedberg used the ultramicroscope to study the size distribution of inorganic colloid particles. As larger particles sediment more quickly than small ones, quantifying the amount of colloid at different levels in a column is a measure of the range of particle sizes. Svedberg found that coarse inorganic colloids would sediment under the gravitational field of the earth, but finer ones were kept in suspension by the thermal motion of the solvent. For metal particles, this imposed a lower size limit of about 200 nm – too large for Svedberg to study the formation of colloids. During a 1923 sabbatical at the University of Wisconsin, Madison, he drew up plans for a centrifuge that could be used to study the sedimentation of smaller particles.

Back in Uppsala, Svedberg started work on the construction of his new centrifuge. The key features were a sector-shaped 'cell' in which the sample was placed (to minimize convection forces) and an optical window for viewing the cell as it rotated. In a paper published at the end of 1924, Svedberg proposed the name 'ultracentrifuge' for this instrument; not because it was ultra-fast – his first machine could generate a centrifugal force of only 5000 times gravity (g) – but rather because of the ultramicroscope that provided the detection system.

Svedberg was not just a machine builder. He was also a talented theoretician who derived the relationship between sedimentation rate and molecular weight. This relationship was based on the realization that at constant centrifugal force a sedimenting particle would eventually reach an equilibrium at which sedimentation was balanced by diffusion. By measuring the concentration of the sedimenting species at two points in the ultracentrifuge cell at a known rate of rotation, the molecular weight could be calculated. In recognition of this contribution, the unit of sedimentation rate was officially named the Svedberg unit (S).

Use of the ultracentrifuge allowed metal particles as small as 5 nm to be studied, bringing almost the entire 'world of neglected dimensions' into the realm of experimental investigation. In 1924, Svedberg conceived of the idea of using the sedimentation technique to characterize organic colloids – the proteins. Ultracentrifugation of bovine milk casein gave a broad distribution (10–70 nm) of particles – exactly as expected from a colloid. Robin Fåhraeus, a pathologist at Uppsala whose wife and Svedberg's were old friends, then asked him to study hemoglobin. When the experiment was set up, Svedberg went home, but Fåhraeus stayed in the laboratory to watch. In the middle of the night Svedberg was awakened by a telephone call. It was Fåhraeus, excitedly telling him: 'The, I see a dawn!' The hemoglobin was sedimenting, leaving a clear zone in the top of the cell.

Svedberg's optical system measured the concentration of hemoglobin at ten positions in the ultracentrifuge cell. These concentrations were converted to molecular weights and averaged, giving (for different experimental conditions) values ranging from 62 000 to 71 000. Svedberg and Fåhraeus concluded that the molecular weight of hemoglobin was 66 800, corresponding to four 16 700 subunits.

The sedimentation equilibrium method could also yield information about the heterogeneity of the sample. If the sample contained molecules of different sizes, the molecular weights measured in the zones closer to the top of the ultracentrifuge cell would be lower than those measured near the bottom of the cell. For hemoglobin, no such variation in molecular weight was found. The ultracentrifugation of hemoglobin had shown not only that this protein was a large molecule, but also that it was monodisperse. The properties of hemoglobin therefore did not appear to be consistent with any version of the colloid theory: it was soluble, diffused slowly, formed crystals and had a uniform particle size.

Perhaps hemoglobin was a rare exception to the colloidal protein rule.

Svedberg wished to study a variety of proteins in his ultracentrifuge, but most proteins lacked the convenient pigmentation of hemoglobin. They could, however, be detected by ultraviolet light. Svedberg used a mercury lamp and halogen filters to produce ultraviolet light of the required wavelength, and an optical window made of quartz rather than glass, which absorbs ultraviolet radiation. By the end of 1926, this system had been used to measure the sedimentation equilibrium of ovalbumin. The results obtained confirmed the molecular weight measured in 1917 by Sørensen, and showed that ovalbumin, like hemoglobin, was monodisperse: 'it may be concluded with confidence that egg albumin in salt-free conditions is composed for the most part, perhaps almost exclusively, of molecules of weight 34 500 ± 1000 for dilute solutions.' In September 1926, Svedberg wrote in his laboratory notebook: 'Not the slightest indication of the presence of differently sized molecules could be noticed.' Two months later the Swedish Academy of Sciences announced that the Nobel prize in chemistry for that year would be awarded to Svedberg for his work on 'dispersed systems'. Ironically, Svedberg was honored for his work on inorganic colloids just as he was disproving the colloidal nature of proteins.

A much more accurate method of determining protein homogeneity would be to measure the rate of sedimentation of a protein through the ultracentrifugation cell. This 'sedimentation velocity' technique would, however, require much higher centrifugal forces. The sum of 25 000 Swedish crowns – a huge amount at that time – was obtained from a medical foundation for the development of an oil turbine ultracentrifuge. By mid-1926, Svedberg had built a machine with a maximum speed of 42 000 revolutions per minute and a maximum centrifugal force of 104 000 g.

Use of the sedimentation velocity technique greatly increased the efficiency of Svedberg's laboratory, as each run took 2–6 hours instead of the 2–6 *days* required for sedimentation equilibrium. By 1930, twenty-eight proteins had been studied in the ultracentrifuge. Of these, exactly half were monodisperse. By this time Svedberg had been completely converted to the polypeptide theory of proteins, and he attributed the polydisperse behavior exhibited by the remaining fourteen proteins to partial degradation of an originally monodisperse population. In the end, the only protein apart from casein to exhibit true colloidal behavior was gelatin, the analysis of which had first prompted Thomas Graham to propose the existence of the colloidal condition of matter.

Svedberg continued with the development of the ultracentrifuge for many years. It was frustrating and dangerous work, as about half the experimental rotors exploded, but eventually resulted in commercial machines that became standard laboratory equipment. Svedberg claimed that every new rotor tested took a year off his life. Despite this, he managed to live until the age of eighty-seven, fathering twelve children by four wives. For an eightieth birthday present, he asked for and received a botanical expedition to Greenland. His own words make a fitting epitaph: 'Both as a scientist and as a botanist, I have chosen the open tundra.'

'A Molecular Weight of 100 000 is Somewhat Terrifying'

By the late 1920s, both the X-ray diffraction of fibrous proteins and the ultracentrifugation of soluble proteins were being interpreted in terms of large molecules rather than colloidal aggregates. A third major blow to the colloidal theory of protein structure came from the work of Hermann Staudinger, professor of organic chemistry at the University of Zürich. In the oddly circuitous way in which science often proceeds, the substance that involved Staudinger in the colloid debate was rubber. By 1910, it had been realized that rubber contained small molecules known as isoprene units which, by virtue of their chemical structure, could react with one another to form chains or rings. In keeping with the colloid orthodoxy, rubber was thought to consist of isoprene rings of a few units each, interacting by secondary valences. When stretched, however, rubber gave a fiber diffraction pattern. For this reason, Staudinger believed that it must consist of long molecular chains held together by 'normal valence bonds'.

To test his conception of polymers, Staudinger devised a means of saturating rubber with hydrogen. If it were a colloid, saturation should break the secondary valences and result in the formation of a volatile hydrocarbon. Instead, a paraffin-like substance was produced. Several of the new plastics, such as polyoxymethylene and polystyrene, were also shown to behave like high-molecular-weight polymers rather than colloidal aggregates. Various chemical modifications of these plastics failed to affect their apparent degree of polymerization. In 1922, Staudinger introduced the term *Makromolekül* (macromolecule) for substances like rubber. Two years later, he defined macromolecules as 'colloidal particles in which the molecule is identical with the primary particle and in which the individual atoms of the colloidal molecule are linked by normal valences'.

In addition to his observations that chemical modification of polymers did not cause them to 'disaggregate', Staudinger used two other arguments to support the macromolecule theory. The first was that densities, melting points and X-ray diffraction patterns of his polymers were extrapolations of those of the corresponding monomer. The second was that there was a direct relationship between viscosity and polymer molecular weight (Staudinger's law).

Staudinger was an organic chemist who mainly worked on synthetic polymers. However, he believed that his macromolecule theory applied equally to natural polymers. Staudinger considered polystyrene analogous to rubber, polyoxymethylene to cellulose and polyacrylic acid to proteins.

The macromolecule theory encountered tremendous resistance. A 'stormy' lecture Staudinger gave to the Zürich Chemical Society in 1925 ended with him shouting the words of Martin Luther: '*Hier stehe ich, ich kann nicht anders*' ('Here I stand, I cannot do otherwise'). According to Herman Mark, Staudinger's critics were of three types: those who found the macromolecule theory incredible, because they had only ever encountered molecules of a few hundred daltons; those who felt it was unnecessary, because the compounds Staudinger studied were in fact colloids; and those who felt it was impossible, because the crystallographic unit cell cannot be smaller than the molecule. However, even Mark could not accept the existence of giant molecules. His and Meyer's 1928 structure of cellulose consisted of chains of thirty to fifty glucose molecules, forty to sixty of these chains aggregating to form a 'micelle'. Not until 1935 did Meyer and Mark fully accept the macromolecule theory.

In 1926, a debate on macromolecules was held in Düsseldorf as part of the annual meeting of the Society for Natural Science and Medicine – the same forum at which Fischer and Franz Hofmeister had debated the polypeptide structure of proteins in 1902 (see Chapter 5). Hans Pringsheim and Max Bergmann presented the arguments in favor of a colloid structure for Staudinger's 'macromolecules', such as the evidence for secondary valences and the fact that many inorganic colloids had viscosities as high as Staudinger's polymers. Staudinger himself then presented his data on rubber and artificial polymers, noting that proteins and cellulose were also probably macromolecular. The final speaker, Herman Mark, concentrated on the evidence from X-ray diffraction, pointing out that situations in which the crystallographic unit cell was smaller than the molecule were now well established. The structure of graphite, which recently had been solved by Bernal, consisted of sheets of carbon atoms with primary valences linking atoms in adjacent unit cells. Diamond, whose structure had earlier been solved by the Braggs, exhibited the same phenomenon in three dimensions. If these carbon compounds consisted of unit cells covalently bonded into giant networks, why should not carbon-containing fibers exhibit the same phenomenon in one dimension? According to Mark's account, the chair of the debate, Richard Willstätter, closed the session with these words: 'For me, as an organic chemist, the concept that a molecule can have a molecular weight of 100 000 is somewhat terrifying, but, on the basis of what we have heard today, it seems that I shall have to slowly adjust to this thought.'

'I Decided to Take a "Long Shot"'

Richard Willstätter was indeed slow in adjusting to the thought of proteins as giant molecules. Even after the 1926 symposium in Düsseldorf, he was a leading proponent of the idea that enzymes were colloids. This belief was rooted in his finding that enzymes could be purified to the point that no protein was detectable. In his 1927 Faraday lecture, Willstätter wrote: 'It seems that we must consider an enzyme to be composed of a specifically active group and a colloidal carrier. To this, other substances of high molecular weight cling in various ways.' The colloidal carrier was not inert, however – Willstätter considered it to be responsible for the temperature- and pH-dependence of the enzyme-catalyzed reaction, as well as the effects of inhibitors and activators. The 'enzyme' was only responsible for catalysis.

Willstätter's theory of colloidal enzymes is an excellent example of a scientist being wrong for perfectly good reasons. He was quite logical in arguing that the stepwise purification of an activity should result in a relative increase in the chemical entity responsible for the activity and a relative decrease in contaminants. One can hardly fault him for failing to realize that his apparent preparation of protein-free enzymes was merely because the assays of enzyme activity were more sensitive than those for protein. It was a strange idea for an organic chemist that a substance could be purified to the point that its action could be detected, but not its physical presence. It should also be said in Willstätter's defense that his idea of enzymes consisting of an 'active group' attached to a 'colloidal character' was in some ways prescient – if one substitutes 'active site' and 'protein molecule' for the terms he used, one arrives at a conception of enzymes very close to the contemporary one. From this point of view, Willstätter's physical separation of catalytic activity from the effects of inhibitors and activators anticipated the finding that enzymes have separate catalytic and regulatory sites (or subunits).

The same year as Willstätter's Faraday lecture, however, came the news that enzymes could be crystallized. James Batcheller Sumner of the Department of Biochemistry at Cornell University, then thirty years old and lacking an arm from a hunting accident, had decided in 1917 to isolate an enzyme in pure form: 'At that time I had little time for research, not much apparatus, research money or assistance. I decided to accomplish something of real importance. In other words, I decided to take a "long shot". A number of persons advised me that my attempt to isolate an enzyme was foolish, but this advice made me feel all the more certain that if successful the quest would be worthwhile.'[54]

Sumner chose to work with the enzyme urease from jack beans because it was abundant and its activity could easily be measured. Lacking a proper apparatus to grind up the beans, he used a coffee mill. The enzyme activity could be extracted from the ground beans with 30% ethanol, and then precipitated by lowering the temperature. However, Sumner did not have an 'ice chest', either, so he had to put his solutions on the window ledge and 'pray for cold weather'. When he tried using acetone instead of ethanol, no visible precipitate formed. However, the acetone contained tiny crystals that tested positive for both urease activity and protein – apparently, the elusive crystalline enzyme.

Sumner's evidence for the crystallinity of urease was not universally accepted. Willstätter thought that he had merely co-crystallized the enzyme with a protein 'carrier'. Three years later, however, John Northrop of the Rockefeller Institute for Medical Sciences, a former associate of Jacques Loeb, obtained crystals of the intestinal enzyme pepsin. Northrop realized that the isolation of biological molecules would require new criteria of purity. In organic chemistry, a compound was considered pure if its elemental composition and melting point remained constant through repeated crystallization steps. For enzymes, these tests were useless – the elemental composition of a protein was so complex as to be almost meaningless, and protein crystals decompose instead of melting on heating.

Northrop had read Sørensen's papers showing that proteins obeyed the Gibbs' phase rule. This states that the equilibrium between a solid phase and a solution will be changed by the addition of another solid substance but not by the addition of the same substance. By this extremely sensitive test of chemical homogeneity, Northrop's crystalline pepsin was pure. It also behaved as a monodisperse macromolecule in Svedberg's ultracentrifuge. By this time, Svedberg's student Arne Tiselius had developed a method of separating proteins on the basis of electrical charge by allowing them to move through a gel towards the positive or negative electrode. Northrop's pepsin migrated as a single band in the Tiselius electrophoresis apparatus.

If it had been possible to quibble about the nature of Sumner's crystalline urease, there did not seem to be much room for doubt about Northrop's crystalline pepsin. Based on the most sensitive chemical tests and the most modern biochemical methods, the latter appeared to be homogeneous. Enzymes, like many other proteins, were crystalloids.

The tide had turned decisively against the colloid theory of protein structure and, by inference, the colloid theory of life. By 1930, enzymes had been shown to be proteins; many proteins had been shown to be crystalline, others (the fibrous proteins) para-crystalline; the great majority of proteins studied in the ultracentrifuge were homogeneous. Little more would be heard of colloidal aggregates in protein chemistry. From this point onwards, the emphasis would be on establishing the patterns of folding of the polypeptide chain. In 1946, Sumner and Northrop shared the Nobel prize in chemistry for the preparation of crystalline enzymes. Staudinger would have to wait longer for vindication; it was not until 1953, when he was seventy-two years old, that the true significance of his macromolecule hypothesis was recognized by the award of the Nobel prize for chemistry.

'The Molecules of the Proteins are Formed According to a Common Building Plan'

The state of protein chemistry at the end of the decade was summed up in a 1928 review article by Hubert Vickery and Thomas Osborne. They listed twenty amino acids found in proteins, exactly the number now known to be used by the cellular biosynthetic mechanism. However, their list included four amino acids ('oxyglutaminic acid', 'oxyproline', 'iodogorgoic acid' and thyroxine) that do not occur in proteins or are modified forms of others listed. Of the four amino acids missing from Vickery's and Osborne's list that do occur in proteins, methionine was mentioned in a footnote. Completely absent were glutamine and asparagine, which are converted to glutamic acid and aspartic acid, respectively, during acid hydrolysis of proteins, and threonine, which is often destroyed. Until the solving of the genetic code in the 1960s, it was not entirely clear which amino acids were incorporated into proteins and which were formed by post-synthetic modification of other residues.

For Vickery and Osborne, the evidence for the macromolecular nature of proteins was now unequivocal: 'there are several independent and trustworthy lines of evidence that point unmistakably to a molecular weight of the order 34 000 for [ovalbumin] . . . A careful consideration of the iron and sulphur contents of hemoglobins from different animal species indicates . . . a figure of the order 68 000.' These values represented giant amino acid polymers, not colloids: 'the peptide hypothesis of Hofmeister is still, in spite of certain shortcomings, the foundation stone of protein chemistry . . . The future development of protein structure theory may be most confidently looked for in an expansion of the peptide hypothesis.'[55]

At least for these authors, the colloid theory of protein structure could be laid to rest. Many other questions remained unanswered, though, including what happened to proteins when they were denatured by heating, and why fibrous proteins such as keratin were insoluble. Vickery and Osborne also noted that: 'It seems highly probable that some configuration must be present which acts in such a way as to render the molecule far more compact than would be supposed upon the simplest and most obvious premises.' This question of how a linear polypeptide could be folded into a roughly spherical or cylindrical shape clearly motivated proposals such as the diketopiperazine structure of Abderhalden, and was to dominate protein chemistry for many decades to come.

As noted above, the best evidence in support of the view that proteins were giant polymers of amino acids rather than colloidal aggregates came from the techniques of ultracentrifugation and fiber X-ray diffraction. By the late 1920s, some intriguing parallels were being drawn between the results obtained from these two systems. In 1929, Svedberg noted that proteins studied in the ultracentrifuge tended to have molecular weights of 34500 or multiples of this value by the factors two, three or six. Ovalbumin was 34500; hemoglobin was two times, serum globulin three times and edestin four times that value. To Svedberg, this algebraic regularity suggested that 'one must assume that the molecules of the proteins are formed according to a common building plan.'

One person who enthusiastically adopted Svedberg's 'common building plan' for proteins was William Astbury. In fact, he went further than Svedberg by proposing in 1931 a theory of why 34500 daltons represented a maximum size for polypeptides. According to Astbury, 'vibrational instability' and 'disruptive resonance' of larger polypeptides results in 'spontaneous decomposition into shorter chains'. This is an early example of the weakness for wildly speculative models based on flimsy experimental evidence that would plague Astbury's attempts to understand the structures of proteins and nucleic acids.

However, Astbury was more interested in the similarity between Svedberg's molecular weight factors of two, three and six and the rotational symmetries of crystals: two-, three-, four- and six-fold symmetries were the only ones possible. This suggested to Astbury that the proteins of Svedberg's higher-molecular-weight groups may be composed of 34 500 dalton polypeptides organized in a symmetrical fashion. Therefore, Astbury generated models of multi-subunit proteins with the kinds of symmetries suggested by ultracentrifugation. To achieve two-fold symmetry, Astbury proposed a structure composed of two extended polypeptide chains running in opposite directions with 'secondary valences' between the C=O and N–H groups of the two chains (Figure 7.5). For three-fold symmetry, he proposed a structure composed of three polypeptides all running in the same direction but staggered such that bonds between C=O and N–H groups could occur perpendicular to the chain axes.

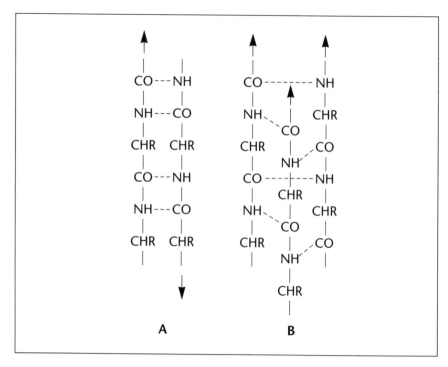

Figure 7.5: William Astbury's 1931 structures of proteins consisting of two (A) or three (B) polypeptide chains held together by 'secondary valences'. Note that the polypeptides in (A) run in opposite directions, whereas those in (B) run in the same direction. The two-chain protein corresponds to crystallographic space group C2, the three-chain protein to space group C3. Adapted from Astbury, W. T. and Woods, H. J. (1931) *Nature* **127**, 664

In this, Astbury was well ahead of his time. Basing macromolecular models on crystallographic space groups would later be used in determining the three-dimensional structures of proteins and nucleic acids. Astbury's 'secondary valences' between C=O and N–H groups, later recognized as hydrogen bonds, would be shown to be key determinants of protein conformation. Furthermore, his two-chain structure can be recognized as the earliest precursor of the pattern of protein folding later known as the antiparallel β-pleated sheet (see Chapter 11). Unfortunately for Astbury, the credit for these insights would go to others.

'A Polymer of the Tetranucleotide'

In 1910, Albrecht Kossel was awarded the Nobel prize in chemistry for his 'contributions to our knowledge of cell chemistry made through his work on proteins, including the nucleic substances'. In his presentation speech, Count Mörner of the Karolinska Institute mentioned that, through lack of time, Kossel's work on the nucleic acids 'must be passed over on this occasion'. The Nobel laureate, however, knew on which side his bread was buttered, and devoted more than half his lecture to the nucleic acids.

Largely thanks to the work of Kossel's laboratory, it was known that 'yeast nucleic acid' and 'thymus nucleic acid' both contained four bases (two each of the purine and pyrimidine types), sugar and phosphate (see Chapter 5). Thanks to Phoebus Levene, it was known that the sugar in yeast nucleic acid was D-ribose. Now that the *Bausteine* of the nucleic acids were established, Kossel noted: '. . . two new questions arise: What are the relative amounts of each block, and how are they mutually arranged?'

Regarding the first question, Kossel, like Levene, accepted Hermann Steudel's view that the bases were present in equimolar amounts. However, this was by no means universally accepted. Other nucleic acid chemists reported that there was no uracil present in yeast nucleic acid, or that only one uracil was present for every two of the other bases.

Identifying the chemical structure of the nucleic acids was a much more difficult problem. In proteins, the building blocks were all chemically similar – each amino acid had a carboxylate group and an amino group. If all the amino acids were to be joined in the same manner, the only real possibility was to form amide bonds between the carboxylate (COOH) group of one amino acid and the amino (NH_2) group of the next, as envisaged in the polypeptide structure of Hofmeister and Fischer. There were simply no other reactive groups common to all amino acids. For the nucleic acids, the picture was much more complicated. The sugars had several hydroxyl (OH) groups that were all equally likely to be involved in bond formation (see Figure 4.3), the rings of the bases had several carbon and nitrogen atoms that could be used as points of attachment (see Figure 5.1), the phosphate ion had three reactive oxygen atoms – the possible ways to combine these components was dauntingly high.

In the decade following Kossel's Nobel lecture, various structures were proposed for the yeast and thymus nucleic acids (Figure 7.6). The only feature these had in common was the presence of Levene's nucleotide units of phosphate–sugar–base. Levene's own 1912 structure of thymus nucleic acid contained sugar–phosphate and sugar–sugar bonds; Siegfried Thannhauser proposed in 1917 a structure of yeast nucleic acid containing similar linkages; in Walter Jones's 1917 structure of this acid, the nucleotides were joined via the sugar groups; Robert Feulgen's 1918 structure of thymus nucleic acid had the nucleotides joined by a mixture of sugar–sugar, phosphate–phosphate and sugar–phosphate bonds.

For several years after 1912, Levene published almost nothing on nucleic

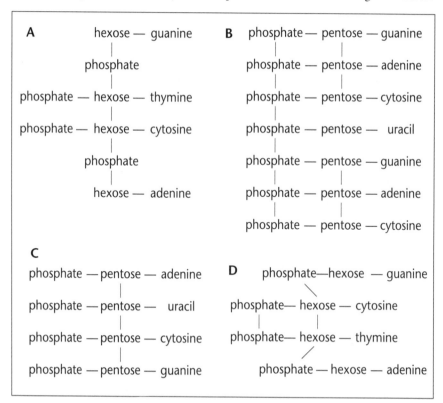

Figure 7.6: Proposed structures of the nucleic acids. (A) Phoebus Levene's 1912 structure of thymus nucleic acid. (B) Siegfried Thannhauser's 1917 structure of yeast nucleic acid. (C) Walter Jones' 1917 structure of yeast nucleic acid. (D) Robert Feulgen's 1918 structure of thymus nucleic acid

acids, in part because World War I interrupted his supply of yeast nucleic acid. When he returned to the field in 1917, Levene decided upon a fresh start to the problem of nucleic acid structure. From first principles, he noted, there were six possible ways in which nucleotides could be linked: phosphate–phosphate, phosphate–sugar, phosphate–base, sugar–sugar, base–sugar, and base–base. Because of the different chemistries of the bases, the three forms of union involving the bases seemed unlikely.

Levene's thinking about nucleic acid structure was heavily influenced by his now-considerable experience with different methods of hydrolysis. In particular, he was struck by the ease with which yeast nucleic acid could be broken down with alkali. As little as 2.5% ammonium hydroxide at 100°C for 1 hour was sufficient to convert this nucleic acid to mononucleotides. It was inconceivable that such mild conditions would cleave sugar–sugar or phosphate–phosphate bonds. However, mild alkali could hydrolyze sugar–phosphate bonds. Accordingly, Levene proposed in 1919 a structure of yeast nucleic acid in which all the nucleotides were linked by ester bonds between the phosphoric acid group of one and the ribose group of the next (Figure 7.7). Two years later he was sufficiently confident in this sugar–

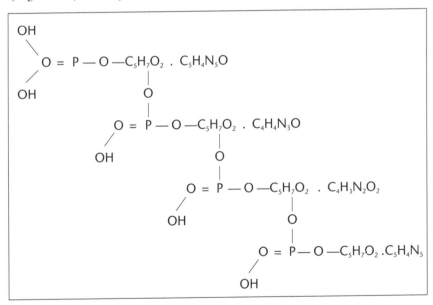

Figure 7.7: Phoebus Levene's 1919 structure of yeast nucleic acid. As in his 1909 structure (Figure 5.3), the molecule can be dissociated into nucleotide (phosphate–sugar–base) units. However, the backbone of the molecule now consists of alternating phosphate and pentose ($C_5H_7O_2$) groups rather than just phosphates. Note that the hydroxyl (OH) groups should be attached to the phosphorus (P) atoms rather than to oxygen (O) atoms. Adapted from Levene, P. A. (1919) *Journal of Biological Chemistry* **40**, 420, with permission

phosphate structure to extend it to thymus nucleic acid, even though that substance was not susceptible to mild alkaline hydrolysis.

If Levene were correct in thinking that yeast nucleic acid consisted of a phosphate–ribose 'backbone' to which the bases were attached via the ribose molecules, only minor aspects of the structure remained to be resolved, such as the exact carbon atoms of ribose to which the base and the phosphates were attached. For thymus nucleic acid, however, a more fundamental question remained: the nature of the sugar. For over thirty years, this had been believed to be a hexose, based largely on the fact that acid hydrolysis of thymus nucleic acid produced levulinic acid ($C_5H_8O_3$). The hexose itself had never been isolated, because any conditions that released it seemed to convert it to levulinic acid.

In 1923, the great digestive physiologist Ivan Pavlov, then seventy-five years old, was visiting New York. When he had his pocket picked at Grand Central Station, Pavlov turned to his former student, Levene, for help in getting money and travel documents. When Pavlov learned of Levene's difficulty in isolating the sugar from thymus nucleic acid, he suggested that he try using enzymatic digestion. Pure enzymes were not available, but it was known that the intestinal mucosa contained activities that broke down nucleic acids. The following year, Levene visited Pavlov's laboratory in Leningrad. The attempt to use intestinal extracts to produce small fragments of thymus nucleic acid – possibly free sugars – that could be analyzed chemically was unsuccessful.

In 1929, Pavlov's colleague Efim London came to Levene's laboratory to try a different approach. Gastric and intestinal fistulas were made in dogs so that thymus nucleic acid could be added through the stomach and recovered from the intestine. It was found that the resulting intestinal fluid contained the four nucleoside components of the nucleic acid. Hydrolysis of the purine nucleosides with very mild acid produced a free sugar that Levene and London provisionally named 'thyminose'. Analysis of its elemental composition and chemical properties showed that thyminose was in fact 2-D-deoxyribose, a derivative of D-ribose in which a hydrogen atom, rather than a hydroxyl group, is attached to carbon 2 of the sugar ring (Figure 7.8).

The isolation of deoxyribose greatly simplified the analysis of the structure of thymus nucleic acid. Whereas hexoses have five hydroxyl groups, deoxypentoses like 2-D-deoxyribose only have three. In Levene's sugar–phosphate structure of nucleic acids, one hydroxyl group of the sugar was required for attachment to the base, and two more for attachment to the

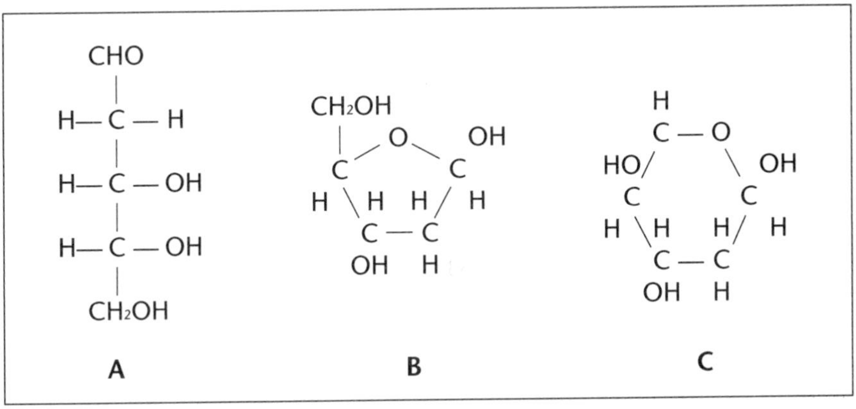

Figure 7.8: Structures of 2-D-deoxyribose. (A) Straight chain. (B) Furanose ring. (C) Pyranose ring

phosphates. Once Levene had established that the base was attached to the hydroxyl group on carbon 1, it was clear that the phosphate groups must be attached to the remaining two positions. However, there was a complication: pentoses can adopt two ring conformations, known as the pyranose and furanose forms. In the pyranose (six-membered ring) form, the hydroxyl groups are on carbons 1, 3 and 4; in the furanose (five-membered ring) form, the hydroxyl groups are on carbons 1, 3 and 5 (Figure 7.8).

The demonstration that the 2-D-deoxyribose of nucleotides is in the furanose conformation removed this last ambiguity in the chemical structure of thymus nucleic acid. In 1935, Levene published his final structure of this nucleic acid, which he was now referring to as deoxyribose nucleic acid, or DNA. This structure consisted of deoxypentose nucleosides connected by phosphate groups between the carbon 5 hydroxyl group of one sugar and the carbon 3 hydroxyl of the next, with the bases attached to carbon 1 (Figure 7.9).

Levene's 1935 paper also showed a new structure for yeast nucleic acid, which was now referred to as ribosenucleic acid, or RNA. Ribose has four hydroxyl groups, which in the furanose form are on carbon atoms 1, 2, 3 and 5. Unlike in the case of DNA, therefore, there was one hydroxyl group in excess of those required for bonding to the other components. The simplest assumption would have been that the phosphate groups were attached to carbons 3 and 5, as in DNA. However, Levene still had to explain the difference in alkaline hydrolysis between the two nucleic acids. For this reason, he proposed – incorrectly – that the phosphoester bonds in RNA were between carbons 2 and 3.

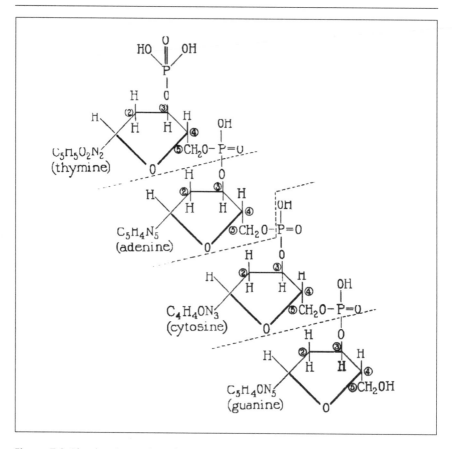

Figure 7.9: Phoebus Levene's and Stuart Tipson's 1935 structure of deoxyribose nucleic acid. The component bases, sugars and phosphate groups are attached in the same general way as in Levene's 1919 structure of yeast nucleic acid (Figure 7.7), but the conformation of the sugar (furanose ring; Figure 7.8) and the point of attachment of the phosphates to deoxyribose (carbons 3 and 5) are now specified. The point of attachment of the sugars to the bases is not specified, but the structure is otherwise complete and correct. Reproduced from Levene, P. A. and Tipson, R. S. (1935) *Journal of Biological Chemistry* 109, 625, with permission

The 1935 structures of the nucleic acids were the culmination of Levene's career, resulting from thirty-five years of work by his laboratory. The achievement of an entirely correct (although incomplete) structure of DNA and a slightly incorrect structure of RNA represents one of the most outstanding contributions of organic chemistry to biology. As noted by Alexander Todd, who fifteen years later was able to confirm Levene's DNA structure, the analysis of the nucleic acids took chemistry into uncharted terrain: 'As water-soluble polar compounds with no proper melting points [the nucleotides] were extremely difficult to handle by the classical techniques of organic chemistry.' Even more impressive is the fact that, when Levene isolated 2-deoxyribose, this sugar was unknown to science.

Levene's and Tipson's structure of DNA was drawn as a tetranucleotide, containing one unit each of adenine, cytosine, guanine and thymine, and four units each of 2-deoxyribose and phosphate. By this time, however, Levene was aware that this was merely the minimum unit of structure. In his 1931 monograph, *The Nucleic Acids*, he addressed the question of nucleic acid size for the first time: 'it must be borne in mind that the true molecular weight of nucleic acids is as yet not known. *The tetranucleotide theory is the minimum molecular weight and the nucleic acid may as well be a multiple of it*' [author's emphasis]. Levene must have been aware that the 'energetic' methods used to isolate nucleic acids from tissue were hardly guaranteed to preserve their native structure. The standard Kossel–Neumann method, for example, involved boiling tissues in sodium hydroxide for up to two hours. Levene may also have realized that the techniques used to characterize nucleic acids rarely gave information about the size of the molecule.

In 1938, two years before his death, Levene published a study in which a 'ribonucleo-depolymerase' was used to cleave nucleic acid from yeast, which he now referred to as ribonucleic acid. From an analysis of the products obtained, he concluded: 'Native ribonucleic acid is a polymer of the tetranucleotide.' That same year, Levene reported that 'native' DNA sedimented in the ultracentrifuge with an apparent molecular weight of 200000–1000000. He was therefore one of the first to propose a macromolecular structure of the nucleic acids – it was only in 1938 that the physical studies of Rudolf Signer and Astbury suggested that DNA was a large molecule (see Chapter 8).

Levene was a brilliant organic chemist with an encyclopedic knowledge of the discipline and a rare intuition for the correct alternative. As described in Chapter 5, his initial belief that nucleic acids were physiologically important, tissue-specific molecules was abandoned as early as 1907. Thereafter, he appeared to regard nucleic acids purely as a problem of structural chemistry. His solutions to this problem advanced nucleic acid chemistry from the elemental analyses of Kossel to detailed structural characterizations of the DNA and RNA molecules.

Unfortunately for Levene, the most significant impact of his work was to convince biologists that nucleic acids could not function as genes. The techniques available to him were not capable of detecting the subtle differences

in base content that occurred between different species of organism. As a
result, Levene believed that the four bases were present in equal amounts.
On that basis, he made the conventional assumption that the molecule contained only one unit of its least abundant components – that is, that nucleic
acids were tetranucleotides.

When, at the end of his career, Levene realized that nucleic acids were
macromolecules, he made another simplifying assumption – that DNA and
RNA were 'polymers of the tetranucleotide'. Thus, the base sequence of
DNA, shown in the 1935 structure as thymine-adenine-cytosine-guanine,
or T-A-C-G, would become in the polymeric form T-A-C-G-T-A-C-G-
T-A-C-G etc. This reinterpretation did not change the opinion of biologists that nucleic acids were uninteresting molecules. A redundant polymer
of identical tetranucleotide units could not match the chemical diversity of
the proteins, whose variations in amino acid content had long been
recognized. Such a molecule could act only as a structural element – 'the
wooden stretcher behind the Rembrandt', as Horace Judson put it. Such a
molecule could not be the gene.

Chapter 8
The Giant Molecules of the Living Cell

'A Flamboyant Don Juan'

As described in Chapter 7, John Desmond Bernal left William Bragg's group at the Royal Institution in 1927 to become lecturer in structural crystallography at Cambridge University. When he arrived in Cambridge, however, Bernal might have wished that William Astbury had been appointed instead. The 'hut' in which the crystallographers were housed was so cold in winter that their organic solvents froze. His student Max Perutz recalled 'a few ill-lit and dirty rooms on the ground floor of a stark, dilapidated grey building'. According to his research assistant, Dorothy Crowfoot, overhead wires running from transformers to X-ray tubes created such strong fields in Bernal's laboratory that 'when you went into the room, your hair used to stand on end, literally'. Bernal and Perutz were nearly electrocuted in one accident, and a student survived a massive dose of radiation in another. Things only got worse in 1935 when Bernal failed to achieve the Jacksonian chair in natural philosophy and a university reorganization shifted crystallography from the mineralogy to the physics department. This reorganization made the crystallographers responsible to Ernest Rutherford, the Cavendish professor of physics, who disapproved of Bernal's theoretical approach, politics and lifestyle.

By the standards of 1930s Britain, there was plenty to disapprove of in Bernal's politics and lifestyle. He was married to the same woman from 1922 until his death in 1971, but fathered children by two others. Perutz described Bernal as 'a bohemian, a flamboyant Don Juan'; the historian Pnina Abir-Am listed Bernal's marital status as 'libertine'! He joined the Communist Party of Great Britain in 1923 and remained 'as red as the flames of hell' until he died. Not even the suppression of Soviet genetics by Lysenko in the late 1940s, which was denounced by such left-wing scientists as Hermann Muller, J. B. S. Haldane and Jean Brachet, could shake Bernal's faith.

Despite the inhospitable environment of Cambridge, Bernal managed to set up a research program that was to have far-reaching influence on the field of biological structure. Just as William Bragg had concentrated on organic crystals while his son Lawrence worked on inorganic ones, Bernal reached a 'gentlemen's agreement' with Astbury that the fibrous substances would be studied in Leeds and the crystalline ones in Cambridge. As he wrote in 1968: 'I took the crystalline substances and he [Astbury] the amorphous or messy ones. At first it seemed that I must have the best of it but it was to prove otherwise . . .'

This arrangement left Bernal free to work with the classical low-molecular-weight crystalloids, such as amino acids. It also allowed him to study the soluble, non-fibrous proteins. Most such proteins that had been studied by physical techniques were thought to be of approximately spherical shape, and hence were known as globular, or 'corpuscular'. By 1930, many globular proteins and enzymes were known to be crystalline (see Chapter 7), and thus, in principle, amenable to analysis by X-ray methods.

However, Bernal did not attempt the X-ray diffraction of a protein until 1934. That year, his friend Glen Millikan was visiting The Svedberg's laboratory in Uppsala and discovered in a refrigerator a solution of pepsin that had crystallized while its owner was on a ski trip. The crystals were beautiful hexagonal bipyramids, 2 mm in length, and Millikan thought that Bernal might be interested in studying them. It was fortunate that Millikan took the pepsin crystals back to Cambridge in their 'mother liquor' (the solution in which they had formed), as it subsequently turned out that the crystal structure was lost on drying.

By diffracting the pepsin crystals in a capillary tube containing some mother liquor, Bernal and Crowfoot were able to obtain excellent X-ray photographs. These showed that 'the arrangement of atoms inside the protein molecule is also of a perfectly definite kind, although without the periodicities characterising the fibrous proteins.' It was also possible to determine the molecular weight of the unit cell, which appeared to be about six times that of the pepsin molecule. From the external symmetry of the pepsin crystals and the apparent presence of six molecules per unit cell, Bernal inferred that the pepsin molecules lay along a six-fold screw axis of symmetry – that is, the operation of going from one molecule to the next within the crystal involved a rotation of 60° and a translation along the rotation axis.

The big question was whether the ability of protein crystals to diffract X-rays meant that the protein molecule was itself symmetrical, or merely that several asymmetrical protein molecules could be packed into a symmetrical unit cell. Only in the former case would the crystal structure of a protein yield information about its molecular structure. Understandably, Bernal inclined towards a symmetrical protein molecule: 'Peptide chains in the ordinary sense may exist only in the more highly condensed or fibrous proteins, while the molecules of the primary soluble proteins may have their constituent parts grouped more symmetrically around a prosthetic nucleus.' In the hope that protein structure would be solvable by X-ray methods, Bernal was prepared to discard the polypeptide theory in favour of a structure that sounded much like Verworn's 'biogen' structure of the 1890s (see Chapter 5).

The next steps in the X-ray analysis of globular proteins were taken by Dorothy Crowfoot, who moved to Oxford University later in 1934. Her new laboratory, in a back room of the Ruskin Science Museum, had already earned a place in the history of science – it was the same room in which Thomas Huxley had debated evolution with Bishop Samuel Wilberforce in 1860 (see Chapter 3). It was not equipped for X-ray diffraction analysis, however. Robert Robinson, Waynflete professor of chemistry at Oxford, provided Crowfoot with some basic apparatus. More importantly, Robinson gave her some crystals of the hormone insulin. This protein would prove to be a good choice for crystallographic analysis – not only was it being purified in large quantities for the treatment of diabetes, but it also turned out to be a relatively small molecule.

Crowfoot published a preliminary X-ray analysis of insulin in 1935. Insulin crystals were of the rhombohedral class, which meant that the unit cell must have a three-fold axis of rotation. From the unit cell dimensions and crystal density, a unit cell molecular weight of 37 200 was calculated. This value was very close to the molecular weight determined for insulin by ultracentrifugation, suggesting that there was only one insulin molecule per unit cell. If this were the case, the crystal symmetry could not be explained by the arrangement of protein molecules in the unit cell – the insulin molecule itself must have three-fold symmetry. However, Crowfoot was less willing than Bernal to reach this conclusion: 'it is possible also that the crystal attains apparent trigonal [three-fold] symmetry by a statistical regularity of arrangement of molecules about the lattice points . . .' This special pleading was founded upon a sound instinct – the true molecular weight of bovine insulin turned out to be 5700, corresponding to six molecules per unit cell.

At the same time Crowfoot was performing her first studies on insulin, a technique was being developed that would eventually revolutionize the solving of protein structures by X-ray diffraction. William Bragg had suggested as early as 1915 that the periodicity of atoms in a crystal could be described by a mathematical expression known as a Fourier series, which was the sum of a series of sine waves. This idea was adopted by others to determine the density of electrons[a] along certain crystal edges and diagonals. Lawrence Bragg was the first to express crystallographic data as a two-dimensional Fourier series, producing contour maps of electron density in planes of the crystal. This allowed the younger Bragg to determine the positions of individual atoms in mineral crystals, but was impractical for the complicated diffraction patterns of proteins because it required that the 'phases' (positive or negative amplitudes) of the reflected X-rays be known.

Arthur Patterson, a former student of William Bragg's, believed that Fourier methods could be used to solve the crystal structures of complex molecules, and in 1931 took a two-year leave of absence from his job at the Johnson Foundation to investigate this further. Patterson came up with a method of expressing diffraction data in the form of 'vectors' that did not require determination of the phases. The vectors, which represented the major spacings between X-ray-scattering centers in the crystal, could be plotted on various crystal planes, as Lawrence Bragg had done for inorganic crystals, to give contour maps of sections through the three-dimensional structure. The resolution was low and atomic locations were not unambiguously established, but the 'Patterson maps' pointed the way to solving complex biological structures by X-ray analysis.

Crowfoot[b] was the first person to use the Patterson method (as subsequently modified by David Harker) on a protein crystal. The Patterson–Harker analysis of insulin yielded only fifty-nine vectors. None of these represented interatomic spacings below 7 Å, so no information was obtained about spacings between neighboring atoms. The contour maps therefore represented only a vague outline of the insulin molecule (Figure 8.1). Nonetheless, such maps were eventually to produce atomic-level details of protein structure; as Crowfoot wrote in 1979: 'No protein crystal structures could have been solved without their use.'

'As Promising as a Journey to the Moon'

When Herman Mark was working at the I. G. Farben plant in Ludwigshafen, his colleagues sometimes asked him why he bought the Nazi newspaper, the *Völkischer Beobachter*. Mark, who was half-Jewish, replied: 'If I want to know what is happening in Germany today, I shall read your newspapers,

---

a   It is mainly the electrons, not the nuclei, of crystals that diffract X-rays.
b   Crowfoot married in 1937 and subsequently published under her married name of Hodgkin. For the sake of simplicity, she is referred to here by her maiden name.

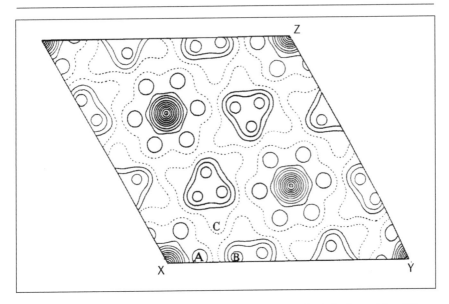

Figure 8.1: One of Dorothy Crowfoot's 1938 contour maps of relative intensities of X-ray scattering by the insulin crystal. Note the presence of features with three-fold and six-fold symmetries. Reproduced from Crowfoot, D. M. (1938) *Proceedings of the Royal Society of London A* **164**, 594, with permission

but I want to learn what will happen in Germany four or five years from now. Therefore, I read and believe the *Völkischer Beobachter*.' Mark could not have been entirely surprised when, in the summer of 1932, the managing director of the Farben plant informed him that the impending Nazi takeover would put his career in jeopardy. He applied for and was awarded the position of professor of physical chemistry at the University of Vienna.

In Vienna, one of Mark's students was Max Perutz. When Mark visited Britain in 1934, Perutz asked him to arrange a PhD position for him with Frederick Gowland Hopkins, professor of biochemistry at Cambridge. Mark went to see Bernal first and was so excited about the X-ray pictures of pepsin that he 'forgot' to talk to Hopkins. Instead, he arranged for Perutz to work with Bernal. When Perutz objected that he knew no crystallography, Mark breezily replied: 'You will learn it, my dear boy!'

Perutz arrived in Cambridge in 1936 and spent a year doing just that. In the summer of 1937, he visited his wife's cousin, Felix Haurowitz, in Prague. Haurowitz showed Perutz under the microscope how trigonal crystals of hemoglobin changed to monoclinic crystals when oxygen was added, and suggested that this protein would make a good PhD project. Little did either man suspect that this casual suggestion would determine the life's work of

Max Perutz. On his return to Cambridge, Perutz was able to obtain beautiful crystals of horse hemoglobin from Gilbert Adair of the physiology department. He then began his research towards a PhD, commuting by bicycle between the biochemistry laboratory at the Molteno Institute and the crystallography laboratory at the Cavendish.

At that time, the largest molecule whose structure had been solved by X-ray methods was phytocyanine, a pigment containing 58 atoms. Hemoglobin contains 11 000 atoms. In attempting the X-ray analysis of hemoglobin, Perutz and Bernal were motivated by the hope that there would be a regularity to the structure of the molecule. If not, any prospect of solving the hemoglobin structure was, as Perutz later put it, 'as promising as a journey to the moon'. In fact, it would take almost as long to achieve; a structure of hemoglobin at atomic resolution would not be published until 1968.

'The New Biology'

Another chemist who became interested in hemoglobin in the mid-1930s was Linus Pauling. Pauling was born in 1901 in Condon, Oregon, and studied chemistry at the Oregon Agricultural College (now Oregon State University) from 1917 to 1922. He then went to the California Institute of Technology (Cal Tech) to work with Roscoe Dickinson on the crystallography of minerals. On graduating from Cal Tech with a PhD in 1926, Pauling spent eighteen months in Europe, on a kind of scientific Grand Tour that took him to the Institute of Theoretical Physics in Munich, where fourteen years earlier Laue had discovered X-ray diffraction, as well as to Copenhagen, Göttingen and Zürich. In Munich, Pauling worked on a quantum interpretation of the hydrogen chloride bond. However, his aspirations to become a theoretical physicist were dashed when the brilliant but abrasive Wolfgang Pauli brusquely dismissed his ideas in two words: 'not interesting'.

When he returned to Cal Tech in 1927 with the unusual title of 'assistant professor of theoretical chemistry and mathematical physics', Pauling attempted to apply his knowledge of physics to the understanding of chemical bonds. Four years later, this resulted in the proposal of the valence bond theory, a quantum-mechanical description of the covalent (shared-electron) bond that explained the tetravalent carbon atom and numerous other chemical structures. On the experimental side, Pauling became the acknowledged master of what he described as structural chemistry – a rigorous application of crystallography and chemical bond theory to the analysis of chemical structure. By 1935, he was the rising star of American chemistry – only 34 years old, he was already a full professor at Cal Tech, recipient of the Langmuir prize of the American Chemical Society and a member of the prestigious National Academy of Sciences.

In many ways, Pauling was a twentieth-century Louis Pasteur. Both were men of humble origin who started their research careers in crystallography and ended them in medicine. Like Pasteur, Pauling gained tremendous honors, including the creation of an institute named after him. However, both men were also very sensitive to criticism, burying scientific adversaries under masses of data and using their powerful positions in the scientific establishment to promote their own views.

In 1929, Pauling formulated a series of rules for predicting the structure of a crystal from the chemical characteristics of its lattice ions. This work brought Pauling for the first time to the attention of Lawrence Bragg, who had developed but never published a similar set of principles. Bragg was understandably upset when 'his' principles of crystal building became known as 'Pauling's rules'. Unfortunately for Bragg, this was not to be the only time his scientific aspirations were thwarted by Pauling.

The conversion of Linus Pauling from chemical physics to biological chemistry was largely instigated by the Rockefeller Foundation, which had been set up to support the sciences and humanities with the vast bequest of oil baron John D. Rockefeller. In the mid-1930s, the president of the Rockefeller Foundation was the mathematical physicist Max Mason; Warren Weaver, a former student of Mason's at the University of Wisconsin, headed its natural sciences division.[c] Both men believed that rapid progress in biology, and therefore the betterment of humanity, could be made possible by the application of techniques from physics, chemistry and mathematics. Weaver used a variety of terms to describe the application of physical techniques to biology, including 'new biology', 'vital processes' and 'experimental biology', but in 1938 he invented a term that stuck – 'molecular biology'.

It is difficult to overestimate the influence of the Rockefeller Foundation on biological science during the period between the World Wars. From 1932 to 1959, its natural sciences division spent over US$90 million, most of it on biology. Grants from the Foundation supported the research of Svedberg, Astbury, Bernal, John Randall and George Beadle, among others, and

---

c   In 1924, Mason and Weaver developed a theory of sedimentation equilibrium that was later used by The Svedberg.

recipients of its fellowships or other forms of salary support included the future
Nobel laureates Arne Tiselius, Jacques Monod, Max Delbrück, Crowfoot,
Perutz and Hermann Muller. Weaver realized that one of the most important
effects of the Rockefeller Foundation on biology was to reorient the research
programs of some key scientists. Among this group of Rockefeller-influenced
converts to biology, perhaps the most important was Linus Pauling.

Soon after Weaver was appointed, he met with Pauling and told him that the
Foundation would not support his structural chemistry work – funding would
be forthcoming only for biology. Pauling decided to accede to the *Zeitgeist*
and study proteins. Like Perutz, he chose to work with hemoglobin, but for
quite different reasons. It was known that the hemoglobin molecule consisted
of a polypeptide, globin, and a non-polypeptide 'prosthetic group', heme.
Heme, which contained the iron atom and was the oxygen-binding site of
the protein, was known to be a relatively simple molecule of the porphyrin
class, which Pauling reasoned would be more amenable to structural analysis
than a large polypeptide. In addition, the iron atom of heme could be studied
by the techniques of inorganic chemistry. In 1935, Pauling set a graduate
student to work on the chemistry of the iron–heme bond. From that point
on, with the substantial support of the Rockefeller Foundation, Pauling's
research program became more and more directed towards biology. The man
who had learned theoretical physics from Sommerfeld and Bohr would end
his career doing research in nutrition and psychology.

'A Scientific Bloomsbury'

During his later years at Cambridge, Bernal was a member of the scientific
discussion group known as the Biotheoretical Gathering. This 'scientific
Bloomsbury', as Abir-Am described it, began in 1929 when Joseph Needham,
a lecturer in biochemistry at Cambridge University, met Joseph Woodger,
reader in biology at the Middlesex Hospital Medical School in London.
Woodger was interested in theoretical biology and the philosophy of science,
while Needham sought a unification between biochemistry and embryology.
Discussions between Needham and Woodger, later joined by the experimental morphologist Conrad Waddington, resulted in the first formal meeting of
the Biotheoretical Gathering in August 1932, at Woodger's house on Epsom
Downs. Also in attendance were Bernal; Dorothy Wrinch, a mathematician
and philosopher of science then living in Oxford; and the wives of Needham[d]
and Woodger. Subsequent meetings were held on an irregular basis.

---

d   Dorothy Needham was to become a distinguished scientist in the area of muscle biochemistry.

The discussions of the Biotheoretical Gathering focussed on the interrelationships between scientific disciplines and between science and philosophy, politics and ethics. Needham and Waddington were socialists, Bernal a communist. The political philosophy of the Biotheoretical Gathering was a somewhat utopian program of using scientific progress and scientific method for the betterment of society. This was a not uncommon idea in the 1930s, and to some extent it was a similar ethos that motivated the Rockefeller Foundation and the eugenic writings of Hermann Muller. The scientific program that emerged from the discussions of the Biotheoretical Gathering was broadly concerned with biological organization, initially at the level of morphogenesis (pattern formation) and subsequently at the level of chromosomes and proteins.

In 1935, Needham and Wrinch requested support from the Rockefeller Foundation for the establishment of an institute of 'mathematico-physico-chemical morphology' at Cambridge. The proposed institute was to have five divisions: chemical embryology (Needham), experimental morphology (Waddington), crystal physics of biological compounds (Bernal), physico-chemical genetics (Wrinch) and theoretical biology (Woodger). The Institute of Mathematico-Physico-Chemical Morphology should have been a perfect vehicle for the goals of Mason and Weaver, but Cambridge University refused to support it and peer review of the proposal by the Rockefeller Foundation was generally negative. In the end, the Foundation agreed only to support Wrinch, who was given a long-term grant to study 'application of knowledge of physical chemistry to proteins and chromosomes', 'linearity of chromosomes and linearity of proteins chains' and the 'chemical basis of genetics'. The Biotheoretical Gathering continued to meet and even added some new members, including Dorothy Crowfoot and the philosopher Karl Popper, but no collaborative endeavors occurred after 1935. Woodger became a full-time philosopher of biology and Waddington specialized in developmental genetics. Needham turned to the history of Chinese science, becoming 'the greatest sinologist in the West'.

'An Intruder from the Quiet Groves of Mathematics'

The most important scientific idea to come out of the Biotheoretical Gathering was Dorothy Wrinch's 'cyclol' structure of proteins. This was an ambitious, if premature, attempt to apply topology (the geometry of surfaces) to biological macromolecules. In her 1936 paper proposing the cyclol structure, Wrinch noted a number of characteristic features of proteins: there were few (possibly no!) free ends of polypeptide chains; protein crystals were normally of three-fold symmetry; many proteins were globular; many dissociated into subunits at high pH. From this she concluded that there was 'a general uniformity among proteins of widely different chemical constitution which suggests a simple general plan in the arrangement of amino acid residues'. Wrinch then observed that polypeptide chains containing numbers of amino acids in the series 2, 6, 18, 42, 66 . . . $(18 + 24\,n)$ could fold into hexagonal arrays with three-fold central symmetry (Figure 8.2). The hexagons were closed by covalent bonds between the C=O and N–H bonds of amino acids two positions apart in the chain. This created a planar structure that was 'dorsiventral' – all the amino acid side-chains projected from one side.

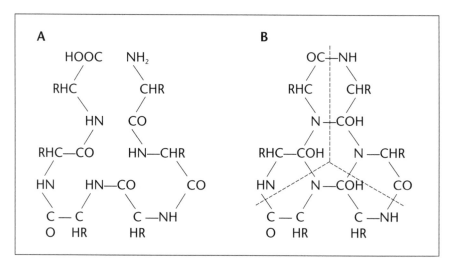

Figure 8.2: Folding of a polypeptide chain consisting of six amino acids into a cyclol. Note the presence of new hydroxyl (OH) groups and three-fold rotation axis (perpendicular to center of molecule) in the cyclol structure. The shape and symmetry of this cyclol are similar to those of the three-fold symmetrical features in the contour map of insulin (Figure 8.1)

Subsequently, Wrinch proposed that a three-dimensional cyclol structure could be present in the globular proteins. Folding the planar cyclol into a three-dimensional 'cage' was possible only with a more restricted number of amino acids, corresponding to the formula $72n^2$. For $n = 2$, the total number of amino acids in the structure was 288, which corresponded to a molecular weight of approximately 36 000 – similar to Svedberg's molecular weight values for several globular proteins: 'The suggestion has therefore been made that these [cyclols] may be appropriate structures for insulin, pepsin, egg albumin, and the other globular proteins which have molecular weights in the neighbourhood of 35 000, and that each globular protein consists of a $72n^2$ residue structure or of an association of such structures.'[56]

The cyclol theory had many attractive features. First, it reconciled the diketopiperazine theory of Emil Abderhalden with the polypeptide theory of Franz Hofmeister and Emil Fischer by suggesting, in essence, that proteins were a hybrid of both. Proteins were synthesized as linear polypeptides, but folded up into cyclols by a reaction similar to the one that produced diketopiperazines from free amino acids. Second, the cyclol structure had similarities to Astbury's 'hexagonal fold' model of α-keratin – in fact, the 'lactam–lactim' bonds that stabilized cyclols had been proposed in 1933 as a modification of Astbury's model – and therefore was generally consistent with the X-ray data. Third, Wrinch's theory explained Svedberg's finding that proteins were multiples of a 34500 dalton unit, as the 288 amino acid 'C₂' cyclol was approximately that size. Fourth, and perhaps most importantly, it addressed the vexing question of how a linear polypeptide could fold up into the compact shape of a globular protein.

The cyclol structure was well received at the Cold Spring Harbor symposium on proteins in 1938, and attracted the support of some influential figures, including Astbury and Lawrence Bragg. Bernal and Crowfoot, fellow members of the Biotheoretical Gathering, encouraged Wrinch in her theorizing; it was Crowfoot who suggested folding the cyclol 'fabric' to account for globular proteins. The eminent physical chemist Irving Langmuir later became a convert. Even Pauling was cautiously supportive of Wrinch's model, writing in 1937 that 'there is a great deal of truth in her general picture'. Many others, however, were extremely skeptical. Needham's colleague Norman Wingate Pirie, has recorded that, 'for ideological, objective and esthetic reasons', the cyclol theory was 'derided in the Cambridge Biochemistry Laboratory'. On more prosaic grounds of chemistry, Felix Haurowitz, who had introduced Max Perutz to hemoglobin, was also highly critical.

By 1938, the Rockefeller Foundation's Warren Weaver was worried enough about this mixed reception that he asked Pauling, then lecturing at Cornell University, to invite Wrinch for a seminar and report back with an opinion of her work. After the seminar, according to Pauling, 'several pertinent questions were raised' and the following day he and Wrinch had a 'lengthy discussion' as a result of which he reported to Weaver that the evidence for cyclols was 'quite weak'.

All these quotations may be taken as understatements. Wrinch, who described herself as 'an intruder from the quiet groves of mathematics', had written of the cyclol theory: 'a considerable part of the argument is concerned wholly with topological considerations and so is independent of any metrical data. This provides an elegant illustration of the important part which topology will inevitably play in the megachemistry of the future.' From Pauling's point of view, nothing could be *less* elegant than Wrinch's idea that her theory was supported by an absence of quantitative data. From this point on he would use his considerable influence to oppose the cyclol theory.

Later in 1938, the stakes in the cyclol debate were raised considerably. In Dorothy Crowfoot's X-ray analysis of insulin using Patterson–Harker sections, she wrote that: 'the patterns calculated do not appear to have any direct relation either to the cyclol or to the various chain structures put forward for the globular proteins.' Nonetheless, Wrinch and Langmuir wrote a paper claiming that Crowfoot's density maps were exactly in agreement with the structure of the C_2 cyclol, concluding that: 'these X-ray data, in giving so perfect a picture of the C_2 structure, provide the experimental basis for the cyclol theory.'

This development forced Bernal and Crowfoot to publicly disavow the cyclol theory. Bernal wrote a scathing letter to *Nature* about the musings of his former Biotheoretical Gathering comrade, accusing Wrinch of reporting data that supported her hypothesis while ignoring a much greater body of data that contradicted it; Bernal concluded that: 'Vector maps of Dr Wrinch's hypothetical cyclol structure bear no resemblance to those which have been derived by Miss Crowfoot from her observations.' In an accompanying letter, Lawrence Bragg, another early supporter of the cyclol theory, echoed Bernal's sentiments in more subtle terms: 'Exaggerated claims as to the novelty of the geometrical method of approach and the certainty with which a proposed detailed model is confirmed are only too likely, at this stage, to bring discredit upon the patient work which has placed the analysis of simpler structures upon a sure foundation.'

Wrinch and Langmuir's paper must also have come as a considerable irritant to Pauling. The support of the great Irving Langmuir meant that the cyclol theory could no longer be dismissed as the uninformed musings of a mathematician. The fact that their paper was published in the official journal of the American Chemical Society brought the controversy right to Pauling's backyard. When Carl Niemann, who had been highly critical of Wrinch's presentation at the Cold Spring Harbor symposium, suggested to Pauling that they publish an article on cyclols, he must have been only too happy to agree.

The resulting paper was a devastating refutation of the cyclol theory. Pauling

Plate 1: Antoine Lavoisier (1743–94) and his wife Marie, by Jacques Louis David (detail). Reproduced with the permission of the Metropolitan Museum of New York

Plate 2: Justus Liebig (1803–73). Reproduced, with permission, from *The Illustrated London News*, May 3, 1873

Plate 3: August Kekulé (1829–96). Reproduced, with permission, from the Edgar Fahs Smith Collection, University of Pennsylvania

Plate 4: Louis Pasteur (1822–95). Reproduced, with permission, from the Edgar Fahs Smith Collection, University of Pennsylvania

Plate 5: Friedrich Miescher (1844–95). Reproduced, with permission, from the portrait collection of the University of Basel Library

Plate 6: Albrecht Kossel (1853–1927). Reproduced, with permission, from the Edgar Fahs Smith Collection, University of Pennsylvania

Plate 7: Phoebus Levene (1869–1940). Reproduced, with permission, from the Edgar Fahs Smith Collection, University of Pennsylvania

Plate 8: Archibald Garrod (1857–1936). Reproduced, with permission, from *Genetics* **56**, p. 1 (1967)

Plate 9: Thomas Hunt Morgan (1866–1945). Reproduced, by permission of the President and Council of the Royal Society, from *Obituary Notices of Fellows of the Royal Society of London* **5**, p. 451 (1947)

Plate 10: Hermann Muller (1890–1967). Reproduced, by permission of the President and Council of the Royal Society, from *Biographical Memoirs of Fellows of the Royal Society of London* **14**, p. 349 (1968)

Plate 11: William Lawrence Bragg (1890–1971). Reproduced, by permission of the President and Council of the Royal Society, from *Biographical Memoirs of Fellows of the Royal Society of London* **25**, p. 75 (1979)

Plate 12: William Astbury (1898–1961). Reproduced, by permission of the President and Council of the Royal Society, from *Biographical Memoirs of Fellows of the Royal Society of London* **9**, p. 1 (1963)

Plate 13: John Desmond Bernal (1901–71). Reproduced, by permission of the President and Council of the Royal Society, from *Biographical Memoirs of Fellows of the Royal Society of London* **26**, p. 17 (1980)

Plate 14: Theodor Svedberg (1884–1971). Reproduced, by permission of the President and Council of the Royal Society, from *Biographical Memoirs of Fellows of the Royal Society of London* **18**, p. 595 (1972)

Plate 15: Max Delbrück (1906–81). Reproduced, by permission of the President and Council of the Royal Society, from *Biographical Memoirs of Fellows of the Royal Society of London* **28**, p. 59 (1982)

Plate 16: George Beadle (1903–89). Reproduced, with permission, from *Genetics* **124**, p. 1 (1990)

Plate 17: Oswald Avery (1877–1955). Reproduced, with permission, from *Genetics* **51**, p. 1 (1965)

Plate 18: John Kendrew (1917–97), with his model of the myoglobin molecule. Reproduced, with permission, from *Methods in Enzymology* **114A**, p. 18 (1985)

Plate 19: John Masson Gulland (1898–1947). Reproduced, by permission of the President and Council of the Royal Society, from *Obituary Notices of Fellows of the Royal Society of London* **6**, p. 67 (1948)

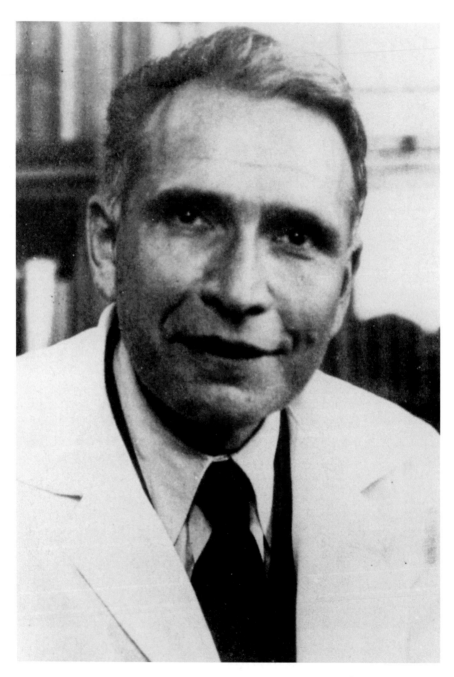

Plate 20: Erwin Chargaff (b. 1905). Reproduced, with permission, from *Annual Review of Biochemistry* **44**, p. 1 (1975), copyright 1975, by Annual Review http://www.Annua Reviews.org

and Niemann pointed out that cyclols were energetically far less stable than open polypeptide chains. The cyclol structure also brought atoms closer together then the sum of their van der Waals radii, which represented the minimum interatomic distance. Another problem was that cyclol formation should produce a large number of new hydroxyl (OH) groups (Figure 8.2), for which there was an absence of chemical or spectroscopic evidence. As for the Patterson–Harker analysis of insulin, to make these data fit a cyclol structure Wrinch and Langmuir had not only assigned arbitrary values to electron density parameters, but also had introduced new arbitrary parameters!

Wrinch then published a rebuttal of Pauling and Niemann's rebuttal, but the cyclol debate was essentially over. Wrinch, who had suffered the personal tragedy of her husband's insanity during this period, moved to the USA in 1939 and withdrew from the study of biological structure. Her only consolation was the eventual discovery of cyclol-like structures in certain plant alkaloids.

The cyclol incident illustrates the difficulties facing the scientist who tries to cross disciplinary boundaries. The bad feelings generated were probably the final blow to the interdisciplinary program of the Biotheoretical Gathering. As Crowfoot later recalled: 'Bernal first brought [the cyclol link] to the attention of Dorothy Wrinch . . . I suggested, I am sorry to say, that she folded the fabrics over a surface . . . We were friends of hers; we helped her to develop her theories; but we did not believe in them, and that was our trouble.'

'The Hypnotic Power of Numerology'

Astbury's 'hexagonal fold' structure of α-keratin and Wrinch's cyclol structure of the globular proteins were both inspired by two observations: that proteins appeared to be folded into a smaller volume than would be occupied by a fully extended polypeptide chain, and that the protein molecule appeared to exhibit some kind of internal periodicity. Neither of these structures depended upon a protein having a particular amino acid composition. The proposed periodicities were patterns of polypeptide folding such that its constituent amino acids formed a lattice in one-, two- or three-dimensional space. However, amino acids occupying equivalent positions in the lattice could be chemically different. It was realized that the position occupied by a lysine in one unit cell could be occupied by a tyrosine in another unit cell, and the structure would still exhibit periodicity (diffract X-rays).

A complete description of the structure of a protein would therefore involve not only determining the pattern of polypeptide chain folding but also the sequence of amino acids along that chain. In the 1930s, there was no way in which amino acid sequence could be determined. However, although it was now clear that different proteins varied widely in amino acid composition, it also seemed likely that certain proteins contained repetitive distributions of amino acids. The X-ray analysis of silk fibroin by Rudolf Brill (see Chapter 7), for example, had assumed a basic unit of glycine–alanine dipeptides. Gelatin (collagen) from many sources always consisted of about one-third glycine, suggesting that this amino acid might occupy every third position in the collagen polypeptide.

There was another, usually unstated, reason to suspect that amino acids were periodically distributed along the polypeptide chain: if proteins did not possess some kind of pattern of amino acids, how could they be produced by the cell? The synthesis of homopolymers – containing only one type of monomer, such as the glucose polymer cellulose – was conceptually quite straightforward. Such macromolecules required only an enzyme capable of putting together the monomer units. The synthesis of a polymer containing two different monomers presumably required two enzymes, one capable of adding A to B, and one capable of adding B to A. A polymer containing equal amounts of n subunits, periodically distributed, could likewise be produced by the action of n different enzymes. However, even by the late nineteenth century it was clear that proteins did not contain equal amounts of their constituent amino acids. This made the problem of protein synthesis more difficult, but still not insuperable – if one assumed that amino acids were distributed periodically along the polypeptide chain, the synthetic mechanism only had to follow relatively simple rules of assembly.

Carl Niemann, scourge of the cyclol hypothesis, was one of the founders of the theory that all proteins contained a periodic arrangement of amino acids. Before moving to Cal Tech in 1938, Niemann had worked with Max Bergmann at the Rockefeller Institute for Medical Research in New York. Bergmann was perhaps the leading protein chemist of the 1930s. He had been an assistant to Emil Fischer until the latter's death in 1919, and then worked at the Kaiser Wilhelm Institute for Fiber Chemistry. In 1926, he had argued against Staudinger's macromolecule hypothesis at the *Naturforscher* symposium in Düsseldorf (see Chapter 7). A Jew, Bergmann left Germany in 1933 and joined the Rockefeller Institute. Three years later he succeeded Phoebus Levene as head of the division of chemistry.

In 1936, Bergmann and Niemann proposed that amino acids were periodically distributed along the polypeptide chains of proteins. By measuring the compositions of several proteins, Bergmann and Niemann found that the total number of amino acids was always of the form $2^n \times 3^m$. For ovalbumin, the total number of amino acids was 288 ($2^5 \times 3^2$); for hemoglobin and fibrin, 576 ($2^6 \times 3^2$); and for silk fibroin, 2592 ($2^5 \times 3^4$). More significantly, the number of times any individual amino acid occurred in these proteins was also of the form $2^n \times 3^m$. This was interpreted to mean that a given amino acid could occur at every second, third, fourth, sixth, eighth, ninth, twelfth, etc., position in the polypeptide chain, but not at every fifth, seventh, tenth, eleventh, etc. In the case of silk fibroin, for example, the measured periodicity of glycine was 2, that of alanine was 4 and that of tyrosine was 16, giving a structure of:

$$(G\text{-}A\text{-}G\text{-}X\text{-}G\text{-}A\text{-}G\text{-}X\text{-}G\text{-}A\text{-}G\text{-}X\text{-}G\text{-}A\text{-}G\text{-}Y\text{-})_{162}$$

where G is glycine, A is alanine, Y is tyrosine and X is any other amino acid.

In 1937, Bergmann and Niemann proposed that the synthesis of each periodic protein required an 'intracellular organizer'. As these organizers had the property of adding specific amino acids together, they could be proteinases, the enzymes that break down proteins, working in reverse. Proteinases were known to cleave polypeptide chains between specific amino acids, and like all enzymes, catalyzed the 'reverse' reaction as well as the 'forward' reaction. Thus, the silk fibroin sequence shown above could be synthesized by a minimum of four enzymes: one that added glycine at every second position, one that added alanine at every fourth position, one that added tyrosine at every sixteenth position, and one or more that added any other amino acid to every fourth position not occupied by tyrosine.

Any periodic distribution of amino acids could be explained by the action of 'organizers' that either added specific amino acids at certain positions in the sequence, or that added specific amino acids to certain existing sequences; but how would the organizers, themselves proteins, be synthesized? Here Bergmann and Niemann had to fall back on the old idea of 'autocatalytic' proteins, which had recently been given new life by Stanley's apparent demonstration in 1935 that tobacco mosaic virus was a crystalline protein (see Chapter 9).

The periodicity hypothesis, like the cyclol hypothesis, was largely inspired by the work of The Svedberg. As described above (see Chapter 7), Svedberg had noted in 1929 that the molecular weights of proteins studied by ultra-centrifugation to that time were 34 500 or multiples of that value. Bergmann and Niemann pointed out that the 'Svedberg unit' of 34500 daltons was approximately the size of a protein containing 288 ($2^5 \times 3^2$) amino acids, assuming an average amino acid molecular weight of 120. Therefore, 288 appeared to be some kind of 'magic number' for proteins: it conformed to Bergmann's and Niemann's $2^n \times 3^m$ series, it was the number of amino acids in Wrinch's C_2 cyclol, and it corresponded to the molecular weight of Svedberg's basic unit of protein structure.

However, a periodic distribution of amino acids would greatly limit the number of possible protein sequences. As described in Chapter 6, the vast number of possible permutations of amino acids helped persuade geneticists and biochemists in the period 1900–20 that proteins must constitute the material of heredity. This kind of thinking had, however, fallen somewhat out of favor by the 1930s, and would not be explicitly revived until the publication of *What Is Life?* in 1944 (see Chapter 10).

As with the cyclol theory, initial reviews of the periodicity hypothesis were positive. Pauling lent his authority to it, and Astbury wrote that 'the basic idea must be right'. Richard Goldschmidt, professor of genetics at the University of California, even saw in it a 'chemical theory of heredity'. However, holes in the argument eventually became apparent. It was pointed out that a modest error of 6% in the determination of amino acids would result in an 80% chance that a completely random polypeptide sequence would be in 'agreement' with the periodicity hypothesis. Another weakness was that many amino acids could not be quantified in the 1930s – Bergmann and Niemann measured no more than nine amino acids for any of the proteins studied; for hemoglobin and ovalbumin, this left unaccounted 73% and 64% of the total amino acids, respectively. A heavy blow came in 1939, when one of Bergmann's colleagues produced an amino acid composition for the enzyme ribonuclease that did not obey the $2^n \times 3^m$ rule. In 1942, Albert Chibnall wrote: 'I think that those interested in proteins would be wise to regard the Bergmann–Niemann hypothesis as still tentative and in any case applicable only to the component peptide chains of the molecule, for much of the evidence brought forward to support it has been based on inadequate experimental data and has demonstrated nothing more than the hypnotic power of numerology.'[57] Although it would not be formally disproved until Frederick Sanger determined the amino acid sequence of the B chain of insulin in 1951 (see Chapter 11), little more would be heard of the periodicity hypothesis.

The story of the Bergmann–Niemann hypothesis is an example of the fine line that exists in science between a prescient insight and a hasty overinterpretation. Its authors were correct, it turns out, in thinking that the one-third proportion of glycine in gelatin meant that this amino acid occupies every third position in the collagen polypeptide. They were partly correct in believing that glycine and alanine alternate in the amino acid sequence of silk fibroin. Unfortunately for Bergmann and Niemann, gelatin and silk fibroin were the exceptions rather than the rule.

'Hydrogen Bridges'

With the possible exception of the use of X-ray diffraction to study globular proteins, the most significant development in protein chemistry in the 1930s was the recognition of the importance of hydrogen bonds in protein structure. Astbury and Woods had proposed in 1931 that 'secondary valences' between carbonyl (C=O) and amino (N–H) groups stabilized the hexagonal folds in α-keratin and could provide interchain bonds between fully extended polypeptide chains. At this time, Astbury did not seem to be aware of the hydrogen bond, which had been proposed in 1920.

The concept of the hydrogen bond was based on the realization that hydrogen in organic compounds is often attached to more electronegative elements such as oxygen or nitrogen. Therefore, the shared electrons forming the bond lie closer to the more electronegative atom, leaving the hydrogen atom with a fractional positive charge, or dipole. This means that weak interactions can develop between such hydrogens and atoms with a negative dipole, usually oxygen or nitrogen.

The importance of hydrogen bonds in protein folding was first made explicit by Pauling and Alfred Mirsky. Mirsky worked at the Rockefeller Institute for Medical Research in New York, but in 1935 Pauling persuaded the Rockefeller Institute to grant him a leave of absence for a couple of years to come to Cal Tech. In Pasadena, Pauling and Mirsky collaborated on a study of protein denaturation. It had been known since antiquity that heating solutions such as egg white or serum produced an insoluble precipitate. As early as the mid-eighteenth century, it was known that proteins could be precipitated by extremes of pH or by organic solvents such as alcohol. This change of state from soluble to insoluble protein became known as denaturation. Mirsky and Pauling pointed out that denaturing conditions were also those that would break hydrogen bonds, and that denaturation was associated with the loss of specific properties of proteins, such as enzyme activity and the ability to crystallize. From this they concluded that the native (non-denatured) protein molecule: 'consists of one polypeptide chain which continues without interruption throughout the molecule . . . this chain is folded into a uniquely defined configuration in which it is held by hydrogen bonds between the peptide nitrogen and oxygen atoms and also between the free amino and carboxyl groups of the diamino and dicarboxyl amino acid residues.[58]

Another early proponent of hydrogen bonds in proteins was Maurice Huggins. He had trained at Cal Tech with Roscoe Dickinson, as had Pauling, and had moved to the Eastman Kodak Company in 1936. There he was attempting to identify the physical state of the gelatin used in photographic emulsion. In 1937, Huggins pointed out that the 'secondary valences' postulated by Astbury to provide intrachain bonds in α-keratin and interchain bonds in the β conformation could well be 'hydrogen bridges'. However, Astbury's structures had assumed that the polypeptide chain was linear, whereas in fact, because of the tetrahedral nature of the carbon atom, the chain had to zigzag. By incorporating the correct bond angles and assuming that all the atoms of the peptide bonds joining adjacent amino acids lay in the same plane, Huggins was able to formulate three-dimensional versions of Astbury's interchain-bonded β structure and intrachain-bonded α structures (Figure 8.3).

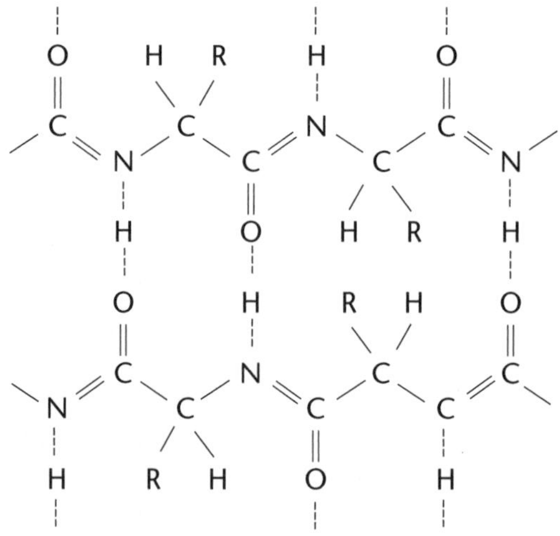

Figure 8.3: Maurice Huggins's 1937 structure of β-keratin. The two sections of polypeptide chain shown run in opposite directions and are held together by hydrogen bonds involving every C=O and N–H group. This structure is similar to Astbury's 1931 two-chain protein (Figure 7.5), but with accurate bond lengths and angles. Adapted from Huggins, M. L. (1937) *The Journal of Organic Chemistry* **1**, 449, with permission. Copyright 1937 American Chemical Society

'The Great Complexity of Proteins'

The prospects for solving protein structures by X-ray methods were discussed in review articles published in 1939 by three leading crystallographers: Astbury, Bernal and Pauling. A major question was whether the globular proteins would, like the fibrous proteins, prove to have some kind of repeating folded structure. Astbury was characteristically optimistic: 'The X-ray interpretation of the structure of the elastic fibrous proteins surely points directly to the solution of the problem of the 'globular' proteins also . . . once having demonstrated the principle of intramolecular folding, if only along one direction, the natural conclusion is that the massive roundish molecules of the globular proteins are but generalized examples of the same principle.'[59] Bernal, the leading authority on the crystallography of globular proteins, took a more cautious position. On the bright side, the presence in several globular proteins of identical reflections corresponding to interatomic spacings as low as 2 Å suggested that many proteins may have similar internal structures at the atomic level. Indeed, Crowfoot's X-ray analysis of insulin had revealed reflections similar to those characterizing the fibrous proteins. Nonetheless, 'It is difficult to imagine any kind of fold or coil by which a single chain can occupy the observable space and at the same time not be so intricate that its formation by any natural process would be enormously improbable.' The usually confident Pauling sounded even more daunted at the task ahead: 'the great complexity of proteins makes it unlikely that a complete structure determination for a protein will ever be made by X-ray methods alone.'

If Astbury were correct, the globular protein molecule would have a lattice structure, and solving that structure would, in principle, be no different from solving those of inorganic crystals. If Bernal and Pauling were right, however, there would be no repeating pattern of folding of the polypeptide chain; globular proteins would not be simply a folded version of Astbury's ribbon chain; and every globular protein might have a different pattern of polypeptide chain folding. If this were the case, the course of the polypeptide would have to be plotted, atom by atom, throughout the three-dimensional space of a huge molecule.

Astbury was to make no more significant contributions to protein structure. Bernal thought that Astbury 'knew more crystallography than either Sir William or Sir Lawrence[e] Bragg'. However, he was handicapped by 'rashness and lack of self-criticism'. According to his student Mansel Davies, Astbury

---

e   Lawrence Bragg was knighted in 1941.

'did not indulge in open collaboration' and as a result left no disciples. As Francis Crick put it, 'Astbury was an adventurous but not a meticulous experimenter, as well as being a sloppy model builder.'

By the late 1930s, it was apparent that there were two ways of using X-ray crystallography to solve the structure of biological macromolecules. One was the painstaking method of using Fourier analysis or the Patterson–Harker method to construct contour maps of sufficiently high resolution that the individual atoms of the molecule could be resolved. The other was the 'stochastic method' that Pauling had devised in the late 1920s, in which the observed unit cell characteristics and symmetry were used to make an educated guess at the structure, and the calculated X-ray intensities of the model were then compared with those actually obtained. Astbury, it seems, mastered neither – not 'meticulous' enough an experimenter for the first, too 'sloppy' a model-builder for the second.

It was perhaps inevitable that Bernal would fail to achieve as much as he seemed capable of, if only because his potential seemed unlimited – probably no scientist has ever so impressed his peers. As a Cambridge undergraduate, Bernal had acquired the nickname 'Sage' because 'he knew so much'. In his letter of recommendation to William Bragg, Hutchinson described Bernal as 'something of a genius'. Perutz considered Bernal 'the most brilliant talker I ever encountered'. Needham described him to Woodger as 'perhaps the most acute mind of my acquaintances here' and Hermann Muller to Edgar Altenburg as 'one of the best, if not the best scientific mind in the world'.

However, Bernal suffered, in Perutz's opinion, from 'a lack of critical judgement'. The novelist C. P. Snow wrote of him that: 'He likes to start something, drop an idea, get the first foot in – and then leave it for someone else to produce the final finished work . . . he has suffered from a certain lack of the obsessiveness which most scientists possess and which makes them want to carry out a piece of creative work to the end.'[60] Bernal's obituary in *Nature* noted that he 'was in some ways a poor experimentalist, being impatient in a field that required long periods of concentration and accurate measurement.' Early in his career, Bernal made important methodological contributions to crystallography, including the technique of single crystal rotation photography and improvements to the goniometer. Also to his credit are the first significant X-ray analyses of amino acids, steroids, globular proteins and viruses. The fruits of these breakthroughs, however, were left to others. Nonetheless, his students Perutz and Crowfoot would dominate the structural analysis of globular proteins, and Bernal's eloquence and enthusiasm were to attract other important players, including John Kendrew and John Randall, to the study of biological macromolecules.

'When I Showed Him My X-ray Pictures of Hemoglobin His Face Lit Up'

Lawrence Bragg was having a difficult time in Manchester. Despite his great talent for oral exposition, Bragg encountered great problems when first required to do undergraduate teaching at Manchester, where many of the students were ex-servicemen. This culminated in an incident in which the mild-mannered Bragg slapped a student who had set off a firework during a lecture. To make things worse, anonymous letters were sent to university administrators, apparently by a junior colleague or his wife, accusing Bragg of incompetence. The happier times included a trip to Sweden in 1922 to receive the Nobel prize (delayed because of the war). New breakthroughs were also being made in the laboratory. Measurement of the absolute, rather than relative, intensity of diffracted X-rays allowed Bragg and his co-workers to study crystals with large numbers of parameters (atomic locations not determined by the symmetry of the crystal). Using this method, the crystal structures of several complex minerals were solved.

Pauling visited Manchester in 1930 and was taken aback when Bragg kept his distance. Bragg may still have been disappointed about Pauling publishing 'his' principles of ionic packing in crystals two years earlier. However, Pauling's reception was more probably a reflection of Bragg's mental condition at the time; his mother had died in 1929 and he suffered a nervous breakdown a few months after Pauling's visit. Bragg finally left Manchester in 1937 to become director of the National Physical Laboratory in London. Shortly before he left Manchester, the Cavendish professorship at Cambridge became vacant with the death of Rutherford. Bragg was offered the opportunity of replacing Rutherford, as he had at Manchester, and moved back to Cambridge the following year.

Lawrence Bragg's departure from Manchester initiated a game of musical chairs in British physics. Patrick Blackett moved from Birkbeck College in London to replace Bragg, and Bernal moved to Birkbeck to replace Blackett. Most of Bernal's biological crystallography group moved to London with him, but Perutz loved Cambridge and wished to stay. Bernal arranged that Bragg would supervise the remainder of Perutz's PhD project. When Bragg arrived at the Cavendish, Perutz waited day after day for him to show up in the crystallography lab. Finally he summoned up the nerve to visit Bragg's office: 'When I showed him my X-ray pictures of hemoglobin his face lit up. He realized at once the challenge of extending X-ray analysis to the giant molecules of the living cell.'

Linus Pauling's interests were also becoming focussed on proteins. He realized that the structural analysis of such huge and complicated molecules would require a great deal of crystallographic groundwork. Specifically, the bond lengths and angles of the polypeptide backbone would need to be determined from simpler 'model' compounds. A stroke of good fortune made this possible. In 1937, Robert Corey arrived at Cal Tech with a year's salary after his supervisor's laboratory at the Rockefeller Institute was eliminated in a restructuring exercise. Corey was a crystallographer who had worked on biological materials such as porcupine quill (keratin) and hemoglobin. Even better, he had already started the crystallographic analysis of glycine, the simplest amino acid. Pauling allowed him to continue this work; by 1939, the crystal structures of glycine and diketopiperazine were completed. Pauling and Corey were on their way to the α-helix.

The Crystallography of DNA

Pauling was less fortunate in another recruitment. In 1938, the Cal Tech division of chemistry and chemical engineering, of which Pauling was now chairman, inaugurated a new building for research in organic chemistry. To establish a strong group in organic chemistry, a world-class scientist was sought. With the help of the Rockefeller Foundation, Pauling tried hard to attract Alexander Todd, even bringing the Scottish chemist to Cal Tech for six months. For once even Pauling's powers of persuasion were not enough, and Todd went to Manchester instead. He moved to Cambridge in 1944, where his studies on the synthesis of nucleotides won him the Nobel prize for chemistry in 1957. Pauling had failed to attract to his division the leading nucleic acid chemist of the day – a failure that would have profound consequences for him when he turned his attention to the structure of DNA.

Bernal and Astbury had already started to study the nucleic acids. In a 1937 letter to Astbury, Bernal mentioned that his X-ray diffraction of DNA suggested that the axial repeat distance was 'some multiple of 3.3 Å'. Astbury replied: 'I was rather amused at your getting a fibre photograph of nucleic acid. I got a similar photograph 3 or 4 years ago . . . and I have been thinking of investigating it in more detail now with regard to the chromosome business.' Bernal then felt honor-bound to leave 'the chromosome business' to the Leeds group: 'Faithful to my gentleman's agreement with Astbury, I turned from the study of the amorphous nucleic acids to their crystalline components, the nucleosides.' Bernal may have got the worst of the bargain, but his studies on the nucleosides were to lead to the solving of the structure of cytidine by Sven Furberg in 1949, an important step on the way to the double helix.

From its beginnings in the late 1930s to its culmination in the early 1950s, the X-ray analysis of DNA was critically dependent upon the state of polymerization of the nucleic acid. As noted in Chapter 7, the methods of the organic chemists were hardly optimal for obtaining undegraded DNA. In the period 1934–8, however, techniques for extracting DNA from cells with water were developed at the Karolinska Institute in Stockholm by Einar Hammarsten and Torbjörn Caspersson. In 1936, Caspersson travelled to Bern to study the optical properties of his highly polymerized DNA preparation using the flow birefringence apparatus built by Rudolf Signer. Signer was the former student and colleague of Hermann Staudinger, the father of macromolecular chemistry, and it is therefore fitting that this collaboration led to a definitive demonstration of the macromolecular nature of DNA.

In a short note published in *Nature* in 1938, Signer, Hammarsten and Caspersson reported that: 'The molecules of sodium thymonucleate [DNA] have in solution the form of thin rods, the length of which is approximately 300 times their width . . . the molecular weight of the preparation studied lies between 500000 and 1000000.' From the fact that DNA exhibited negative flow birefringence, as had previously been observed by Bernal and others, Signer and co-workers concluded that the bases lay perpendicular to the long axis of the molecule.

Astbury had obtained a sample of the high-quality Karolinska DNA, and later in 1938 published an X-ray diffraction analysis of this material. Two strong reflections were observed. One, perpendicular to the axis of the fiber and therefore corresponding to the fiber diameter, was at 20 Å. The other was at 3.34 Å parallel to the fiber axis, which Astbury thought 'corresponds to that of a close succession of flat or flattish nucleotides standing out perpendicularly to the long axis of the molecule' (Figure 8.4). The fact that DNA gave a fiber diffraction pattern showed that it, like keratin, was a one-dimensional crystal, consisting of structural units that repeated along the axis of the fiber. The diffraction pattern obtained by Astbury and Bell was not of sufficient quality for them to measure accurately the length of the axial repeat (the length of the repeating unit along the axis of the fiber), but they estimated that it was 'perhaps seventeen times as great' as the 3.34 Å distance attributed to the spacing between adjacent nucleotides.

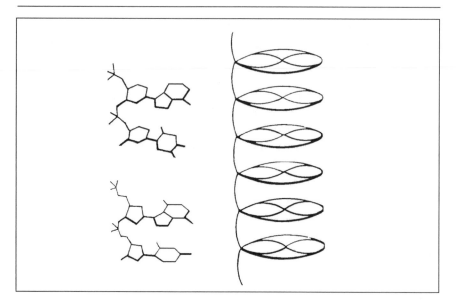

Figure 8.4: William Astbury's 1938 model of DNA structure. The deoxyribose–bases units lie parallel to one another, 3.34 Å apart, and are linked by the phosphate groups. Reproduced from Astbury, W. T. and Bell, F. O. (1938) *Cold Spring Harbor Symposia on Quantitative Biology* **6**, 112

Astbury made particular note of the fact that the 3.34 Å internucleotide distance 'is almost identical with that of a fully extended polypeptide chain system'. It seems clear that Astbury was already thinking of the implications for chromosome structure of interactions between equally spaced nucleotides in DNA and amino acids in proteins. However, the similarity between the X-ray diffraction patterns of DNA and fully extended proteins, like that he had previously noted between the diffraction patterns of cellulose and α-keratin, was to prove nothing more than a coincidence.

These two studies of 1938 laid the foundations of the three-dimensional structure of DNA. It was a giant molecule, hundreds or thousands of times larger than Levene's minimum tetranucleotide. It was extremely long and thin, and contained structural units that repeated along the fiber axis. The nucleotides of which DNA is composed were separated by 3.34 Å, with their bases lying perpendicular to the fiber axis, 'like a great pile of plates'.

In Bern, Signer had realized the importance of extracting DNA from cells in undegraded form, and set out to improve on Hammarsten's and Caspersson's technique. No significant progress in the X-ray diffraction of DNA would occur until 1951–3, when Rosalind Franklin's analysis of Signer's DNA was used to construct the double helix.

Chapter 9
The Chemical Basis of Genetics

'Light and Life'

While he was an undergraduate at Cambridge, Lawrence Bragg wrote to his father: 'I got an awful lot from a Dane who had seen me asking [James] Jeans questions, and after the lecture came up to me and talked over the whole thing. He was awfully sound on it, and most interesting, his name was Böhr or something that sounds like it.'[61]

His name was Niels Bohr and he was indeed 'awfully sound' – he went on to become one of the leading theoretical physicists of his day. Born in 1895, Bohr graduated from Copenhagen University in 1911 with a doctorate in physics and went to work with Rutherford in Manchester. In 1916, he was appointed professor of theoretical physics at Copenhagen and five years later became director of a new Institute for Theoretical Physics. Bohr's major contribution, for which he was awarded the Nobel prize for physics in 1922, was the proposal that electrons occupied orbitals around the nucleus of the atom, between which they could 'jump' with the absorption of a quantum of energy.

Bohr had a life-long interest in philosophy, publishing three volumes of philosophical essays. He was also the son of the distinguished physiologist Christian Bohr, a student of Carl Ludwig, who discovered the cooperative binding of oxygen to hemoglobin. It is therefore not entirely surprising that, when he was asked to give a lecture to the International Congress of Light Therapists in Copenhagen in August 1932, Niels Bohr chose to speak on the philosophy of biology. His lecture, the text of which was subsequently published in *Nature*, was entitled 'Light and life'.

In 'Light and life', Bohr noted that one of the fundamental tenets of quantum mechanics was the principle of complementarity. This stated that although it is possible to determine the location or the velocity of a subatomic particle such as an electron, it is not possible to determine both, because the act of measurement itself perturbs the system. Techniques that analyzed the position of an electron altered its velocity; techniques that measured the

velocity altered the position. Location and velocity are therefore complementary properties of the electron. Not complementary in the ordinary sense
of adding together to make a whole, but rather mutually exclusive, as these
properties can only be measured in different frames of reference.

Bohr saw an analogy between physicists' attempts to characterize the atom
and biologists' attempts to characterize the cell. Living cells were, to be sure,
made of ordinary matter, and therefore amenable to chemical analysis, but
that matter was organized in a complex and particular way. To study the
chemistry, the organization has to be destroyed; to study the organization,
one has to operate at a level at which the chemistry is invisible. Bohr therefore proposed that the chemical basis of an organism and its organizational
hierarchy are complementary properties, just as velocity and location are
complementary properties of an electron.

In stating that the unique characteristic of living systems was their
organization or (in another passage) their teleological (functionally adapted)
properties, Bohr was careful to avoid the implication that he was reviving
the old vitalist doctrine that different physical laws operate in living organisms: 'I think we all agree with Newton that the real basis of science is the
conviction that Nature under the same conditions will always exhibit the
same regularities. If we were able to push the analysis of the mechanism of
living organisms as far as that of atomic phenomena, we should scarcely
expect to find any features differing from the properties of inorganic matter.'[62]

As a result of the 'Light and life' talk, a young theoretical physicist sitting
in the audience was influenced to become a biologist. According to the Bohr
biographer Abraham Pais, his role in turning Max Delbrück into a biologist
was Niels Bohr's 'greatest contribution to biology'.

Max Henning Delbrück was born in 1906 into a family of impeccable lineage
and social standing. His father, Hans, was professor of history at the University of Berlin and served terms in the Prussian parliament and in the
Reichstag. Hans Delbrück's cousins included the professor of German literature at Jena and the chief justice of the Imperial Supreme Court. On his
mother's side, Max Delbrück was the great-grandson of Justus von Liebig.
Neighbors in the Berlin suburb of Grünewald where the Delbrücks lived
included Adolph von Harnack, co-founder and president of the Kaiser
Wilhelm institutes, and Max Planck, the founder of quantum physics, after
whom these institutes were to be renamed.

Max Delbrück's first scientific interest was astronomy, and it was to study this subject that he went to the University of Tübingen in 1924. It was common in Germany for students to move from university to university to find the best teacher for a particular subject, but even by the standards of the time Delbrück's undergraduate career was an itinerant one. From Tübingen he moved to Berlin, then to Bonn, and back to Berlin again before finally graduating from Göttingen. He switched from astronomy to physics at Göttingen, partly because most of the theoretical astronomy literature was in English, a language he did not yet speak. The quantum revolution was then sweeping physics and Göttingen was at its very heart, owing to the presence of Werner Heisenberg, Wolfgang Pauli, Max Born and James Franck. Delbrück stayed at Göttingen to perform research for a doctorate under the supervision of Born, then spent eighteen months in Bristol, England, in the laboratory of John Lennard-Jones. The intellectual environment in Bristol was scarcely less heady than in Göttingen – Delbrück befriended three future Nobel laureates in physics (Paul Dirac, Patrick Blackett and Cecil Powell) and one in chemistry (Gerhard Herzberg). At the end of 1929, he returned to Göttingen to sit his PhD oral examination. Much to Delbrück's embarassment, he failed the examination and had to re-sit the following year. In 1931, he obtained a Rockefeller Foundation fellowship to work with Bohr in Copenhagen and Pauli, by then in Zürich. In Copenhagen, Delbrück roomed with George Gamow, who later also turned to biology (see Chapter 13). Delbrück then spent another six months in Bristol and was on his way to a new position in Berlin when he heard the 'Light and life' lecture.

Max Delbrück was very much his own man. Nonetheless, his contact with Bohr profoundly affected him, and his subsequent career bears the unmistakable imprint of Bohr and the 'Copenhagen spirit'. Like Bohr, Delbrück was interested not only in physics, but also in poetry and philosophy. He published a remarkable book on evolutionary epistemology entitled *Mind or Matter?*, and at the time of his death was preparing a lecture on Rainer Maria Rilke for the Poetry Center in New York. Bohr's most important effect on Delbrück, however, was to awaken his interest in biology. As his student Gunther Stent wrote: 'It was the Copenhagen Spirit that provided [Delbrück] the philosophical infrastructure for navigating latter-day biological thought between the Scylla of crude biochemical reductionism, inspired by 19th century physics, and the Charybdis of obscurantist vitalism, inspired by 19th century romanticism.'

'Light and life' was not the only reason Delbrück left physics. Contact with men like Pauli made him realize his mathematical deficiencies; Pauli had the same intimidating effect even on the redoubtable Linus Pauling (see Chapter 8). However, Bohr's influence on Delbrück was more than just suggesting the easy option of moving into biology ten years before it became fashionable for physicists to do so; Delbrück absorbed the message of 'Light and life' and made it the central aim of his biological research. A lecture given by Delbrück in 1949 eerily echoes the arguments he had heard in Copenhagen seventeen years earlier: 'It may turn out that certain features of the living cell, including perhaps even replication, stand in a mutually exclusive relationship to the strict application of quantum mechanics, and that a new conceptual language has to be developed to embrace this situation . . . we may find features of the living cell which are not reducible to atomic physics but whose appearance stands in a complementary relationship to those of atomic physics.'[63]

Delbrück, however, put quite a different interpretation on Bohr's argument. In fact, he claimed that the published version of 'Light and life' did not correspond to what Bohr actually said in the lecture. To Delbrück, the main reason for physicists to study biology was that analysis of the complementary properties of living organisms might produce a 'paradox' similar to that posed when physicists studied atomic phenomena: 'Instead of aiming from the molecular physics end at the whole of the phenomena exhibited by the living cell, we now expect to find natural limits to this approach, and thereby implicitly new virgin territories on which laws may hold which involve new concepts and which are only loosely related to those of physics.'[63] Biology held the key to 'new physics'. This misunderstanding of Bohr's message was – paradoxically – the most significant effect of 'Light and life.'

The Green Paper

In 1932, however, Delbrück was still a physicist. After his stop in Copenhagen, he traveled to the Kaiser Wilhelm Institute for Chemistry in Berlin, where he worked as an assistant to Lise Meitner. An Austrian physicist who had come to Berlin to study under Delbrück's former neighbor, Max Planck, Meitner had formed a fruitful collaboration with the chemist Otto Hahn. In 1932, Hahn was director of the Institute for Chemistry and Meitner was the head of its physics department. That year the neutron was discovered, leading Hahn and Meitner to study the effects of neutron bombardment on uranium. Delbrück's part in this project was to supply theoretical support to Meitner.

In his spare time, Delbrück attended a discussion group that included the *Drosophila* geneticist Nikolai Timoféeff-Ressovsky and the radiation biologist Karl Zimmer. Timoféeff had come to Berlin in 1924 as part of a bizarre exchange that sent a team of German neurobiologists to Moscow to study the brain of Lenin, who had died earlier that year. According to Max Perutz, he was 'a physical and intellectual giant' whose nickname in the Soviet Union had been 'the wild boar'.

Delbrück's theorizing with Timoféeff and Zimmer led to the joint publication of a paper in 1935. Entitled *Über die Natur der Genmutation und der Genstruktur* (On the nature of mutation and gene structure), it was published in the obscure journal *Gesellschaft der Wissenschaften zu Göttingen*. Because the reprints had bright green covers, this publication became known as the 'Green Paper' or *Dreimännerwerk* (three-man work).

The Green Paper was in four sections: the first by Timoféeff, the second by Zimmer, the third by Delbrück, and the fourth, a summary, by all three. The first two sections represented a detailed discussion of what was known as the 'target theory'. This theory had originated with J. A. Crowther, who in 1924 measured the dose of ionizing radiation required to kill chick embryo cells and tried to derive from this dose the size of the structure damaged by the radiation. When Hermann Muller demonstrated in 1927 that X-rays could induce mutations, it became possible to use the target theory to estimate the size of the gene. A certain dose of ionizing radiation would produce a known number of ionizations per unit volume. By irradiating a population of organisms with that dose of radiation and determining how many of them developed a particular mutation as a result, one could calculate the minimum size of the radiation-sensitive structure whose mutation produced that phenotype – in other words, the size of a gene.

The target theory involved a number of assumptions. One of these was that the mutation had to result from a single 'hit' of ionizing radiation. If this were the case, there would be a linear relationship between radiation dose and the number of mutations. In Muller's studies this was not the case – the mutation number varied as the square of the X-ray energy absorbed. He was therefore skeptical about the utility of the target theory approach to determining gene size: 'Should this lack of exact proportionality be confirmed . . . we should have to conclude that these mutations are not caused directly by single quanta of X-ray energy that happen to be absorbed at some critical spot.'

However, Timoféeff and Zimmer had performed studies on the *eosin* eye color mutation of *Drosophila* which showed a linear relationship between radiation dose and mutation frequency, suggesting a single-hit mechanism. Specifically, with a radiation dose of 6000 roentgens (R), the *eosin* mutation occurred with a frequency of 1 in 7000 flies.[a] From this, Timoféeff and

---

a　1 roentgen = 2.58×10^{-4} Ci/kg.

Zimmer calculated that the radiation dose required to give a mutation rate of unity was $6000 \times 7000 = 42\,000\,000$ R. This amount of radiation would produce ion pairs with an energy of 30 electronvolts (eV) from 1 in every 1000 water molecules. Therefore, the minimum size of the radiation-sensitive structure (presumed to be a gene) corresponded to approximately 1000 atoms. The energy available to create a mutation was 30 eV.

The third section of the Green Paper, written by Delbrück, attempted to use these studies on radiation-induced mutagenesis to deduce not just the size but also the physical properties of the gene. He took as his starting point the observation that the mutation rate of *Drosophila* increased approximately five-fold for every 10°C increase in temperature. From this, he was able to calculate an 'activation energy' for mutation which represented the minimum amount of energy that would have to be absorbed by the gene in order to induce a mutational change. From his calculated value of 1.5 eV, Delbrück suggested that the physical change involved in mutation was an electronic transition from one stable state to another. He was also able to determine a spontaneous mutation rate, at 37°C, of approximately once every sixteen months.

Unfortunately for Timoféeff, Zimmer and Delbrück, each of the observations that underlay the Green Paper proved to be false. Mutation rates turned out to be independent of or inversely related to temperature. Zimmer's later studies showed that the linear relationship between X-ray dose and mutation rate was an artifact resulting from the superposition of two or more non-linear curves. The ion pairs created by X-rays can diffuse comparatively long distances, rendering meaningless the 'sensitive volumes' calculated by target theory. In 1940, Muller wrote that the assumptions underlying the target theory were 'not merely gratuitous but improbable on theoretical grounds and there is in fact strong empirical evidence against each of them'.

Nonetheless, the Green Paper occupies an important place in the history of biology. A copy came into the hands of the physicist Erwin Schrödinger, and it appears to have been this that stimulated him to prepare the lectures that were published as the 1944 book *What is Life?* (see Chapter 10). Delbrück's misinterpretation of 'Light and life' led him to write the Green Paper; Schrödinger's enthusiasm for the erroneous 'Delbrück model' led him to write *What is Life?*

'The Twilight Zone of Life'

As described in Chapter 6, Muller had proposed in 1922 an approach for elucidating the nature of the gene that was quite different from the target theory – the study of viruses. These mysterious infectious agents appeared to cause a variety of diseases in plants and animals. They were distinguished from other infectious agents by their small size, which allowed them to pass through filters that retained micro-organisms such as bacteria, and the fact that they could only reproduce in the presence of the appropriate host organism. By the 1930s, however, viruses had been discovered that were larger than certain bacteria, while other viruses were thought to be as small as protein molecules. Their size range, intermediate between the dimensions of the inanimate and animate, meant that viruses occupied 'the twilight zone of life'.

There were two major theories about the nature of viruses. One was that they were intracellular parasites that required the 'nutrients' of the host cell in order to multiply. The other was that viruses were autocatalytic proteins – enzymes that catalyzed their own replication. The ideas that genes were proteins, that gene replication involved an autocatalytic mechanism and that viruses were 'model genes' were all current in the 1920s (see Chapter 6). The autocatalytic theory of gene replication received apparent experimental support when John Northrop, who had first prepared crystalline pepsin (see Chapter 7), showed that several protein-degrading enzymes occur in a precursor form that can be converted to the active form by a small amount of the active enzyme. The time course for such reactions was of the same S-shape as the growth curve of a population of organisms in a closed system, suggesting to some that genetic replication was likely to involve an auto-catalytic process.

In 1935, Wendell Meredith Stanley, working at the Princeton laboratories of the Rockefeller Institute for Medical Research, succeeded in isolating the infectious agent of mosaic disease of tobacco in crystalline form. This material exhibited constant physical, chemical and biological properties through ten successive crystallizations, following which it was infectious at dilutions as low as 1 in 1 000 000 000. The crystalline material appeared to be pure protein. Stanley concluded that: 'Tobacco-mosaic virus is regarded as an autocatalytic protein which, for the present, may be assumed to require the presence of living cells for multiplication.'

The public response to the crystallization of tobacco mosaic virus was similar to that of Thomas Huxley's protoplasmic theory of life (see Chapter 3). A story appeared on the front page of the *New York Times*; another newspaper described viruses as 'perhaps the final missing link between the dead of the chemical world and the living of the biological'. As Scott Gilbert wrote, the crystallization of tobacco mosaic virus was 'a philosophical as well as a biological breakthrough, similar in its significance and interpretation to Wöhler's synthesis of urea'.

Only a few months later, a group from Cambridge that included the bio-chemist Norman Wingate Pirie and the crystallographer John Desmond Bernal reported that crystalline tobacco mosaic virus was not pure protein, but contained a small amount of RNA. The Cambridge scientists were not convinced that the infectious material precipitated out of solution was even crystalline, preferring to refer to it as a liquid crystal or fiber. Even if Stanley's tobacco mosaic virus preparations were crystalline, that was not necessarily proof of homogeneity, in Pirie's eyes – he had previously found that Northrop's crystalline trypsin was an excellent source of the enzyme ribo-nuclease!

Stanley countered with studies showing that crystalline tobacco mosaic virus was physically, chemically and serologically homogeneous, writing in 1938 that: 'we are unable, at the present time, to conclude other than that the high molecular weight protein under discussion is the virus.' This conclu-sion was supported by observations that treatment of the virus with protein-denaturing agents abolished its infectivity. In retrospect, this only shows that the protein is required for infection of the host cell, not that it is the genetic material of the virus. In any case, Stanley's crystallization of tobacco mosaic virus was quickly accepted as evidence that genes were autocatalytic proteins. In 1938, Dorothy Wrinch stated that: 'the tobacco mosaic virus, under appropriate conditions, is produced by an autocatalytic reaction.' The following year, Delbrück referred to viruses as 'large protein molecules'. As late as 1947, Torbjörn Caspersson described the agent of tobacco mosaic as 'the simplest virus, consisting of one single kind of protein'.

The crystallization of tobacco mosaic virus made a great impression upon Delbrück. In 1937, he prepared a 'memorandum' entitled 'Preliminary Expo-sition on the topic "Riddle of Life"', which included a discussion about the nature of viruses. Was the production of new virus particles by virus–infected cells the result of 'a stimulus which modifies the metabolism of the host in such a way as to produce the foreign virus protein instead of its own normal protein', or was it 'an entirely autonomous accomplishment of the virus'? On this question, Delbrück agreed with Stanley, concluding that 'upon close analysis the first view can be completely excluded . . . Therefore we will look on virus replication as an autonomous accomplishment of the virus, for the general discussion of which we can ignore the host.' Ignoring the host cell, or rather regarding it as a 'nutrient medium' in which the virus grew, had the attraction of making virus replication in principle a very simple process – a process that might yield far more information about the nature of the gene than had the target theory. For Delbrück, the crystallization of tobacco mosaic virus made it 'the hydrogen atom of biology'.

Bacteriophage

Berlin was Delbrück's home town, and he wanted to obtain a permanent position there. In 1933, however, the National Socialists came to power in Germany and instituted new restrictions on civil service employment. One of these restrictions was on the hiring of Jews. Despite having been baptized and raised as a Protestant, Meitner was considered to be Jewish, and, when the *Anschluss* (union) with Austria made her subject to German laws, she escaped to Sweden. Another Nazi regulation required new government employees to obtain a certificate of 'political maturity'. Delbrück unsuccessfully attended *Dozentakademie* in 1934 and again the following year. His life-long attitude towards authority ranged from the irreverent to the subversive, so it is not at all surprising that Delbrück failed to demonstrate the required 'maturity'.

The Rockefeller Foundation saved Delbrück's career – and quite possibly his life – with the unsolicited offer of a second fellowship. This offer was made on the strength of the Green Paper and was part of a program to assist European scholars displaced by politics or racism. The Foundation officers suggested that Delbrück go to London to work with a group of mathematical biologists that included John Burdon Sanderson (J. B. S.) Haldane, but he persuaded them to send him instead to Cal Tech. He wanted to learn genetics, and the best person to teach him was the master of *Drosophila*, Thomas Hunt Morgan, who had moved from Columbia to Cal Tech in 1928. At the end of 1937, Delbrück arrived in Pasadena.[b]

Delbrück should have known what he was getting himself into – he was already familiar with *Drosophila* from his collaboration with Timoféeff. Once he arrived in Morgan's laboratory, however, Delbrück found the fly jargon

---

b   Hahn and Meitner's studies on the bombardment of uranium with neutrons had produced not the expected heavier elements, but rather something that looked like radium, four places lower in the periodic table. From her exile in Stockholm, Meitner pressed Hahn for confirmation of this inexplicable result. Further analysis proved that the 'radium' was in fact barium – an even lighter element. Meitner's discussions with her nephew, Otto Frisch, then in Copenhagen, led to the realization that neutron bombardment was splitting the uranium nucleus – a process Meitner called 'fission'. For the discovery of nuclear fission, Otto Hahn – but not Lise Meitner – was awarded the Nobel prize for chemistry in 1944. And what of Max Delbrück, who was supposed to be interpreting the physics of the Hahn–Meitner studies? Perhaps his obsession with biology caused him to overlook the significance of one of the most important findings in twentieth-century physics.

incomprehensible. Even worse, it quickly became clear that this multicellular organism with a complex life cycle and thousands of genes was far too complicated a system for the rigorous physics approach that Delbrück wanted to bring to biology. The paradox that he was seeking, and for which he had left physics and Germany, was not going to emerge from mapping genes in fruit flies.

However, a more appropriate experimental organism was not far away. Early in 1938, Delbrück discovered the basement laboratory of Emory Ellis. Holder of a fellowship in cancer research, Ellis was interested in animal viruses as possible anticancer agents, but had digressed into a study of bacteriophage. He set up his laboratory with little more than an autoclave, a liter of raw sewage and a culture of the obscure intestinal bacterium *Escherichia coli*.

Twenty years after 'd'Hérelle bodies' had been discovered, bacteriophage were even more mysterious objects than the viruses of plants and animals. Ellis described to Delbrück how the addition of phage to a 'lawn' culture of bacteria overnight produced visible holes – 'plaques' – in the bacterial layer; the plaques could easily be counted to determine the number of infectious particles originally present. It was clear to Delbrück, as it had been to Felix d'Hérelle and Hermann Muller before him, that this was not the behavior of a digestive enzyme, but that of a replicating organism.

With Morgan's blessing, Delbrück abandoned *Drosophila* and worked with Ellis on bacteriophage. Their first contribution was the one-step growth curve, which used the plaque assay to measure the kinetics of phage replication. From this curve, it was clear that no free bacteriophage were present in an infected bacterial culture for about twenty-five minutes after infection. There was then an exponential 'burst' of free virus which rapidly reached a 'plateau' level.

Delbrück saw bacteriophage as the perfect model organism to study the nature of the gene and its replication. As an intracellular parasite of a single-celled organism, bacteriophage were probably the simplest of organisms. The property of lysing bacteria, particularly as quantified by the one-step growth curve, allowed the phage researcher to employ clear-cut experimental design and rigorous statistical analysis. As Delbrück wrote in his fellowship report to the Rockefeller Foundation: 'The leading idea was the belief that the growth of phage was essentially the same process as the growth of viruses and the reproduction of the gene.'

After collaborating with Delbrück for a year, Ellis gave up the phage work and returned to cancer research. However, Delbrück was now committed to the use of bacteriophage for an attack on the fundamental physical nature of the gene and its replication. Unfortunately, the former theoretical physicist had made two major theoretical errors. Contrary to what Delbrück had written in 'Riddle of life', viruses turned out not to be independently replicating organisms, but parasites that usurped the metabolism of the host cell. Viruses did not possess the genetic capabilities for their own reproduction, no matter how 'suitable' the 'nutrient medium'. Delbrück was also wrong in thinking that the replication of the gene would involve a 'paradox' that could only be resolved by 'complementarity'. No new physics would be found in the functioning of the genetic apparatus, only some rather straightforward chemistry. Because of the inability of viruses to reproduce without the help of the host cell machinery, and because no physical paradox lay at the heart of gene replication and function, the secret of life would not be found in bacteriophage.

'The Chromosome Micelle'

In the 1930s the gene remained a purely hypothetical entity. As Delbrück later put it, 'Genes at that time were algebraic units of the combinatorial science of genetics.' That is, they were like the x's and y's that mathematicians used to denote unknown quantities in equations – the relationship of one gene to another could be determined, but nothing was known about the nature of the thing itself. As the geneticist Sewall Wright wrote in 1941, 'genes are merely regions of the chromosomes within which crossing over or other breakage has so far not been observed to occur'; the existence of genes as discrete entities had therefore not been established. Even in 1945, as pointed out by George Beadle, 'it is not possible to recognize a given gene unless it exists in at least two forms.' If the mutant form of a gene did not cause a detectable phenotypic change, or if it was lethal, or if it occurred in an organism that did not reproduce sexually, the existence of that gene might never be suspected.

All that was known with certainty of the physical nature of genes was that they formed linear arrays on chromosomes. Nothing whatsoever was known of the chemical structure of genes, how they replicated or how they controlled the hereditary characteristics of organisms. This experimental void could be approached from a number of directions; some of these approaches were astonishingly fruitful, others proved to be dead ends.

As described above, one method of obtaining information about the gene was to use the target theory to estimate the size of the gene and the physical basis of mutation. Another was the 'topological' approach of the Oxford mathematician Dorothy Wrinch. In 1936, the same year in which Wrinch proposed the cyclol theory of protein structure (see Chapter 8), she also speculated that 'the chromosome itself' would yield up its secrets to 'the mathematical theory of potential' and 'the concepts of pure physical chemistry'. As had been known practically since the time of their discovery, chromosomes were composed principally or entirely of protein and nucleic acid. Proteins were now recognized to be large polypeptides, whereas nucleic acids were thought to be small molecules. Wrinch therefore proposed that 'a gene is simply an annular structure, a set of identical polypeptide molecules in parallel, held together by nucleic acid molecules which form rings outside.' In other words, a bundle of sticks tied with string, in which the proteins were the sticks and the DNA the string (Figure 9.1). This structure, for which Wrinch used Carl von Naegeli's term 'micelle' (see Chapter 6), was stabilized by electrostatic interactions between positively charged amino acids and the negatively charged phosphate groups of the nucleic acid.

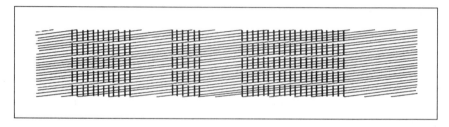

Figure 9.1: Dorothy Wrinch's 1936 model of the chromosome. Nucleic acids (thick vertical lines) interact with proteins (thin diagonal lines) to stabilize the 'micelle'. The distribution of nucleic acid molecules reflects the presence of positively charged amino acids, with which the nucleic acids interact, in the proteins. As a tetranucleotide, the nucleic acid can interact with four protein molecules. Adapted from Wrinch, D. M. (1936) *Protoplasma* **25**, 559

The Wrinch model of the chromosome lacked the geometric elegance of her cyclol structure for proteins, and Wrinch herself made the unusually candid disclaimer that 'the detailed formulation of a possible structure for the chromosome micelle given on this occasion may be of little worth in itself.' Indeed, it was soon pointed out that the micellar structure was inconsistent with the observed negative birefringence of nucleoproteins.

However, it was in discussing the replication of the 'micelle' that the deficiencies of the Wrinch model became most apparent, and its author had to resort to some egregious hand-waving: 'We may expect that the incorporation of new molecules here regarded due, possibly, to changes in the acidity of the medium will lead to an ever growing instability which finds its ultimate solution in the division of the whole.' This replication mechanism may have been inspired, if that is the right word, by the 'vibrational instability' that Astbury claimed prevented polypeptide chains from exceeding a maximum size (see Chapter 7).

In fact, Wrinch's approach to chromosome structure, like Astbury's, was based on the idea of a stereochemical complementarity between the negatively charged phosphate groups of DNA and the positively charged side-chains of basic proteins. Astbury conceived of colinear nucleic acid and polypeptide chains, each characterized by a 3.4 Å spacing between charged groups. Wrinch envisioned parallel polypeptide chains with their basic residues in register, and the electrostatically interacting DNA chains at right angles. Although satisfying the criterion of intermolecular stability, the Wrinch micelle did little to explain gene replication and said nothing at all about gene function.

'Autosynthesis'

The same year that Wrinch's paper appeared, a completely different tack was taken by the geneticist Hermann Muller. As it had been in 1921, Muller's approach was to deduce the physical nature of the gene from what was known about its biological properties. The result was a paper that stands as one of the most remarkably prescient documents in the history of biology.

In 1924, while still at the University of Texas, Muller started to study the effects of ionizing radiation on *Drosophila*. Three years later, he reported that X-rays caused a 150-fold increase in mutation rate, and claimed to have identified one hundred new mutants. The discovery of radiation-induced mutagenesis was to win Muller the Nobel prize for physiology or medicine in 1946. In 1932, however, he was experiencing a number of problems in Austin. His socialist views were unpopular, he was in conflict with his collaborators and the pressures of overwork and the collapse of his marriage led to a suicide attempt. Muller took a leave of absence and went to Berlin, where he worked with Timoféeff-Ressovsky and participated in the type of biophysical discussions that were later joined by Delbrück. In 1933, the rise of National Socialism in Germany led Muller to move to the Soviet Union, where he lived and worked until 1937.

In Moscow in March, 1936, Muller gave a lecture entitled 'Physics in the attack on the fundamental problems of genetics'. He started with a definition: 'genes are particles of sub-microscopic volume, probably on the order of about one twentieth of a micron in length, and considerably less in their other diameters, probably of protein composition, and bound to one another in line, single file, so as to form solid threads.'[64] This was about as much of a physical description as could be given at the time, and even then was wrong on the composition and overestimated the length. However, it was in inferring the structure of the gene from its properties that Muller's 'a priori thinking' came into its own. According to Muller, the gene had two fundamental properties. The first was 'specific auto-attraction of like with like' – the ability of the two sets of chromosomes to pair up side by side that could be inferred from the property of crossing-over. The second fundamental property of the gene was 'autosynthesis':

> each gene, reacting with the complicated surrounding material enveloping all the genes in common, exerts such a selectively organizing effect upon this material as to cause the synthesis, next to itself, of another molecular or sub-molecular structure, quite identical in composition with the given gene itself. The gene is, as it were, a modeller, and forms an image, a copy of itself, next to itself, and since all the genes in the [chromosome] chain do likewise, a duplicate chain is produced next to each original chain, and no doubt lying in contact with a certain face of the latter.[64]

This was not much more than stating the obvious, but Muller quickly jumped to novel intellectual terrain: 'This gene-building is not mere "auto-catalysis", in the ordinary sense of the chemist, since reactions are not merely speeded up that would have happened anyway, but the gene actually initiates just such reactions as are required to form precisely another gene just like itself; it is an active arranger of material and arranges the latter after its own pattern.' With these words, Muller discarded the old gene–enzyme analogy; autocatalysis, the key concept in gene replication for twenty years, was replaced with 'autosynthesis'.

Muller went even further with a brilliant insight. He realized that the gene, like any cellular macromolecule, must be constructed from building blocks – Albrecht Kossel's *Bausteine* (see Chapter 5). In order for the gene to copy itself, these individual building blocks, *not just the gene as a whole*, must have the property of auto-attraction:

> If the attracting principle of like for like, which we already know to be possessed by the gene considered as a whole, extend also to more elementary parts of the gene, to 'blocks' whose difference in arrangement constitute the specific differences in gene pattern whereby one gene differs from another and which form the basis of the mutational changes, then, if we suppose that representatives of these more elementary 'blocks' exist in scattered disorganized form in

the space surrounding the genes, it can be seen that each gene-part or 'block' would tend to attract to itself another, like part, and so a second group of parts would gather next to the original gene in the same pattern as in the latter, in much the same way as, on a still grosser scale, each chromonema ['gene-chain'] as a whole builds up a second chromonema, having its individual genes identical with and arranged in the same order as in the first one. If, then, the auto-attraction holds not merely for genes as a whole but also for gene-parts the auto-synthesis of a gene as a whole would be largely explained in terms of this auto-attraction.[64]

Muller's model may be summarized as follows: genes consisted of a linear series of 'blocks'; the arrangement (sequence) of these blocks constituted the difference between one gene and another; changes in the arrangement of blocks was the basis of mutation; the like-with-like attraction of blocks permitted gene replication (Figure 9.2). In this remarkable synthesis, Muller

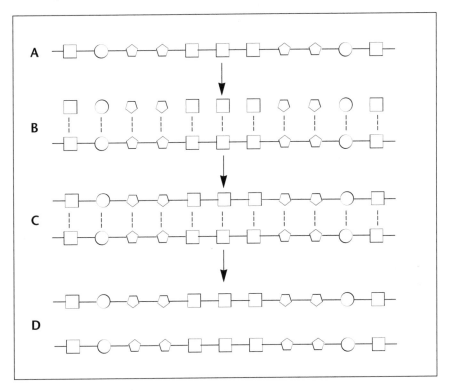

Figure 9.2: Hermann Muller's 1936 model of gene duplication. (A) The gene is a linear macromolecule composed of non-identical subunits. (B) Each subunit attracts an identical free subunit (upper) from solution. (C) Polymerization of the free subunits produces a twinned gene. (D) Separation of the two chains releases the original gene and an identical copy

anticipated the key concepts of Schrödinger's *What is Life?* – the 'heredi-tary codescript' and the 'aperiodic crystal' (see Chapter 10). The idea that the sequence of subunits in a macromolecule determined genetic specificity was not novel (see Chapter 6). Muller made this much more explicit by proposing that changes in the arrangement of subunits constituted 'the basis of the mutational changes'. However, what was truly revolutionary about Muller's 1936 model of the gene was the idea of subunit pairing as the basis of gene replication. The full significance of this concept would only be realized seventeen years later when the structure of DNA was determined.

At the end of his 1921 lecture, Muller had suggested the use of bacterio-phage for the study of genetic mechanisms (see Chapter 6). In his 1936 paper, Muller looked beyond that to a new approach, suggesting that solu-tions to 'the general problems of gene composition' might be found 'through the study of X-ray diffraction patterns' and 'parallel studies on such material carried out by the methods of the chemists'. With his advocacy of X-ray diffraction and chemistry – two years before Astbury published his first X-ray pictures of DNA, long before Chargaff, Todd, Wilkins, Franklin or Crick had any interest in gene structure, at a time when James Watson was six years old – Hermann Muller laid out the strategy that would lead to the double helix.

'A Process Analogous to the Copying of a Gramophone Record'

Muller's model of gene replication was wrong in only one important respect. Not in the belief that the gene was a protein, for Muller's argument was couched in terms of chemical generality, but the belief that the basis of gene replication was the attraction of like with like, rather than, as it turned out, the attraction of opposites. The gene parts that pair during gene copying are complementary, not identical – and this complementarity has only a semantic relationship to that proposed as a characteristic of living systems in Bohr's 'Light and life' lecture. The first suggestion that gene copying involved complementary molecules seems to have been made by J. B. S. Haldane in 1938:

> As the gene is of the dimensions of a protein molecule and does not consist of a number of similar parts, we cannot regard its reproduction as a process of growth by accretion ending in division when a limiting size is reached [as Wrinch had suggested]. It must, on the contrary, be a process of copying. The gene, considered as a molecule, must be spread out one *Baustein* deep. Otherwise it could not be copied. The most likely method of copying is by a process analo-

gous to crystallization, a second similar layer of *Baustein* [*sic*] being laid down on the first. But we could conceive of a process analogous to the copying of a gramophone record by the intermediation of a 'negative' perhaps related to the original as an antibody to an antigen. The process normally stops when one copy has been made, or at least the further copies are not attached to the chromosomes.[65]

The analogy between biological growth and crystal growth was already a hoary one when Haldane wrote these words, but his alternative mechanism of gene replication was novel. A gene could be copied in the way that a gramophone record is pressed from a mold, the way that a photograph is printed from a negative, the way that an antigen-binding antibody is formed in the presence of that antigen. It was this antigen–antibody analogy that was, quite independently, to lead Pauling to a more explicit statement of the gene-copying mechanism as the decade came to a close.

'The Conditions Under Which Complementariness and Identity Might Coincide'

The idea of applying physics to biology was clearly in the air in the 1930s. Niels Bohr had no difficulty convincing the Rockefeller Foundation in 1936 to fund a 'physico-biological conference' in Copenhagen that brought together, for the first time in any formal way, crystallographers, geneticists and cytologists. Similar conferences were held in 1938 in Klampenborg, Denmark, and Spa, Belgium.

At the Klampenborg conference, the quantum physicist Pascual Jordan heard Bernal suggest that long-range non-specific forces might be responsible for chromosomal pairing. This prompted Jordan to propose that a form of quantum-mechanical stabilization between identical molecules might be responsible for such 'autocatalytic' processes as gene replication and even the formation of antibodies. Linus Pauling had become interested in antibodies as a result of a conversation in 1936 with the great Austrian immunologist Karl Landsteiner. When Delbrück ran into Pauling one day in 1940, he mentioned the papers of Jordan, and the two men retired to the Cal Tech library to look at them. It took Pauling only five minutes to pronounce them 'baloney'. A few days later Pauling told Delbrück that he had written a note to *Science* about the Jordan hypothesis, and asked him if he would like to sign it. Although Delbrück's only real contribution to this work was to draw Pauling's attention to the work of Jordan, he agreed to be a co-author.

In the resulting paper, Pauling and Delbrück listed a number of reasons why the resonance forces described by Jordan could not stabilize interactions between identical molecules, and then concluded:

> It is our opinion that the processes of synthesis and folding of highly complex molecules in the living cell involve, in addition to covalent bond formation, only the intermolecular interactions of van der Waals attraction and repulsion, electrostatic interactions, hydrogen-bond formation, etc., which are now rather well understood. These interactions are such as to give stability to a system of two molecules with *complementary* structures in juxtaposition, rather than of two molecules with necessarily identical structures; we accordingly feel that complementariness should be given primary consideration in the discussion of the specific attraction between molecules and the enzymatic synthesis of molecules [emphasis in original].[66]

The main point is intuitively obvious – if molecules are to bind to one another, they must have complementary, not identical, structures. Opposite poles of a magnet attract, like poles repel.

When Delbrück read Pauling's draft manuscript, he must have been struck by the use of the term 'complementarity'. However, Pauling used the term in a quite different way to Bohr. By 'complementary', Pauling meant physically or electrostatically opposite, 'like die and coin', not mutually exclusive, like the quantum properties of an electron. Eight years after 'Light and life', Delbrück had found complementarity in biology – but no paradox.

The Pauling–Delbrück paper concluded with some thoughts about the possible structural basis of complementarity: 'The case might occur in which the two complementary structures happened to be identical; however, in this case also the stability of the complex would be due to their complementariness rather than their identity. When speculating about possible mechanism of autocatalysis it would therefore seem to be most rational from the point of view of the structural chemist to analyze the conditions under which complementariness and identity might coincide.'[66]

This is a classic example of Pauling's remarkable intuition. The forces holding molecules together all involve chemical entities that are opposites. Haldane had realized this and proposed that 'autocatalysis' – the duplication of the gene – requires a complementary intermediate, an 'anti-gene'. Pauling, however, appeared to recognize that there are situations in which molecules could be opposite and also identical – an apparent contradiction. Pauling had no mechanism to suggest for this; there was not even an analogy.

Gramophone records and their molds, antigens and their antibodies, photographic negatives and positives, dies and coins – all are complementary, but not identical. Yet somehow Pauling seems to have foreseen the possibility of complementarity with identity that would be realized thirteen years later in the antiparallel, base-paired structure of DNA.

Chemical Genetics

In the 1938 article that compared the replication of genes to the copying of a gramophone record, Haldane also wrote: 'Just as the biochemist finds hundreds of distinct enzymes in the protoplasm, the geneticist finds his genes in the nucleus, and if the biochemist will only help him, will discover what they are doing. The biochemist should be grateful to the geneticist, because one of his main difficulties is in splitting up metabolic processes into successive stages. The geneticist similarly needs the biochemist if he is to discover what genes are doing, still more if he is to find out what they are.'[65] Haldane practiced what he preached – he was involved in a series of genetic studies on pigment production in plants that represented the most significant advance since Garrod in linking genes to metabolic reactions.

William Bateson, who had interpreted Garrod's 'metabolic sports' in terms of Mendel's laws of inheritance (see Chapter 6), became in 1910 the first director of the John Innes Horticultural Institution at Merton in the southeast of England. Bateson and Muriel Wheldale (Onslow) studied the inheritance of flower color in *Matthiola* (stock), *Antirrhinum* (snapdragon) and *Lathyrus* (sweet pea), and noticed that pigment production appeared to be under genetic control. The elucidation of the chemistry of the plant pigments anthocyanin, anthoxanthin and flavone, largely the work of Richard Willstätter and Robert Robinson, made possible a study of plant coloration by 'chemical genetics'.

Haldane, who joined the Horticultural Institution in 1927, collaborated with the chemist Rose Scott-Moncrieff to determine the chemical basis of the previously isolated color variant plants. These studies led to a complete biochemical understanding of pigment production in a number of plant species. For example, it was shown that eight different colored strains of *Primula sinensis* resulted from the action of three genes: *K*, which controlled the oxidation of an anthocyanin pigment; *B*, which determined the presence of a flavone co-pigment; and *R*, which controlled the pH of the cell sap of the petals. The eight color strains – magenta, blue, red, slaty, tinted coral, white, coral and tinted plum – were shown to correspond to the genotypes *KBR*, *KBr*, *KbR*, *Kbr*, *kBR*, *kBr*, *kbR* and *kbr*, respectively, where *K*, *B* and *R* refer to the dominant and *k*, *b* and *r* to the recessive alleles. The relationship between pigment chemistry and plant genetics was so exact that crossing a recessive red *Primula* strain with a dominant white one produced a plant of the same color as a laboratory mixture of the purified red anthocyanin and white flavone pigments.

To achieve this detailed understanding of the genetics of plant coloration, Scott-Moncrieff not only had to perform some complicated pigment analyses but also liaise between Robinson's chemists and Haldane's geneticists. As is so often the case, this interdisciplinary project was easier in theory than in practice: 'At first, workers in the two disciplines did not appreciate fully the significance of their different approaches and were unaware of the potential of combining their findings. They were sceptical of each other's methods.' Scott-Moncrieff received only grudging recognition for her part in this work, as Cambridge University at this time would give female students only the 'title of a degree': 'I treasure the vivid recollection, while at work at my bench in the Biochemical Laboratory in 1930, of receiving the roll of my long-coveted PhD (Titular) Certificate from the hands of the Vice-Chancellor's proxy, in this case Mrs Onslow's red-headed fifteen-year-old lab. boy in stained overalls, on his round of errands. It had the flavour of being knighted in the field.'[67]

'The Attack on Development'

The man who took the link between genetics and biochemistry to the next level was also a plant geneticist, although he had to change the species he studied in order to do it. George Wells Beadle was born in Wahoo, Nebraska, in 1903 and completed a PhD in maize genetics at Cornell University in 1931. He applied for a National Research Council fellowship to continue this work, but the chairman of the fellowship board felt that Beadle should learn another system. Accordingly, he was given his second choice – Morgan's lab at Cal Tech. Like Delbrück a few years later, Beadle found the *Drosophila* jargon incomprehensible. Delbrück switched to bacteriophage because he felt that *Drosophila* was too complicated an organism with which to get to the heart of the matter – the nature and workings of the gene. For Beadle, however, the problem with *Drosophila* was, in a sense, that it was not complicated enough. The *Drosophila* group treated genes as abstractions, with little regard for what they were or how they acted. Beadle wanted to understand how genes controlled cellular processes such as metabolism and development; for him, genetics and biochemistry were 'two doors leading to the same room'.

Most scientists who make fundamental discoveries happen upon them by accident. This is not surprising – if a discovery is predictable, it can rarely be fundamental. In such cases, the genius of the scientist lies in realizing the significance of his or her finding – which usually involves the jettisoning of much conceptual baggage – rather than in having planned for the breakthrough. Far fewer are those scientists whose fundamental discoveries result from a conscious decision on a long-term program of study. But George Beadle did this twice – first in his decision to work on *Drosophila* development, then in his decision to switch to *Neurospora*.

One way of using flies to study what genes did, rather than where they were, arose from Alfred Sturtevant's discovery of genetic gynandromorphs of *Drosophila simulans*. Flies with this mutation lost one copy of the X-chromosome early in embryonic development, resulting in a 'mosaic' adult in which some organs had developed from cells carrying one copy of the X chromosome and some from cells carrying the other copy. The genetic mosaicism of the gynandromorph was revealed if it were heterozygous for (carried two different alleles of) an X chromosome gene. Some parts of the adult fly would develop from cells that retained the X chromosome carrying the wild-type (normal) gene, other parts from cells that retained the mutant gene. Thus, if a population of gynandromorphs were heterozygous for the *vermilion* eye color trait – that is, one of their X chromosomes carried the *vermilion* mutation and the other carried the wild-type eye color gene – one would expect to find flies that had one eye of normal color (red) and one eye that was vermilion. Such flies were not found, however – all the adults had eyes of the wild-type color. Thus the curious situation existed that a genetically mutant tissue actually expressed the wild-type phenotype. To explain this, Sturtevant proposed that the 'hormone' that controlled eye color diffused into the mutant eye from adjacent wild-type tissue.

These gynandromorphs appear to have languished as a genetic curiosity until the summer of 1934, when Sturtevant was informed by a former member of the Morgan group, Kurt Stern, of a technique for transplanting imaginal disks (pieces of embryonic tissue that give rise to specific adult organs) between *Drosophila* larvae. Sturtevant realized that imaginal disk transplantation offered a much more direct way than his gynandromorphs of determining the link between mutations and embryonic development – a piece of tissue from an embryo carrying a mutation of interest could be grafted onto a specific location on a wild-type embryo, and the effect of the mutation on the adult tissue subsequently determined. He wrote back to Stern: 'If you can develop a grafting technique for *Drosophila* all this gynandromorph stuff will be out of date, and the attack on development can really go ahead.'

Sturtevant's enthusiasm for the embryonic transplantation technique was shared by Beadle and by Boris Ephrussi, who worked in Morgan's laboratory in 1934. When Ephrussi returned to the Rothschild Institute for Physico-Chemical Biology in Paris, Beadle arranged to spend the first half of 1935 there. Morgan was even more accommodating to Beadle than he would be three years later with Delbrück, agreeing not only to let him go but also to pay his living expenses. Beadle lived as cheaply as possible – his room and board cost him $2 per day. It was only much later that Beadle realized that the funds to support his stay in Paris had come from Morgan's own pocket.

Larval tissue transplants in *Drosophila* required enormous patience and dexterity. Beadle and Ephrussi sat at opposite sides of a table, each looking down a binocular microscope: 'One person washed donor larvae and dissected out the disks while his partner washed the host, etherized it – at just the right rate to get the larva to extend – and placed it in position for injection. The dissector then sucked the disk slowly into the micro-pipette [a capillary tube drawn down to 0.01 mm] and injected into the side or ventral surface of the host larva, which his partner held in position with a blunt needle.'[68]

An additional problem was that it was almost impossible to tell the imaginal disks apart, so Beadle and Ephrussi could not be sure which tissue they were transplanting. However, one morning in June they came to the laboratory to find a newly hatched fly with a third eye in its abdomen, and realized that they had successfully transplanted an eye disk. When news of this finding reached Cal Tech, Morgan described it as being as important as Muller's production of mutations by X-rays.

Having succeeded in transplanting a piece of embryonic tissue and showing that it developed normally at an abnormal location, the next step was to repeat the experiment using a mutant donor and a wild-type host. Specifically, Beadle and Ephrussi decided to reproduce the combination of traits that had previously been studied in the gynandromorphs by transplanting eye disks from *vermilion* donors into wild-type hosts. As expected, an imaginal disk from a *vermilion* larva grafted to a wild-type host produced an eye of wild-type color. Beadle and Ephrussi now started a series of transplants between *Drosophila* embryos carrying two different eye color mutations. They knew that a wild-type host could supply the 'hormone' that restored the normal color to a mutant eye, but which color would result when both donor and host were mutant?

Once the transplantation technique was optimized, Beadle and Ephrussi were able to do between one hundred and two hundred transplantations a day, with a success rate of 30–60%. However, it was still frustrating work. Many eyes developed within the abdomen or attached to other organs. Worse, the anatomical location affected the eye color. Worst of all, the great majority of transplants gave a negative result – the grafted tissue developed into an eye of the donor color, unaffected by the 'hormones' of the surrounding host tissue.

Of twenty-six eye color mutants tested, only two, *vermilion* and *cinnabar*, developed the color of the host larva. Results from a third, *claret*, were ambiguous. However, transplants between these three strains of fly produced a completely unexpected result. When Beadle and Ephrussi transplanted an imaginal disk from a *cinnabar* donor into a *vermilion* host, they thought that one of two things would happen. If the product of the *cinnabar* gene, which they called cn^+, was the same as the product of the *vermilion* gene, which they called v^+, the transplanted disk should develop into an eye of the donor color. If the cn^+ and v^+ substances were different, the transplanted disks should develop the wild-type color, as the recipient tissue should be able to supply the missing substance. What happened, however, was that a *cinnabar* disk transplanted into a *vermilion* recipient produced a *cinnabar* eye, whereas a *vermilion* disk transplanted into a *cinnabar* recipient gave a wild-type eye.

The only plausible explanation seemed to be that the v^+ substance was a precursor of the cn^+ substance in a sequence of reactions producing the normal red eye-color pigment (Figure 9.3). Similar results involving the *claret* mutant indicated that the product of the *claret* gene (ca^+) occurred even earlier in the pathway of eye pigment synthesis. This allowed Beadle and Ephrussi to construct the following pathway for the synthesis of the *Drosophila* eye pigment:

$$ca^+ \rightarrow v^+ \rightarrow cn^+$$

Once again there was unstinting praise from Pasadena. Jack Schultz claimed: 'The analysis of the behaviour of genes in combination with each other forms a bridge between the description of development in the different mutant races – what may be called Mendelian embryology – and the study of gene action proper.' Not everyone was impressed, however. Richard Goldschmidt wrote to Schultz in 1942: '. . . genetically it's just a parallel to what we know rather thoroughly from plant pigments.' It is difficult to disagree with Goldschmidt's view – Beadle and Ephrussi's studies did not advance embryology at all and took the link between biochemistry and genetics only

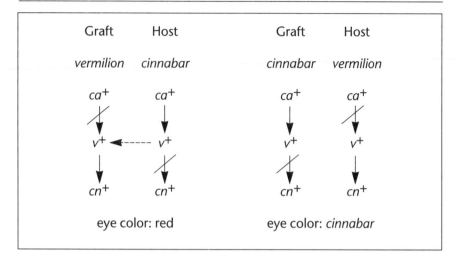

Figure 9.3: George Beadle's and Boris Ephrussi's interpretation of imaginal disk transplants between mutant strains of *Drosophila*. Both the vermilion and cinnabar mutations result from defective genes in the pathway for synthesis of eye color pigment, but the gene affected by the vermilion mutation occurs earlier in the pathway. Thus, in the case of a vermilion graft into a cinnabar host, the product of the vermilion gene (v^+) can diffuse from host to graft and be further metabolized by the wild-type cinnabar gene of the graft. In the reciprocal case (cinnabar graft into vermilion host), this does not happen, as the host cannot form v^+.

as far as the studies of Scott-Moncrieff, Haldane, Robinson and Willstätter had already gone. In fact, the interpretation that Beadle and Ephrussi put upon their results was a step short of the work of Garrod thirty years earlier, in that they did not seem to realize that genes acted through enzymes. The historical importance of Beadle and Ephrussi's *Drosophila* studies was not that they provided any dramatic new insights, but that they made possible Beadle and Tatum's *Neurospora* studies.

'To Invent a Genetics of Fungi'

The success of the *Drosophila* transplantation work led to Beadle being offered a position at Harvard University in 1936. He accepted, but a year later moved to Stanford University. The award of a grant from the Rockefeller Foundation allowed Beadle to hire the biochemist Edward Tatum. With Tatum's help, Beadle hoped to be able to identify the intermediates of the *Drosophila* eye pigment pathway. They were close to identifying the v^+ substance in 1940 when a German group reported that it was the tryptophan derivative kynurenine.

Beadle did not dwell upon this disappointment. Instead, he formulated a new strategic plan – one that would bring genetics and biochemistry even closer together. His idea was an ingenious one: 'rather than starting with visible mutations and laboriously working out their complex chemistry, why not start with biochemical mutations and work out the genetics?' However, this was easier said than done. Mutations could easily be induced with X-rays, but how could mutations of a defined type be identified? The easiest way, Beadle thought, was to look for mutants unable to grow in a medium that would support the growth of the wild-type. Tatum was already doing nutritional studies on *Drosophila*, but Beadle realized that this was too complicated an organism for ready identification of nutritional mutants. Instead he decided to use the Indonesian bread mold *Neurospora crassa*. As Robert Kohler put it, 'Beadle understandably thought it would be easier to invent a genetics of fungi than biochemistry of insects.'

For this new model system, Beadle once more had to thank Thomas Hunt Morgan. While Morgan was still at Columbia, Bernard Dodge of the Brooklyn Botanical Garden tried to interest him in *Neurospora*, which he claimed was more suitable for genetic studies than *Drosophila*. *Neurospora* grows asexually, but occurs in two sexes which will mate if they come into contact. Morgan was interested enough to bring *Neurospora* with him to Cal Tech and have a student work out its basic genetics.

Beadle now had Tatum determine the nutritional requirements of *Neurospora*. It needed only carbon, nitrogen, salts and the vitamin biotin. Here Beadle and Tatum had a stroke of good luck, as biotin had only recently become commercially available. The next step was to develop a technique to measure the growth of the fungus. For this, Beadle and Tatum developed 'race tubes' – graduated glass cylinders along which the fungus was persuaded to grow. To search for nutritional mutants, Beadle and Tatum irradiated *Neurospora* conidia (growth filaments) which were then allowed to fuse with unirradiated conidia of the opposite sex. This fusion produced a spore pod, from which spores were isolated and grown in separate cultures. Nutritional mutants would fail to grow in minimal medium, and the nature of the deficiency could be identified by adding various nutrients not required by the wild-type organism.

Beadle and Tatum had no idea what the frequency of nutritional mutants would be, but decided to isolate a thousand cultures before testing for mutations. Number 299 was the first culture to give a recognizable mutant – one requiring pyridoxine (vitamin B_6). By the time they wrote their work up for publication, Beadle and Tatum had studied two thousand cultures and identified three mutants: in addition to the pyridoxine-requiring strain, they found mutants unable to make thiamin (vitamin B₁) and *para*-aminobenzoic acid.

In the resulting paper, Beadle and Tatum noted that classical genetic studies attempt to identify the bases of already known hereditary traits. However, such studies were limited to 'non-lethal hereditary characters', that is, those involved in 'more or less non-essential "terminal" reactions'. This approach was also limited to visible characteristics, many of which were likely to be biochemically complex: 'Considerations such as those just outlined have led us to investigate the general problem of the genetic control of developmental and metabolic reactions by reversing the ordinary procedure and, instead of attempting to work out the chemical bases of known genetic characters, to set out to determine if and how genes control known biochemical reactions.'[69]

The 1941 paper contained the details of another crucial experiment. To study the mode of inheritance of the pyridoxine-requiring mutation, this strain was mated with a wild-type strain. Seven spore bodies, each containing eight spores, were obtained. As expected from the genetics of *Neurospora*, about half of these fifty-six spores were mutant. Beadle and Tatum concluded from this that: 'this inability to synthesize vitamin B₆ is transmitted as it should be if it were differentiated from normal by a single gene.' The link between genes and enzymes now seemed conclusively established: 'It is entirely tenable to suppose that these genes which are themselves a part of the system, control or regulate specific reactions in the system either by acting directly as enzymes or by determining the specificities of enzymes.'

'One Gene – One Enzyme'

Around the time that the *Neurospora* work appeared, the almost forgotten writings of Archibald Garrod were resurrected by Haldane and the American geneticist Sewall Wright. Beadle, who had been completely ignorant of this work, was generous in admitting Garrod's priority: 'On learning of this long-neglected work it was immediately clear to us that in principle we had merely rediscovered what Garrod had so clearly shown forty years before.' Like Mendel, Garrod was dead by the time the significance of his work was realized. He had finally left his beloved 'Bart's' in 1920 to become regius professor of physic (medicine) at Oxford. Garrod retired in 1927 and died nine years later.

Beadle's studies with Tatum, unlike his previous ones with Ephrussi, were conceptually more significant than those of Garrod (or those of the plant pigment group). Garrod had inferred that genes made enzymes; Beadle had all but proved it. Garrod had been able to identify the causes of a few human diseases; Beadle had devised a method by which entire metabolic pathways could be mapped. In recognition of this advance, Beadle and Tatum shared with Joshua Lederberg the 1958 Nobel prize for physiology or medicine.

In a 1945 review article, Beadle listed twenty-five 'chemical reactions known to be gene-controlled', including pigment biosynthesis in plants, carbohydrate breakdown in yeast, tyrosine and tryptophan metabolism in humans and six reactions in *Neurospora*. Based on these findings, Beadle proposed the 'one gene–one enzyme' concept, defined as 'the thesis that to every gene it is possible to assign one primary action and that, conversely, every enzymatically controlled chemical transformation is under the immediate supervision of one gene, and in general only one.'

The 'one gene–one enzyme' hypothesis was, of course, merely the latest version of Claude Bernard's distinction between the legislative and the executive. The legislative force had been identified with genes, and the executive with enzymes, in the very early years of the twentieth century. Quite apart from the forgotten work of Garrod, William Bateson had in 1909 described genes as having the power to produce enzymes (see Chapter 6). Muller had written in 1910: 'New cytoplasm is made from raw materials through the agency of an enzyme produced by nuclein'; likewise, Beadle's mentor, Morgan, had written in 1926 that: 'the genes may be protein bodies, one of whose activities is to produce enzymes which, being set free, act in each cell, and take part in catalytic reactions in the cytoplasm.' What was new in Beadle's hypothesis was the idea that genes and enzymes existed in a one-to-one correspondence, thus providing a direct linkage between genotype and phenotype.

Nonetheless, the elaboration of Bernard's concept remained unfinished. It was clear that enzymes were proteins, and that they acted by catalyzing metabolic reactions. The physical nature and mode of action of the gene, however, remained mysterious. Was it a molecule capable of existing in different stable electronic configurations, as envisaged by Delbrück? Was it a nucleoprotein micelle undergoing spontaneous division when it reached a certain size, as suggested by Wrinch? Was it an autocatalytic protein, as Stanley and Northrop believed? Was it a polymer composed of auto-attracting subunits, as proposed by Muller and Haldane? Even as Beadle and Tatum were working out the relationship between genes and enzymes, studies at the Rockefeller Institute were providing a first glimpse of the chemical nature of the genetic material.

Chapter 10
The Hereditary Code-script

World War II

The changing political winds of the 1930s that sent Max Delbrück into exile in the USA in 1937 and Lise Meitner to Sweden the following year (see Chapter 9) also buffeted Hermann Muller. His public opposition to Lysenkoism made it dangerous for him to remain in the Soviet Union, and in 1937 he joined a Canadian medical unit serving on the republican side in the Spanish Civil War. On leaving Spain, Muller worked briefly in Edinburgh before returning to the USA in 1940.

The outbreak of general war in September 1939 meant the cessation of most civilian research in Europe. The increasingly technological nature of warfare meant that the scientific resources of combatant nations were called upon to contribute to the war effort to a far greater extent than in previous conflicts. Physicists were recruited for the Manhattan Project and the development of radar, chemists to develop alternative sources of materials made unavailable by the hostilities, and biologists to produce new drugs and ensure adequate nutrition for soldiers and civilians. Lawrence Bragg returned to the field of research he had pioneered in World War I, artillery range-finding; John Gulland worked on the production of fibers from seaweed; J. B. S. Haldane volunteered for studies on the biological effects of high pressure. Even in neutral Sweden, half of The Svedberg's laboratory was turned over to military purposes.

No scientist had a more interesting war than Max Perutz. His Austrian citizenship made him an enemy alien in Britain; he was arrested in May, 1940, and taken to a detention camp near Liverpool. In July, Perutz and others were shipped to Canada and imprisoned on the Plains of Abraham, near Quebec City. Thanks to the intervention of Lawrence Bragg, Perutz was returned to England and released in January 1941. He resumed his work on the crystallography of hemoglobin, but in the fall of 1942 was asked by the eccentric inventor Geoffrey Pyke to work on a top-secret military operation, code-named 'Habakkuk'. Pyke had managed to interest the authorities in the idea of using huge platforms of ice as floating landing strips for military aircraft. Perutz was an obvious choice to work on the design of these 'berg-ships'; he had been conducting research on glaciers since 1938 as an excuse to make mountaineering trips to Switzerland.

Also involved in Habakkuk were two of Perutz's mentors. Desmond Bernal's complaints to the Home Office about the state of Britain's preparedness for air raids had earned him a meeting with the minister, Sir John Anderson. Bernal's presentation of the problem so impressed Anderson that he was asked, in spite of his communist beliefs, to serve on a Home Office com-mittee. From there, Bernal was seconded to Combined Operations, where he became scientific adviser to Lord Louis Mountbatten and landed with the Allied troops on D-Day.[a]

An American contributor to Habakkuk was Perutz's former physical chem-istry teacher, Herman Mark. Mark, another keen mountaineer, had shown in the mid-1930s that glacier ice was enriched in the heavy hydrogen isotope deuterium. Like Meitner, Mark found himself with unwelcome German citizenship when the Nazis invaded Austria in 1938. He had his passport impounded and was interrogated by the Gestapo. However, he bribed an official to get his passport back and set off with his family for Switzerland. From there, Mark went via France and Britain to Canada, where he worked in a pulp mill for two years. He became adjunct professor of chemistry at the Polytechnic Institute of Brooklyn in 1940. When the USA joined the war, Mark worked on the development of an armored snow vehicle and an amphibious landing craft.

Mark's main contribution to Habakkuk was the suggestion of increasing the strength of the floating platform by reinforcing the ice with wood-pulp. Perutz and his colleagues made a block of this material, which they nick-named 'pykrete', for a demonstration to senior British military officials. The demonstration was only a qualified success. The block of pykrete success-fully deflected a bullet fired into it, but the ricochet hit the Chief of the Imperial General Staff in the shoulder! Perutz was ordered to Washington, DC, to plan the construction of a prototype bergship. However, the US embassy refused to issue a visa to a citizen of a hostile country, and the man who had been incarcerated as a possible fifth columnist had to be given expe-dited British citizenship. When he arrived in Washington, Perutz found that the production of pykrete bergships was no longer a strategic priority. The

---

a   At one point in his military work, Bernal requested the assistance of a former colleague. Mount-
    batten was told that the request had been denied as the man had been associated with a known com-
    munist. The communist associate was, of course, Bernal himself!

improved range of land-based aircraft meant that the floating platforms were now not needed, and the US Navy cancelled Habakkuk in September, 1942.

Delbrück had left Germany for California in 1937 (see Chapter 9). The fates of his Green Paper co-authors showed how fortunate Delbrück was to escape Europe. Both Nikolai Timoféeff-Ressovsky and Karl Zimmer were arrested when the Red Army took Berlin in 1945. Zimmer was sent to a uranium factory, and was not released to West Germany until 1955. Timoféeff was sentenced to hard labor in a prison camp, where he shared a cell with Alexander Solzhenitsyn, and was subsequently portrayed in *The Gulag Archipelago*. In 1947, Timoféeff was ordered to set up a secret prison institute of radiation biology; he was only allowed to go free in 1964.

The Phage Group

Although Delbrück was undoubtedly lucky to be out of Germany, the outbreak of war put him in a difficult position. His Rockefeller Foundation fellowship had ended, and he was, like Perutz in Britain, an enemy alien. Thanks to the intervention of Thomas Hunt Morgan, the Rockefeller Foundation came to Delbrück's rescue again, arranging a position for him as instructor in physics at Vanderbilt University in Nashville, Tennessee. The Foundation paid a portion of Delbrück's salary on condition that he be allowed to continue his research on bacteriophage. He moved to Nashville in January 1940, and was *en route* to Mexico for immigration purposes when he had the meeting with Linus Pauling at Cal Tech that led to their joint paper on complementarity (see Chapter 9).

In December 1940, Delbrück met the Italian physician turned biophysicist Salvador Luria at a meeting of the Physical Society in Philadelphia. Luria later recalled: 'After a few hours of conversation (and a dinner with W. Pauli and G. Placzek, during which the talk was mostly in German, mostly about theoretical physics, mostly above my head), Delbrück and I adjourned to New York for a 48-hour bout of experimentation in my laboratory at the College of Physicians and Surgeons.'[70] They arranged to work together again at the Cold Spring Harbor Laboratory on Long Island the following summer.

This collaboration quickly produced a vindication of Delbrück's faith that genetic questions could be answered using a mathematical approach. In a 1943 paper to which Delbrück contributed the theory and Luria the experimentation, it was shown that bacteria resistant to bacteriophage arose spontaneously, not in response to the presence of the virus. To do this, Luria and Delbrück exploited the fact that a mutational mechanism would produce much more variation in numbers of resistant cells between different cultures. As William Hayes noted, 'This paper . . . provided the first real evidence that bacterial inheritance, like that of the cells of higher organisms, is mediated by genes and not by some Lamarckian mechanism of adaptation as was widely held at the time.'

That same year, Delbrück met Alfred Hershey, a phage researcher at Washington University in St Louis. In a letter to Luria, Delbrück described Hershey thus: 'Drinks whiskey but not tea. Simple and to the point. Likes living in a sailboat for three months, likes independence.' These three men – 'two enemy aliens and another misfit in society' – formed the core of what was to become known as the Phage Group. This was a new type of scientific research group that shared an experimental philosophy but not a physical location. Its unquestioned leader was Delbrück. In 1944, he negotiated the 'phage treaty', an agreement that phage workers would study certain standard strains of bacteria and viruses. The following year, the first course on bacteriophage was held at Cold Spring Harbor. The phage course was held every summer into the 1960s and represented the principal mechanism for recruitment of new phage workers and coordination of research efforts.

What is Life?

Delbrück conceived of the bacteriophage system as a means of solving the 'riddle of life' that had fascinated him since Niels Bohr's 'Light and life' lecture (see Chapter 9). At the same time as the Phage Group was becoming established, however, the pragmatic program of experimentation on bacterial viruses was complemented by a potent theoretical advance that resulted from Delbrück's influence on the physicist Erwin Schrödinger.

Schrödinger was born in Vienna in 1887, graduating with a doctorate in physics from the University of Vienna in 1910. He held academic positions in Stuttgart and Breslau before becoming professor of theoretical physics at the University of Zürich in 1921. Five years later, at the relatively late age of thirty-eight, Schrödinger's discovery of wave mechanics made him one of the leading physicists of the day. In 1927, he replaced Max Planck as professor of theoretical physics at the University of Berlin.

Like Delbrück, Schrödinger made little attempt to conceal his disdain for Nazi policies. In 1933, the year he shared the Nobel prize in physics, Schrödinger resigned his position in Berlin to move to Oxford. However, like Erwin Chargaff six years earlier (see Chapter 11), Schrödinger made the mistake of returning to the sinking ship, becoming professor of physics at the University of Graz in 1936. Two years later, the *Anschluss* between Austria and Germany left Schrödinger, like Lise Meitner and Herman Mark, at the mercy of the Nazis. On being dismissed from his chair at Graz, he was offered the directorship of the new Dublin Institute for Advanced Studies by the *Taoiseach* (prime minister) of the Irish Republic, the former mathematician Eamon de Valera. Schrödinger spent the next seventeen years in Dublin.

One of the duties of Schrödinger's position at the Institute for Advanced Studies was the delivery of an annual series of public lectures on a scientific topic. For 1943, he decided to discuss the relationship between physics and biology. The series of three lectures, entitled 'What is life?', was extremely popular, drawing overflow audiences at two readings. Some of the ideas presented were derived from a lecture Schrödinger had given at the Prussian Academy of Sciences in Berlin eleven years earlier. He had later been given a copy of the Green Paper of Timoféeff-Ressovsky, Zimmer and Delbrück. Impressed by Delbrück's argument that genetic mutation and quantum phenomena both represented transitions between different stable states, Schrödinger incorporated the 'Delbrück model' of the gene into his Dublin lectures. These lectures were published in late 1944 as a book, also entitled *What is Life?*

The problem that Schrödinger addressed was this: 'How can the events *in space and time* which take place within the spatial boundary of a living organism be accounted for by physics and chemistry?' [emphasis in original]. For Schrödinger, this was a question about the 'lawfulness' of biological systems. Quantum physics was based on the average behavior of large numbers of molecules. The behavior of single atoms or molecules could not be predicted – indeed, according to Heisenberg's uncertainty principle, it could not even be measured. However, given a large enough population of molecules, the average behavior of the population over time could be predicted by statistical laws. The genes of living organisms occurred in only two copies per cell. Yet this absurdly small number of molecules somehow ensured the stable reproduction of heritable traits over many generations – for example, the 'Habsburg lip' that characterized the appearance of Austrian emperors from the seventeenth to the twentieth centuries.

Schrödinger noted that liquids and gases tend to be best described by the statistical laws of quantum mechanics. This he described as 'order-from-disorder', because, while the behavior of any individual molecule was completely unpredictable, the average behavior of a large number of molecules could be accurately calculated. Solids, in contrast, were best described by the clock-like laws of classical Newtonian mechanics. This Schrödinger described as 'order-from-order', because the strong forces that held molecules together in solids made them behave quite deterministically, like the pendulum of a clock. From the stability of phenotypes like the Habsburg lip, Schrödinger concluded that the Heitler–London (covalent) bonds that occur in biological macromolecules must be stable enough to confer upon these macromolecules the properties of solids, as opposed to the thermal disorder that characterizes liquids and gases.

Because of this distinction between the clock-like behavior of the large covalently bonded molecules of the gene and the random, statistical behavior of large numbers of small molecules of gases and liquids, Schrödinger believed that a different approach was needed:

> from all that we have learned about the structure of living matter, we must be prepared to find it working in a manner that cannot be reduced to the ordinary laws of physics. And that not on the ground that there is any 'new force' or what not, directing the behaviour of the single atoms within a living organism, but because the construction is different from anything we have yet tested in the physical laboratory.[71]

Schrödinger, like Bohr and Delbrück before him, believed that the properties of living matter could not be reduced to physics and chemistry. For Bohr and Delbrück, this was because the 'organization' of the living cell was somehow 'complementary' to the quantum-mechanical behavior of its components. Schrödinger, however, held the much more specific view that the size and stability of biological macromolecules protected them from the fluctuations of quantum-mechanical 'gas molecules in a bottle' systems. In saying this, Schrödinger articulated a view that has not been seriously challenged since.

Although Schrödinger's principal intention was to explore the workings of the hereditary mechanism at a philosophical level, *What is Life?* also contained two profound ideas about the nature of the genetic material. These ideas, rather than the concept of 'order-from-order' in biological systems, were to account for the remarkable impact of Schrödinger's book. The first of these ideas was the 'aperiodic crystal':

> A small molecule might be called 'the germ of a solid'. Starting from such a small solid germ, there seem to be two different ways of building up larger and

larger associations. One is the comparatively dull way of repeating the same structure in three directions again and again. That is the way followed in a growing crystal. Once a periodicity is established, there is no definite limit to the size of the aggregate. The other way is that of building up a more and more extended aggregate without the dull device of repetition. That is the case of the more and more complicated organic molecule in which every atom, and every group of atoms, plays an individual role, not entirely equivalent to that of many others (as is the case in a periodic structure). We might quite properly call that an aperiodic crystal or solid and express our hypothesis by saying: We believe a gene – or perhaps the whole chromosome fibre – to be an aperiodic solid.[71]

How could a crystal be aperiodic? As noted in Chapter 7, crystals were defined as structures in which a subunit, the unit cell, repeats in space throughout the three-dimensional structure. Aperiodic *solids* could, of course, exist, but these were amorphous – lacking a repeating lattice structure – not crystalline.

What Schrödinger had in mind for the gene was not, in fact, an amorphous structure, but rather a polymer composed of non-identical subunits that nonetheless repeated in space. Such a structure had also been conceived by prewar fiber crystallographers such as William Astbury, when they recognized that proteins giving an X-ray fiber diffraction pattern, and therefore representing one-dimensional crystals, had an amino acid subunit at each structurally equivalent position in the fiber, but did not necessarily have the same amino acid at each equivalent position. Had Schrödinger used a term such as 'aperiodic polymer' to distinguish the 'chromosome fiber' from a true crystal, his meaning would have been much clearer. Nonetheless, he was the first person to make explicit the idea that genes consist of ordered arrays of non-identical, non-repeating subunits.

The second idea, introduced in a passage immediately following the one quoted above, was the 'hereditary code-script'. Schrödinger realized that genes must contain not only the capacity for self-replication and the catalysis of specific reactions, but the entire plan to re-create the organism from a single cell:

It has often been asked how this tiny speck of material, the nucleus of the fertilized egg, could contain an elaborate code-script involving all the future development of the organism. A well-ordered association of atoms, endowed with sufficient resistivity to keep its order permanently, appears to be the only conceivable material structure that offers a variety of possible ('isomeric') arrangements, sufficiently large to embody a complicated system of 'determinations' within a small spatial boundary. Indeed, the number of atoms in such a structure need not be very large to produce an almost unlimited number of possible arrangements. For illustration, think of the Morse code. The two different signs of dot and dash in well-ordered groups of not more than four allow of thirty different specifications. Now, if you allowed yourself the use of a third sign, in addition to dot and dash, and used groups of not more than ten, you could form 88 572 different 'letters'; with five signs and groups of up to 25, the number is 372 529 029 846 191 405.[71]

This conception of the gene as a source of developmental information may have underlain Schrödinger's apparent indifference to its chemical nature. If the gene was just a code, it could have almost any physical form and still contain the same information.

What is Life? was reviewed in at least ten scientific journals, the reviewers including Haldane, Delbrück and Muller. It quickly became a highly influential work, attracting to biology a number of scientists, including the three individuals who would share the Nobel prize for determining the molecular structure of the gene. James Watson wrote: 'from the moment I read Schrödinger's *What is Life?* I became polarized toward finding out the secret of the gene.' His colleague, Francis Crick, was also influenced by Schrödinger's book: 'the book was extremely well written and conveyed in an exciting way the idea that, in biology, molecular explanations would not only be extremely important but also that they were just around the corner . . .' Maurice Wilkins, like Crick trained as a physicist, wrote: 'I had lost some interest in physics. I was therefore very interested when I read Schrödinger's book *What is Life?*'. Other scientists influenced by the views expressed in *What is Life?* included the 'cell chemist' Erwin Chargaff and the microbiologist François Jacob.

It is no exaggeration to say that *What is Life?* is one of the most important scientific books of all time. To many historians and historically minded scientists, however, the defects of the book are more apparent than its merits. Schrödinger was either unaware of or deliberately ignored the chemistry and crystallography of macromolecules – no reference is made to the work of Hermann Staudinger, Astbury, Pauling, Bernal or Phoebus Levene. Reading *What is Life?*, one has to remind oneself that it was written in 1943, when a great deal was known about the structures of the biological macromolecules that Schrödinger refers to in such vague and abstract fashion. Although he apparently did some background reading in preparation for giving the lectures, the vast part of *What is Life?* appears to have been based solely on the Green Paper and the formidable intellect of Erwin Schrödinger.

Nor was the genetics much better than the chemistry. Muller's 1936 model of the gene was far more sophisticated than Schrödinger's, with his prophetic descriptions of genes as chains of subunits replicating by 'autosynthesis' and mutating by changes in subunit sequence (see Chapter 9). In contrast, Schrödinger provided no insight into gene replication and his – or, rather, Delbrück's – mechanism of mutation turned out to be misguided.

However, Schrödinger was clearly not interested in the details of gene function. His was the philosophical approach of asking how physical laws were applied in biological systems. Accordingly, his analysis painted with a broad brush.

The combination of philosophical profundity and chemical naïvety that characterizes *What is Life?* may well represent the secret of its phenomenal impact on the scientific community. Arguably, the success of *What is Life?* was due to the fact that it made explicit two ideas – the aperiodic polymer and the distinction between legislative and executive – that had been subconsciously motivating biology for many decades. The idea of the legislative and the executive had, since it was first proposed by Claude Bernard, mainly manifested itself in biological thought as a series of metaphors – in 1944, the current incarnation was the 'one gene–one enzyme' dogma of George Beadle. Schrödinger, however, clearly distinguished between the gene as a blueprint of the organism and the tools by which this plan was put into effect: 'the miniature code should be in one-to-one correspondence with a highly complicated and specified plan of development and should somehow contain the means of putting it into operation.' Schrödinger was thus the first to state explicitly that genes were purely informational macromolecules. In the postwar period, and particularly after the three-dimensional structure of DNA was elucidated, this concept of the gene as information would become the dominant metaphor of molecular biology.

Schrödinger made even more explicit the idea of the gene as an aperiodic polymer. In order to be capable of 'putting into operation' a 'highly complicated and specified plan', the genetic material must have all the flexibility of a language. In order to do this, the gene must, like a text, consist of a non-redundant linear sequence of a finite number of different parts. The chemistry of this linear molecule did not matter, except insofar as it must be highly stable. It could be a protein, a linear polymer of amino acids; equally well, it could be a nucleic acid, a linear polymer of nucleotides.

Schrödinger's success in putting this latter idea across was largely due to the choice of the term 'aperiodic crystal'. As pointed out by Thomas Hall, the analogy between living organisms and crystals is one of the most recurrent themes in biological thought. Schrödinger's concept of the gene as an aperiodic crystal, however, pointed to a subtle and important difference between inorganic and organic systems. Like a crystal, the gene is composed of an array of subunits; unlike a crystal, the gene has no long-range order. Crystals can grow because they consist of identical subunits; genes can direct the assembly of cells because they consist of non-identical subunits.

In *What is Life?*, Schrödinger wrote: 'If the Delbrück picture should fail, we would have to give up further attempts.' In fact, the 'Delbrück picture' was already obsolete. In June, 1944, Joseph Weiss had shown that the mutagenic effects of ionizing radiation were due to an interaction with water that produced reactive hydroxyl radicals, removing the most important pillar of the target theory. Muller had already realized that mutation had a chemical basis, the replacement of one *Baustein* with another, rather than a quantum-mechanical one – a view highly consistent with Schrödinger's Morse code analogy. The general conclusions of *What is Life?* would therefore withstand the failure of the Delbrück model.

'A Giant Bridge Between Chemistry and Genetics'

Schrödinger, echoing the prevailing sentiment of the time, stated that the gene was 'probably a large protein molecule'. However, in the same year that W*hat is Life?* was published, there appeared a study that seemed to show that some genes did not contain any protein. At least for the process of bacterial transformation, the active ingredient appeared to be that humble 'structural' polymer, DNA.

Studies performed at the Lister Institute in London in the 1920s had shown that the pneumococcus (*Streptococcus pneumoniae*, the bacterium that caused lobar pneumonia) could occur in two forms, R (rough) and S (smooth). These differed in that bacteria of the S form were surrounded by a carbohydrate-containing capsule, whereas in the R form this capsule was absent. Only the S form was virulent (caused infection), suggesting that the capsule was somehow involved in the ability of the bacterium to colonize the lungs.

To study how rough bacteria could be converted to smooth ones, Frederick Griffith, a medical officer with the Ministry of Health in London, inoculated mice with mixtures of S and R strains, one of which had been heat-killed. After two to three weeks, he killed the surviving mice and attempted to culture pneumococci from their blood. When Griffith inoculated a mixture of heat-killed S (virulent) bacteria and living R (non-virulent) ones, he expected to obtain no infection and recover no bacteria. Instead, in some experiments he recovered from the animals something he had not inoculated them with – live bacteria of the S type. Either the presence of live non-virulent bacteria had brought dead virulent ones back to life, or else the presence of dead virulent bacteria had conferred upon non-virulent ones the ability to produce the capsule needed for infection.

A clue to what might be happening came from experiments in which the R and S bacteria were of different strains. These strains were known as serotypes, as they could be distinguished by precipitation with specific antisera. Four such serotypes of pneumococci were known. Griffith found that the generation of live S bacteria in a mouse inoculated with a mixture of dead S and live R bacteria was also associated with a change in type. For example, if a mixture of heat-killed S bacteria of group IV and live R bacteria of type II was used, the organisms recovered were S bacteria of type II. Therefore, it appeared that the second possibility was correct – something present in dead bacteria could endow live bacteria with the ability to produce the capsule, and also to convert them from one serotype to another. This phenomenon became known as 'transformation', and its unknown agent as the 'transforming principle'.

Griffith found that when he tried to concentrate the transforming principle by centrifuging the heat-killed cells and resuspending them in a smaller volume, the activity was in fact decreased. He concluded: 'Although the experiment is not decisive, it rather indicates that the essential material for the building up of a virulent form from the R form may be associated with the capsule of the pneumococcus and may be to some extent washed off.'

It seemed reasonable to implicate the capsule in transformation, as it was the presence of the capsule that distinguished the R and S forms. However, Griffith believed that it was not the capsular carbohydrate itself that was transmitted from dead cells to living ones, but rather the ability to make such a carbohydrate. He proposed that the transforming principle consisted of the 's antigenic substance' – an enzyme that catalyzed the formation of the capsular carbohydrate: 'By S substance I mean that specific protein structure of the virulent pneumococcus which enables it to manufacture a specific soluble carbohydrate.'

Co-inoculation into a mouse was a cumbersome means of achieving bacterial transformation. Griffith therefore tried to produce transformation *in vitro* by simply mixing heat-killed and live bacteria. However, he was unable to produce either conversion of R to S or transformation of type by this means.

Griffith's work was of intense interest to another pneumococcus researcher, Oswald Theodore Avery of the Rockefeller Institute for Medical Research in New York. Avery was born in Halifax, Nova Scotia, in 1877. When he was ten, his family moved to New York, and Avery was educated at Colgate University (BA,1900) and Columbia University (MD, 1904). After working in the Hoagland Laboratory in Brooklyn, Avery joined the Rockefeller in 1913 and worked there until 1947. He spent most of his research career trying to produce a vaccine for pneumonia, which killed 50000 Americans annually in the early twentieth century, and had claimed the life of his mother. This ambition came to naught, as the use of penicillin and other antibiotics made pneumonia a treatable disease by the late 1940s.

Avery, who had earlier shown that the different pneumococcus serotypes differed in capsular polysaccharide, was initially skeptical that transformation of types could occur. However, Martin Dawson of Avery's group soon replicated Griffith's findings. After moving to Columbia University, Dawson managed to achieve *in vitro* transformation of the pneumococcus. Lionel Alloway, working in Avery's laboratory, took this work a further step forward in 1933 by producing a cell-free extract of S bacteria that could be used to transform living R cells. He dissolved the bacteria by heating them in the detergent desoxycholate. On adding alcohol, 'A thick, stringy precipitate formed which slowly settled out on standing.' Rollin Hotchkiss, later a participant in this work, wrote that: 'Alloway was probably the first person to see fibers of precipitated biologically active crude DNA.' Alloway also showed that the 'transforming extract' was unusually heat-stable for a protein, as it was only slowly inactivated by heating at 80°C. According to Hotchkiss, 'Avery outlined to me that the transforming agent could hardly be carbohydrate, did not match very well with protein, and wistfully suggested that it might be a nucleic acid!'

Avery's laboratory now became the leading center for research on the transforming principle. Alloway's isolation of a cell-free transforming extract seemed to open the way to a complete chemical characterization. However, in 1934 or 1935 Avery underwent thyroidectomy for Graves' disease, followed by a long convalescence. Alloway's work was continued by Colin MacLeod, but nothing publishable was achieved and the project was set aside in 1938.

The fact that it took eleven years for Avery to prove that Alloway's 'stringy precipitate' was DNA was largely because research on the transforming principle was for many years a side-line in Avery's laboratory. Even when the development of sulfonamide drugs in the mid-1930s rendered obsolete his search for an antiserum to pneumonia, Avery did not make transformation his primary focus.

A new attempt at characterizing the transforming principle began only in October 1940, when Avery wrote 'Exp. 1 (T.P.)' on the first page of a new notebook. One of the problems in Alloway's studies was the extreme variation in transforming activity from batch to batch of the pneumococcus extract. Avery and MacLeod discovered that a much higher transforming activity was obtained by heating the extracts before treating with desoxycholate. Changes in the conditions used for growing the bacteria also helped reduce the experimental variability. MacLeod left the Rockefeller in July 1941, and that September his place in Avery's 'pneumocosm' was taken by Maclyn McCarty. By November, McCarty had shown that degradation of the capsular polysaccharide with specific enzymes had no effect on the transforming activity. By early 1942, the transforming extract had been purified to the point that no protein could be detected. That spring, analysis of the purified transforming principle in the ultracentrifuge showed that it had a very high molecular weight; the only known substance of such size was DNA.

Avery, McCarty and MacLeod's paper on the chemical characterization of the pneumococcal transforming principle was published in the *Journal of Experimental Medicine* in February 1944. By this time, the man who had inspired this work was dead. Griffith's decision to stay at work during a German air raid cost him his life when a bomb hit the laboratory.

Oswald Avery was sixty-seven years old in 1944; he had officially retired three years earlier. His achievement of a major scientific breakthrough is therefore as remarkable as Guiseppe Verdi composing *Otello* at the age of seventy-three. Perhaps because he knew that the transforming principle paper was to be his swan-song, Avery appears to have taken unusual care with it. The paper, which was written with great clarity, described an extraordinarily rigorous study in which a large number of techniques were used to establish the conclusions.

To prove that the transforming principle was DNA, Avery, MacLeod and McCarty showed that its elemental composition was very close to that of the theoretical tetranucleotide; its ultraviolet absorption spectrum was similar to that of DNA; in the analytical ultracentrifuge, it gave a single sharp band of approximately 500 000 daltons; in the Tiselius electrophoresis apparatus, it gave a single high-mobility component. The crystalline proteolytic enzymes trypsin and chymotrypsin, and also the RNA-degrading enzyme ribonuclease, did not affect the transforming activity.

Crystalline deoxyribonuclease was not yet available, so Avery used the same extract of dog intestinal mucosa with which Phoebus Levene had ten years earlier attempted to isolate the sugar of DNA (see Chapter 7). This crude deoxyribonuclease destroyed the transforming activity of the pneumococcus extract. In order to eliminate the possibility that other enzymes present in the intestinal extract might have been responsible for inactivating the transforming principle, Avery and co-workers performed the elegant experiment of showing that elevated temperature had the same effect on the deoxyribonuclease activity and the ability to destroy the transforming principle.

Apart from elemental analysis, all of the techniques used to characterize the transforming principle were newly developed. Ultracentrifuges were installed in only a handful of laboratories at the time; the air-driven one used by Avery had been developed at the Rockefeller. The Tiselius apparatus was not yet commercially available. The crystalline enzymes Avery used were the best available, given to him by his Rockefeller colleagues John Northrop and Moses Kunitz.

Given that the transforming principle was DNA, could it be said to be a gene? Avery had one powerful indication that it was – the transforming principle had the property of extensibility, which had been identified as a key feature of life by Lionel Beale, Joseph Henry (see Chapter 3) and Theodor Schwann (see Chapter 4), among others. A gene should be able to produce many copies of itself, and this was the case for the transforming principle: 'from the transformed cells themselves, a substance of identical activity can again be recovered in amounts far in excess of that originally added.' Perhaps for this reason, the Avery group, as McCarty has recorded, 'thought it likely that its [DNA's] property of transferring hereditary information could not possibly be limited to the bacterial world.' In the 1944 paper, however, Avery was extremely cautious in his conclusions, stating only that: 'nucleic acids of this type must be regarded not merely as structurally important but as functionally active in determining the biochemical activities and specific characteristics of pneumococcal cells.'

Avery's brother Roy was a bacteriologist at Vanderbilt University. To him Oswald Avery wrote in May 1943 a famous letter that described the history of bacterial transformation and his evidence that the transforming principle was DNA:

> If we prove to be right – and of course that is a big if – then it means that both the chemical nature of the inducing stimulus is known and the chemical structure of the substance produced is also known, the former being thymus nucleic

acid, the latter Type III polysaccharide, and both are thereafter reduplicated in the daughter cells and after innumerable transfers without further addition of the inducing agent and the same active and specific transforming substance can be recovered far in excess of the amount originally used to induce the reaction. Sounds like a virus – may be a gene. But with mechanism I am not now concerned. One step at a time and the first step is what is the chemical nature of the transforming principle? Some one else can work out the rest. Of course the problem bristles with implications. It touches the biochemistry of the thymus type of nucleic acids which are known to constitute the major part of chromosomes but have been thought to be alike regardless of origin and species. It touches genetics, enzyme chemistry, cell metabolism and carbohydrate synthesis. But today it takes a lot of well documented evidence to convince anyone that the sodium salt of deoxyribose nucleic acid, protein free, could possibly be endowed with such biologically active and specific properties and that is the evidence we are now trying to get. It is lots of fun to blow bubbles but it is wiser to prick them yourself before someone else tries to.[72]

Roy Avery showed this letter to his Vanderbilt colleague, Max Delbrück. Thus the 'pope' of the Phage Group received advance notice that genes might be composed of nucleic acid.

The 1944 paper by Avery, MacLeod and McCarty is in every sense a classic of the scientific literature. Max Perutz wrote: 'The paper is absolutely rigorous and leaves no shadow of doubt that the transforming factor consists of DNA and nothing but DNA.' However, according to Rollin Hotchkiss, who led the chemical analysis of the transforming principle in the post-Avery period, few people were convinced that DNA was the genetic material. George Beadle, for example, described the transforming principle in 1948 as a 'transmuting agent' (i.e., something that causes genetic change rather than the gene itself); it was not until 1956 that he unambiguously described it as 'the primary genetic substance'. At a 1950 symposium on Genetics in the Twentieth Century, Gunther Stent noted, 'Only one of the 26 essayists saw fit to make more than a passing reference to Avery's discovery.' Hotchkiss was asked by a classical (Mendelian) geneticist in 1951 how he could be sure that he was working with genes when he 'couldn't show a hybrid cross', and by a bacteriologist whether the identification of genes as DNA was 'based on evidence or was merely a voting agreement'.

There were some reasons for skepticism. It was impossible to eliminate the possibility that an undetectable residue of protein was responsible for the transforming activity of the pneumococcal extract. As the eminent biochemist Joshua Lederberg pointed out, exaggerated claims of homogeneity had been discredited in the cases of Richard Willstätter's 'protein-free enzymes' (see Chapter 7) and Wendell Stanley's 'nucleic acid-free virus' (see Chapter 9). Also, DNA was thought to lack the structural variety necessary to encode genetic information – it was not an 'aperiodic crystal'. Perhaps most importantly, transformation with DNA had been shown only in bacteria – in fact, only in one species of bacterium. If genes were composed of DNA, why was it not possible to transform *all* bacteria, or even all cells, with this substance? It is difficult to disagree with Lederberg's conclusion that 'the critical reception initially given to Avery *et al.* (1944) exemplifies the critical scientific method at its most functional.'

However, another factor that hindered the acceptance of Avery's work reflects less creditable upon the 'critical scientific method'. Alfred Mirsky of the Rockefeller Institute, who had co-authored with Linus Pauling the classic 1936 study on protein denaturation (see Chapter 8), subsequently became interested in chromosomal structure and collaborated briefly with Avery on the transforming principle. Salvador Luria wrote in 1986: 'Mirsky later became one of the severest critics of the Avery group and of their conclusion that TP [transforming principle] consisted of DNA . . . partly, perhaps, because of a grudge over his abortive participation in the research.' Perutz thought that Mirsky's 'smear campaign' may have cost Avery the Nobel prize.[b]

For reasons that are easily understandable, scientists who struggle to win acceptance for revolutionary ideas often tend to exaggerate the hostility of the mainstream scientific establishment. The idea that the transforming principle was DNA was certainly a radical one, but some minds, at least, were prepared for it. As will be discussed in Chapter 12, several prominent scientists had suggested prior to 1944 that DNA had at least a supporting role in the transmission and expression of genetic information. Avery's paper was favorably cited by these and some others.

One scientist who realized the importance of Avery's discovery was Erwin Chargaff, an emigré Austrian biochemist working at Columbia University. For Chargaff, the 1944 paper represented 'the sudden appearance of a giant

---

b   In fairness to Mirsky, he had professional reasons, as well as any personal ones, for doubting the genetic role of DNA. In a 1943 article, he had written, with reference to his own unpublished work: 'All of the nucleic acids so far considered, including those from animal sources as well as that from wheat germ, agree in having equimolar quantities of purines and pyrimidines – a definite restriction in possible variations among the desoxyribose nucleic acids.' It should also be noted that this was probably the first indication of the base composition regularities later known as 'Chargaff's rules' (see Chapter 12).

bridge between chemistry and genetics'. Chargaff decided to drop his work on lipoproteins and apply his expertise in 'cell chemistry' to finding differences between the DNA of different tissues and organisms: 'Avery gave us the first text of a new language, or rather he showed us where to look for it. I resolved to search for this text.'

The identification of the transforming principle as DNA unveiled the last of the major players in the central processes of biology. The work of Gregor Mendel had shown that heredity had a material basis; Morgan showed that these genes existed in linear arrays on chromosomes. A series of discoveries, from Jöns Jacob Berzelius to Eduard Buchner, had shown that biological processes were directed by specific catalytic proteins, the enzymes. The work of Archibald Garrod and Beadle had provided the link between biochemistry and genetics by showing that the function of genes was to make enzymes. With the identification of DNA as the genetic material, the final piece of the puzzle was on the board. DNA mediated Bernard's 'legislative force', enzymes mediated his 'executive force'.

In the spring of 1944, a year after he had learned of Avery's work, Max Delbrück delivered a series of lectures at Vanderbilt on 'Problems of modern biology in relation to atomic physics'. In these, he listed 'three magnificent problems' genetics had created for biochemistry. These were: what do genes consist of; how do they reproduce; and how do they act? Although Delbrück was clearly not yet convinced, Avery's work had suggested the answer to the first question. Answering the second and third questions was to be the main preoccupation of molecular biology in the postwar period.

'The Chemical Physics of Biology'

The end of hostilities in the European and Pacific theaters in May and August, respectively, of 1945 heralded the resumption of peacetime research activities and the return of scientists from wartime duty. In the immediate postwar period, several changes in the organization or personnel of research groups occurred that were to be of great significance in the immediate future. In particular, three groups inspired by the application of physics methods to biology came into being: the Medical Research Council Unit for the Study of the Molecular Structure of Biological Systems at Cambridge; the Medical Research Council biophysics unit at King's College, London; and the Phage Group.

In Cambridge, Bragg's protein crystallography group – consisting solely of Max Perutz – was strengthened by the addition of a new PhD student. John

Kendrew was a chemistry graduate who had been persuaded by Desmond Bernal to take up protein work when the two men were both serving in Burma. In 1947, Bragg requested funding from the Medical Research Council (MRC) for a research unit that would apply X-ray methods to the structures of biological macromolecules. The sum of £2550 per annum for five years was awarded to support Perutz, Kendrew and two assistants.

Two years later, Francis Crick joined the Cambridge group as a PhD student with Perutz. Crick was born in 1916 in Northampton and studied physics at University College, London, graduating in 1937 with a second-class degree. He then commenced graduate studies at University College, but these were interrupted by the outbreak of war. Crick joined the Admiralty research laboratory at Teddington, where he worked on the design of mines. At the end of the war, he joined the Naval Intelligence Division in London. Largely in reaction to *What is Life?*, it seems, Crick conceived the idea of doing research in biophysics, and applied for a studentship from the MRC. His application revealed that Crick had a very definite idea of how physics could profitably be applied to biology:

> The particular field which excites my interest is the division between the living and the non-living, as typified by, say, proteins, viruses, bacteria and the structure of chromosomes. The eventual goal, which is somewhat remote, is the description of these activities in terms of their structure, i.e., the spatial distribution of their constituent atoms, in so far as this may prove possible. This might be called the chemical physics of biology.[73]

Crick went to talk to Bernal, but was fobbed off by his secretary. The MRC then decided that he should go to work with Dame Honor Fell at the Strangeways Laboratory in Cambridge. Crick went to the Strangeways in September 1947, and started work on protoplasmic streaming in fibroblasts. On meeting Perutz, he became much more interested in protein structure than he was in protoplasm. With the support of Bragg and Perutz, he obtained a three-year fellowship from the MRC and moved to the Cavendish in June 1949.

At the end of World War II, William Astbury was appointed chair of a new department of biomolecular structure at Leeds University. In early 1946, he made a verbal proposal to the MRC for support for a group in biophysics. Another physicist seeking support for research on biological tissues was John Randall, who in the fall of 1946 took up the Wheatstone chair of physics at King's College, London. Randall had become interested in biophysics in 1941, when he attended a lecture on protein structure in Birmingham –

given by the ubiquitous Bernal. He asked the Royal Society for funding for a biophysics research unit at King's, and the request was passed to the MRC. Presented with two similar proposals, the MRC decided to support Randall. According to his student Mansel Davies, Astbury attributed his failure to the influence of Bragg, with whom he did not have a good relationship. There may well have been other reasons. Astbury's proposal apparently focussed on artificial fibers rather than proteins or nucleic acids. Also, Randall was a scientific war hero. He was co-inventor of the cavity magnetron, a key component of radar that has been described as 'the most valuable cargo ever brought to the shores of America'.

One member of Randall's new biophysics unit was the physicist Maurice Wilkins. Born in New Zealand in 1916, Wilkins was brought to England at the age of six. He graduated with a BA degree from St John's College, Cambridge, in 1938, and studied for his PhD under Randall at the University of Birmingham. During the war, Wilkins worked in Berkeley on the separation of uranium isotopes, where, as noted above, he 'lost some interest in physics' until reading *What is Life?* He rejoined Randall at St Andrew's University in 1945 and moved with him to London the following year.

Another British biophysics group could have arisen in the immediate postwar period. In 1946, Max Delbrück seriously considered the offer of a chair in biophysics at the University of Manchester. Instead, he joined Beadle's division of biology at Cal Tech. Shortly before leaving Nashville, Delbrück hosted the first 'official' meeting of the Phage Group. François Jacob wrote:

> Research on the phage, the genetics of bacteria, were the business of a small number of people. Some twenty or thirty persons scattered all over the world . . . A sort of rugby team where the ball went from hand to hand . . . Membership in the club gave one the right to a series of privileges: hearing the latest news well before its publication; knowing who was doing what; having access to certain strains of bacteria and viruses or to certain rare substances; the benefit of incessant criticism that obviated blunders.[74]

The Phage Group was characterized by its philosophy as much as it was by its T series bacteriophage and *Escherichia coli* strains. Judson has written: 'The phage group worked with a self-consciously fastidious rigour; they restricted themselves to the simplest biological systems; they distrusted biochemists and earlier microbiologists; the younger of them marked the book off from the others in a manner frankly snobbish.' To a remarkable extent, these workers tried to separate themselves from their scientific forebears. Gunther Stent, a student of Delbrück at Cal Tech, wrote that 'one of the main characteristics of the American Phage Group was that they didn't believe *anything* that anyone had said or done before and insisted on working out everything for themselves' [emphasis in original].

The Phage Group was also a reflection of the attitudes of Max Delbrück. Like his mentor, Niels Bohr, Delbrück saw biochemistry as a reductionist approach that destroyed what was of most interest in living systems. If anyone in his group suggested an experiment that Delbrück felt was reductionist, he would drily 'refer it to the subcommittee on biochemistry'. Jacob described Delbrück as:

> . . . a tall, dry athlete with a thick head of hair and steel-rimmed glasses . . . Delbrück's rigor, his frankness, his way of going to the heart of a problem were combined with his surprising youthfulness of mind as of body . . . He spoke rather slowly and a bit haltingly, with a slight German accent. He would often halt in mid-sentence, his eyes wandering across the ceiling, his hand on his mouth, searching for the right word.[74]

In his Harvey Society lecture of 1946, Delbrück described 'an imaginary theoretical physicist' who was shown how a bacterial cell was lysed in twenty minutes by a bacteriophage and resolved to find out how this occurred:

> Perhaps you would like to see this childish young man after eight years, and ask him, just offhand, whether he has solved the riddle of life yet? This will embarrass him, as he has not got anywhere in the problem he set out to solve. But being quick to rationalize his failure, this is what he may answer, if he is pressed for an answer: 'Well, I made a slight mistake. I could not do it in a few months. Perhaps it will take a few decades, and perhaps it will take the help of a few dozen other people. But listen to what I have found, perhaps you will be interested to join me.[75]

A 'few dozen other people' did answer this siren song, and a 'few decades' were spent on the search. The answer to the 'riddle of life', however, was to come mainly from other directions.

Chapter 11
The Ubiquitous Spiral

The Pauling Giant

In his 1947 application to the Medical Research Council (see Chapter 10), Lawrence Bragg noted that the development of the electron microscope, which illuminated a specimen with a beam of electrons rather than the visible light used by an optical microscope, had resulted in a great increase in resolving power. However, even electron microscopy was not of high enough resolution to determine the structures of biological macromolecules. Chemical methods, on the other hand, were of most use for studying small molecules:

> We are trying to fill this gap by methods of X-ray analysis, which have been so successful in analysing the atomic architecture of inorganic and organic compounds. The resolving power is amply sufficient, interatomic distances of the order of one Angstrom [unit] can be measured with precision by X-rays. The formidable difficulty of applying this method arises from the complexity of the protein molecules which we are studying. It has been a great triumph of the X-ray method that such molecules as penicillins, sterols, or sugars have been completely determined, but these contain less than one hundred atoms. A protein like haemoglobin contains thousands of atoms.[76]

A similar point was made by Linus Pauling in a 1946 article entitled 'Molecular architecture and biological reactions'. To illustrate the abilities of the techniques available to determine molecular structure, Pauling used a brilliant heuristic device. Imagine a human being increased in size by a factor of 250 000 000, the scale factor normally used in molecular models. The height of this giant would be equivalent to the distance between the earth and the moon; to him, the earth would appear to be the size of a billiard ball, and New York City would be barely visible. Using a light microscope, the giant could resolve objects about a thousand feet (300 m) apart. He would thus be able to distinguish Central Park and the Rockefeller Center, but not individual buildings. Using an electron microscope, which would have a resolution of about ten feet (3 m), the giant could determine the shapes of buildings but not much of their interior structure, and automobiles would be mere dots. By X-ray diffraction, detailed information about objects less than one foot (30 cm) in diameter could be obtained. Anything between one and ten feet in diameter – including human beings – would be effectively invisible.[a]

The Pauling giant was like a short-sighted man who was given a powerful magnifying glass instead of a pair of spectacles. The problem for scientists interested in macromolecular structure was that the sizes of proteins and nucleic acids fell within the 'blind spot' between techniques with atomic resolution and those with subcellular resolution – a new 'world of neglected dimensions' (see Chapter 7). However, the solutions proposed by Bragg and Pauling were subtly different: Bragg suggested an expansion of X-ray techniques to the macromolecular level, whereas Pauling believed that the coordinated use of many approaches would be necessary to fill the gap.

Because of the disruptions caused by World War II, the prospects for progress in the area of biological structure could not have seemed bright. In 1946, much of Germany lay in ruins, Britain would suffer shortages for many years, and even in neutral countries research laboratories had been turned over to the war effort. And yet the decade following the war's end was one of unprecedented success in elucidating the molecular mechanisms of life. Of the 'three magnificent problems' Max Delbrück had posed in 1944, two of these – the nature and mode of reproduction of the gene – would be solved by 1955, and the solution to the third – the mode of action of the gene – would be well on its way.

Perhaps the most important reason for the rapid progress of biochemistry in the postwar period was the introduction into standard laboratory practice of a large number of new analytical techniques that filled the gap Pauling had perceived between the atomic and the subcellular levels. Some of these techniques were the products of wartime research. These included the use of reactor-generated radioactive isotopes, first produced by the Oak Ridge National Laboratory in 1946. Radioisotopes such as hydrogen 3 (tritium), carbon 14, phosphorus 32 and sulfur 35 could be used to radiolabel biological macromolecules, and were particularly important in metabolic studies. Another spin-off from military research was the digital computer, which largely arose from attempts to break German codes and would prove essential for the solving of protein structures by X-ray crystallography.

Other postwar innovations, however, were simply the result of earlier techniques becoming widely available. Arne Tiselius had built his first

---

a   Reading Pauling's 1946 paper influenced Francis Crick in his decision to become a biophysicist.

electrophoresis apparatus at Uppsala in 1937, and used it to separate the serum globulins; Cal Tech obtained a Tiselius apparatus in 1943. Also in 1937, the first US ultracentrifuge was installed, at the Rockefeller Institute. Three years later, an American-designed air-driven ultracentrifuge became available. The electron microscope had been invented in Germany in the early 1930s, but was developed into a commercial instrument by the RCA Company in 1940. By 1943, electron microscopes were in operation or under construction at the Massachusetts Institute of Technology, Stanford University and Cal Tech. The best resolving power achieved by then was 2.2 nm, a hundred times better than the theoretical limit of optical microscopy. Chromatographic techniques had been available since the early years of the century. However, these were not used widely for the separation of organic molecules until the development of paper chromatography by Archer Martin and his colleagues at the Chester Beatty Cancer Research Institute in London in 1941. This technique would play a crucial role in the fine-structure analysis of proteins and nucleic acids. Enzymes had also long been used as biochemical reagents. In the postwar period, largely due to the efforts of John Northrop and Moses Kunitz at the Rockefeller Institute, highly purified enzymes were becoming standard laboratory reagents.

Ultracentrifugation, electrophoresis, chromatography, electron microscopy, radioisotopes – virtually all the techniques that were to characterize biochemical research for the next generation were in place, at least in some laboratories, by 1945. Not until the development of recombinant DNA technology in the early 1980s would a major technological change occur in biochemistry.

It was not only the technology of biochemistry that had changed, but also its geographical center of gravity. The air-driven ultracentrifuge, radioisotopes, the commercial electron microscope and crystalline enzymes were all American innovations. Untouched by war and with its research efforts coordinated by the National Institutes of Health, the USA was increasingly to dominate the biomedical sciences in the postwar period.

The Primary Structure of Proteins

The new biochemistry allowed rapid progress to be made in protein chemistry in the late 1940s. Of particular importance was paper chromatography, which by 1944 could separate all the amino acids formed by hydrolysis of proteins. Techniques for the quantitative analysis of amino acids arose from the wartime work of George Beadle's laboratory at Stanford. The nutritional mutants of *Neurospora* that Beadle and Edward Tatum had isolated for mapping metabolic pathways (see Chapter 9) had been used to develop sensitive microassays for nutrients, including several amino acids.

A milestone in the analysis of protein structure was reached in 1945 when a group at Columbia University led by Erwin Brand, a former student of Max Bergmann, published the first complete analysis of the amino acid composition of a protein. Using twenty-eight different analytical methods – eighteen chemical assays, one enzymatic assay and nine nutritional microassays – Brand and co-workers were able to measure the contents of eighteen amino acids in the bovine milk protein β-lactoglobulin. The amide amino acids, asparagine and glutamine, could not be distinguished, so all the amide

Table 11.1: Amino acid compositions of β-lactoglobulin

The amino acid composition of bovine β-lactoglobulin measured by Brand and co-workers in 1945 is compared with the modern values, derived by DNA sequencing. Except for the fact that they underestimated the molecular weight of the protein by a factor of 2, the values of Brand and co-workers are very similar to the modern ones. In the final column, the number of molecules of each amino acid is expressed as being consistent (Y) or inconsistent (N) with the formula $2^m \times 3^n$. Note that the values for eight amino acids are incompatible with the Bergmann and Niemann periodicity hypothesis of protein structure.

Amino acid	Brand et al., 1945	Alexander et al., 1989	Bergmann-Niemann
Methionine (Met)	5	9	Y
Lysine (Lys)	16	33	N
Cysteine (Cys)	7	12	Y
Leucine (Leu)	27	50	N
Alanine (Ala)	18	29	N
Threonine (Thr)	9	21	N
Glycine (Gly)	4	8	Y
Glutamine (Gln) + asparagine (Asn)	15	32	–
Isoleucine (Ile)	10	27	Y
Valine (Val)	11	21	N
Aspartic acid (Asp)	11	36	Y
Tryptophan (Trp)	2	4	Y
Tyrosine (Tyr)	4	9	Y
Serine (Ser)	7	20	N
Proline (Pro)	8	15	N
Arginine (Arg)	3	7	N
Glutamic acid (Glu)	16	24	Y
Phenylananine (Phe)	3	9	Y
Histidine (His)	2	4	Y
Total	178	370	

was arbitrarily assigned to glutamine. As shown in Table 11.1, Brand *et al.* overestimated the total molecular weight of β-lactoglobulin by a factor of two. Apart from that, the amino acid contents measured in 1944 are remarkably close to the present-day values obtained by sequencing the β-lactoglobulin gene.

The availability of a complete amino acid composition for β-lactoglobulin represented a test of the periodicity hypothesis of Max Bergmann and Carl Niemann (see Chapter 8). Of the eighteen amino acids that could be unambiguously measured, the values for ten were consistent with the periodicity hypothesis, while eight were not.

The composition determined by Brand *et al.* – $Gly_8Ala_{29}Val_{21}Leu_{50}Ile_{27}$ $Pro_{15}Phe_9Cys_8Met_9Trp_4Arg_7His_4Lys_{33}Asp_{36}Glu_{24}Gln_{32}Ser_{20}Thr_{21}Tyr_9$ – represented a kind of empirical formula of β-lactoglobulin.[b] Just as an empirical formula gives no clue as to how atoms are arranged in a molecule, an amino acid composition provides no information on the order of these amino acids in the polypeptide chain. For that, techniques would be required that could determine the sequence of amino acids. Because of the linear nature of polypeptides, the amino acid sequence of a protein would correspond to its structural formula.

The accomplishment of an amino acid sequence was first achieved by Frederick Sanger. Born in Rendcombe, Gloucestershire, in 1918, Sanger studied biochemistry at Cambridge University, graduating with a BA in natural science in 1939. Exempted from military service as a conscientious objector, he became a PhD student in the biochemistry department at Cambridge under Norman Pirie, the man who had demonstrated that tobacco mosaic virus contains nucleic acid as well as protein (see Chapter 9). On finishing his PhD in 1943, Sanger joined the research group of Albert Chibnall, the new professor of biochemistry, and started work on the amino acid analysis of insulin. In 1945, the same year that Brand produced the complete amino acid analysis of β-lactoglobulin, Sanger studied insulin with fluorodinitrobenzene (FDNB), which reacted with the free amino groups that occurred at one end (the N-terminus) of polypeptide chains. Reaction with FDNB followed by acid hydrolysis should have resulted in the N-terminal amino acid being recovered as its dinitrobenzene (DNB) derivative. From insulin, however, Sanger recovered two moles of DNB-glycine and two moles of

---

b   Brand and co-workers also reported a true empirical (elemental) formula for β-lactoglobulin: $C_{1864}H_{3012}N_{468}S_{21}O_{576}$. Ignoring the sulfur, this corresponds to $C_{400}H_{646}N_{100}O_{124}$, in excellent agreement with Gerrit Mulder's 1838 formula for the 'protein radical', $C_{400}H_{620}N_{100}O_{120}$ (see Chapter 2).

DNB-phenylalanine per 12 000 daltons of protein. This unexpected finding meant that the true molecular weight of insulin must be 6000 rather than 12 000.[c] It also showed that insulin consisted of two different polypeptide chains, one starting with glycine and one starting with phenylalanine.

Sanger showed that the two polypeptides of insulin, which he named the A (or glycine) and B (or phenylalanine) chains, were joined by a disulfide (S–S) bond between cysteine residues. By oxidizing the disulfide bonds, he was able to separate the A and B chains, which were of different sizes.

In 1947, Sanger made a trip to Tiselius' laboratory at Uppsala, where he learned electrophoresis techniques and met one of the pioneers of paper chromatography. Back at Cambridge, in his 'laboratory in the basement next to the experimental rats', Sanger tried to devise means of determining the sequence of amino acids in proteins. In his initial experiments, insulin was subjected to partial acid hydrolysis, which cleaved it randomly to form short peptides. These peptides were reacted with FDNB and separated by chromatography on silica columns. The peptides were then hydrolyzed to free amino acids, whose nature was determined by paper chromatography. Using this technique, short sequences could be reconstructed. For example, peptide B1 of the insulin B chain contained only phenylalanine (Phe), B2 contained phenylalanine and valine (Val), B3 contained phenylalanine, valine and aspartic acid (Asp), B4 contained phenylalanine, valine, aspartic acid and glutamic acid (Glu). Therefore, the B chain sequence started with: Phe-Val-Asp-Glu.

Determination of longer sequences required the preliminary step of digestion of the polypeptide with specific proteolytic enzymes. As these enzymes cleaved at specific residues – trypsin, for example, cleaves polypeptides at the C-terminal side of lysine and arginine residues – the peptides produced did not overlap. However, the order of peptides within the polypeptide sequence could be obtained by using more than one proteolytic enzyme.

Using this method, Sanger was in 1951 able to publish the complete amino acid sequence of the insulin B chain:

Phe-Val-Asp-Glu-His-Leu-Cys-Gly-Ser-His-Leu-Val-Glu-Ala-Leu-
Tyr-Leu-Val-Cys-Gly-Glu-Arg-Gly-Phe-Phe-Tyr-Thr-Pro-Lys-Ala

---

c   Dorothy Crowfoot had shown in 1938 that the unit cell of insulin corresponded to approximately 37 600 daltons but had predicted that the molecular weight of the insulin molecule might be a submultiple of that value (see Chapter 8).

No trace of a periodic distribution of amino acids was found, and Sanger was able to pronounce the death sentence of the Bergmann–Niemann hypothesis:

> In recent years, several hypotheses of protein structure have been advanced which are based on the assumption of some type of periodic arrangement of amino-acids along the polypeptide chains. It is tempting to assume that this arrangement, which is determined by the unknown mechanisms of protein synthesis, may be rather simple and may follow certain easily discernible principles. An examination of the structure of fraction B, however, fails to reveal any simple periodic arrangement of the residues, nor is it possible to formulate any general principles which might govern the order of amino-acids along the protein chains.[77]

Techniques for sequencing proteins were improved upon only slowly; the rate of generation of sequence data, in amino acids sequenced per researcher-year, was estimated by Brian Hartley to be 10–20 in 1951–8, and only approximately 80 by 1968. However, the large number of researchers attracted to this field meant that 7000–8000 residues of sequence per year were being generated by the late 1960s. Among other things, these sequence data were to be of crucial importance in the X-ray diffraction analysis of proteins.

The Secondary Structure of Proteins

The elucidation of the amino acid sequence of the B chain of insulin – and, two years later, of the A chain – established that proteins consist of an irregular sequence of amino acids. This amino acid sequence became known as the primary structure of the protein. The other unresolved aspect of protein chemistry from the 1930s was the question of how the polypeptide chain folded; these patterns of folding became known as the secondary structure of the protein. It will be recalled that the studies of William Astbury indicated that the conversion of keratin from the stretched (β) form to the unstretched (α) form decreased the length of a keratin fiber by 50% and replaced a prominent X-ray reflection at 3.32 Å with one at 5.1 Å. Astbury believed that folding the polypeptide chain into a series of hexagons stabilized by hydrogen bonds between adjacent amino acids could explain these observations (see Chapter 7). The 'hexagonal fold' model of α-keratin had been inspired by crystallographic analyses of cellulose; it was also this polysaccharide that inspired Astbury in 1939 to muse about 'the ubiquitous spiral' that underlay 'the great fundamental questions in molecular biology'.

In 1943, Maurice Huggins of the Kodak Research Laboratories took a more rigorous approach to protein folding which would be the model for subsequent attempts. Huggins assumed that atoms would pack together as closely as possible and that any folded structure would be stabilized by hydrogen bonds between N–H and C=O groups of the polypeptide 'backbone'. The values for bond lengths and angles were those derived by Pauling's colleague Robert Corey for the three amino acid structures thus far solved by X-ray diffraction: glycine, alanine and diketopiperazine.

For β-keratin, Huggins developed more fully the idea of H-bonded colinear chains initially proposed by Astbury and Woods and in his own 1937 paper. The colinear chains could either be parallel (running in the same direction) or antiparallel (running in opposite directions), but Huggins pointed out that the latter would be more stable. These are the structures that later became known as the parallel and antiparallel β-pleated sheets, common motifs in many globular proteins and the predominant folding pattern of silk fibroin.

To Huggins also goes the credit for being the first person to suggest specific spiral configurations for proteins. For fibrous proteins, Huggins assumed, 'Polypeptide chains extending through the crystalline regions [of the protein] must each have a screw axis of symmetry, or else two or more chains must be grouped around screw axes or other symmetry elements.' What this meant was that the operation of going from one amino acid to the next involved a rotation of the polypeptide chain around a central axis and a translation along the axis. The only structures having such screw axes of symmetry were helices and, in the case of the rotation angle being 180°, a flat ribbon (see Figure 7.3).

Protein helices could be defined by three properties. The first was the 'handedness' of the helix, left or right – most spiral staircases are left-handed helices, corkscrews are right-handed. The second was the number of amino acids per 360° turn of the helix. This corresponded to the rotational symmetry of the molecule, so a helix with two amino acids per turn would have two-fold symmetry, one with three amino acids per turn three-fold symmetry, etc. The third property was the distance (translation) along the rotation axis between adjacent amino acids. Multiplying the residues per turn by the translation per amino acid gave the distance along the rotation axis of the crystallographic repeating unit, or 'pitch' of the helix.

Huggins's 1943 paper showed structures with two and three amino acids per turn of the polypeptide chain. The former was a ribbon structure with the side-chains projecting from alternate sides of the plane; the latter was a right-handed helix with a three-fold rotation axis. In the helical structure, each N–H group was hydrogen-bonded to the C=O group of an amino acid three positions distant in the polypeptide chain, and the side-chains radiated out from the helix (Figure 11.1). However, Huggins's preferred structure for α-keratin was a 'zigzag ribbon' structure with two residues per turn, as this would accommodate the 100% increase in length with stretching required by Astbury's data, and also would be consistent with the X-ray diffraction data.

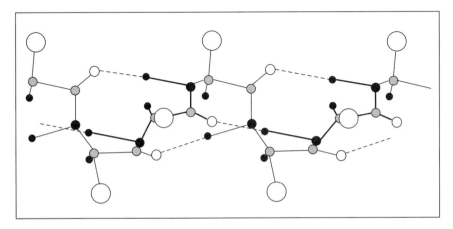

Figure 11.1: One of Maurice Huggins's helical structures of the polypeptide chain, viewed perpendicular to the rotation axis. Grey circles are carbon atoms, small black circles are hydrogen atoms, large black circles are nitrogen atoms, small white circles are oxygen atoms and large white circles are side-chains. There are three amino acids per 360° turn of the helix, giving the molecule a three-fold axis of symmetry (note that every third side-chain occupies an equivalent position). The carbonyl (C=O) group of each amino acid is hydrogen-bonded (dotted lines) to the amino (N–H) group of the amino acid three positions away in the polypeptide chain. Huggins required that the peptide bonds be planar, but did not require that the hydrogen bonds be colinear (i.e., the three bonds in the system C=O . . . H–N all lie on a straight line). Adapted from Huggins, M. L. (1943) *Chemical Reviews* **32**, 211, with permission. Copyright 1943 American Chemical Society

Huggins also made an important point that indicated the sureness of his grasp of protein stereochemistry: 'It may be noted that there is nothing about this [three-fold symmetrical] structure which requires exactly three residues per turn of the spiral. In fact, it would seem, from the models that have been made, that the bond distance and angle requirements are best satisfied by a slightly smaller number of residues per turn.'[78] In other words, the most stable helical conformations of polypeptides appeared to be 'irrational' ones, in which the number of amino acids per turn of the helix is not an integer. Like the β-pleated sheet, the concept of irrational helices is an idea of Huggins' that is often attributed to Pauling.

Huggins's 1943 study was the first attempt to build three-dimensional models of folded proteins. He used all the available data and a clearly stated set of initial premises. Pauling must have read this paper carefully, as the approach he later used was very similar to that of Huggins. Lawrence Bragg read Huggins's paper, too – at least, he later referred to it – but, in the light of later events, obviously not carefully enough.

In 1948, Pauling spent six months as a visiting professor at Oxford University. It was a triumphal tour for the man now widely hailed as the world's leading chemist – he was elected a foreign member of the Royal Society and awarded honorary degrees by the universities of Oxford, Cambridge and London. Pauling visited the Cavendish Laboratory during his sabbatical and found that the X-ray analysis of hemoglobin and myoglobin had progressed to the point that the contours of the molecules could be distinguished. Both the Cambridge and Cal Tech groups were intensely interested in determining the structures of ordered regions of polypeptide chains, but the experimental philosophies being used were quite different. Pauling's and Corey's approach was the 'bottom up' one of basing models on detailed analysis of simpler 'model compounds'; Bragg, Max Perutz and John Kendrew were using the 'top down' approach of trying to obtain Patterson (electron density) maps of sufficiently high quality that the dimensions of ordered regions could be measured.

Pauling may have been the world's leading chemist, but Bragg was the world's leading crystallographer. It was time to return to the problem of protein folding before Bragg and his group solved it. In bed with a cold one day, Pauling decided to try ways of folding a polypeptide chain into a hydrogen-bonded helix.

The way in which this was attempted provides a revealing insight into the genius of Linus Pauling: 'By making a drawing of a polypeptide chain on a sheet of paper and folding the paper on parallel lines passing through the alpha-carbon atoms, I tried to bring the N–H group and the O=C group into the proper orientation and distance from one another to correspond to the formation of an acceptable hydrogen bond, with the N–H . . . O distance about 2.8 Å. It took me a couple of hours to find this structure and to make calculations about the repeat distance.'[79]

Using nothing more complicated than paper and pen, Pauling found two helical structures that appeared to maximize hydrogen bonding without violating any other structural parameters. One of these had 3.7 residues per turn, the other 5.1. The former appeared to be capable of doubling in length if stretched, like α-keratin. However, when Pauling returned to Cal Tech and had his helices checked out by Corey, there was a problem – the 3.7 helix had an axial repeat distance (pitch) of 5.4 Å, whereas the most prominent X-ray reflection of α-keratin was 5.1 Å. The difference between these values – 0.3 Å – was only half the diameter of a hydrogen atom, but Pauling knew that it was too much to be explained by experimental error. He had no interest in publishing a folded polypeptide structure that did not occur in proteins. He decided to sit on the 3.7 and 5.1 helices.

Pauling knew that it was only a matter of time before the Cavendish group came up with a structure for α-keratin. They did so in 1950. In a long paper entitled 'Polypeptide chain configurations in crystalline proteins', Bragg, Kendrew and Perutz constructed models of polypeptide structures with two, three- or four-fold screw axes of symmetry and hydrogen bonds between amino acids 1, 2, 3 or 4 positions apart. Of these, ten were considered in detail, including the three-fold symmetrical helix discussed in the Huggins 1943 paper (Figure 11.1) and the Astbury and Woods hexagonal fold structure of α-keratin (see Figure 7.4). Preference was given to structures in which the maximum number of hydrogen bonds was formed and in which this bond system (C=O . . . H–N) was approximately linear. Because of Astbury's evidence that the 5.1 Å reflection of α-keratin corresponded to three amino acids, structures which involved a three-amino acid 'fold' were also favored. The Cavendish group hoped that by comparing the theoretical X-ray diffraction patterns of these structures with the diffraction patterns of hemoglobin and myoglobin, one or more of the models would be strongly supported. The hexagonal fold structure, which Bragg and co-workers referred to as a $2_{13}\frac{1}{3}$ helix, was in good agreement with the two-dimensional Patterson map of myoglobin, but not with the three-dimensional map of hemoglobin (Figure 11.2). It also had only one-third of the possible number of hydrogen bonds. The other structures examined were even less satisfactory:

> The problem is very complex, and one is forced to rely on a number of items of evidence each of which is very slight. The conclusion that the chains are of a folded coplanar form resembling the $2_{13}\frac{1}{3}$ [Astbury and Woods] or $2_{14}\frac{1}{3}$ type would be more convincing if there were any indication that the form has obvious advantages over others. In X-ray analysis in general, when a crystal structure has been successfully analyzed and a model of it is built, it presents so neat a solution of the requirements of packing and interplay of atomic forces that it carries conviction as to its essential correctness. In the present case the models to which we have been led have no obvious advantages over their alternatives. Much more evidence must be accumulated before conclusions can be safely drawn.[80]

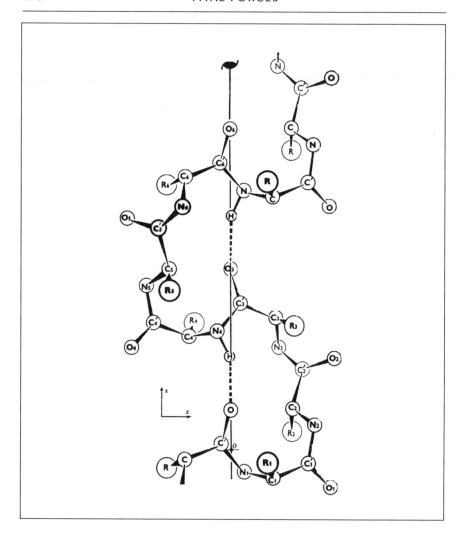

Figure 11.2: Lawrence Bragg's '$2_{13}\,{}^{1}\!/_{3}$' helical structure of the polypeptide chain, viewed perpendicular to the rotation axis (vertical line). There are six amino acids per 360° turn of the helix, with the carbonyl (C=O) group of every third amino acid hydrogen-bonded (dotted lines) to the amino (N–H) group of the amino acid four positions away in the polypeptide chain. The hydrogen bonds (C=O . . . H–N) are approximately colinear, but the peptide bonds are not planar (e.g., atoms O_4, C_4', N_5 and C_5 should all lie in the same plane). This structure is a three-dimensional version of Astbury's 'hexagonal fold' model of α-keratin (Figure 7.4). Because the polypeptide chain winds around the rotation axis, however, the repeat unit is six amino acids rather than three. Reproduced from Bragg, W. L., Kendrew, J. C. and Perutz, M. F. (1950) *Proceedings of the Royal Society of London A* **203**, 336, with permission

In modeling molecular structures, the initial assumptions were crucial. If the assumptions were not restrictive enough, a large number of plausible structures would be generated, with nothing to choose between them. If the assumptions were too restrictive, no plausible structures would be generated. The Cavendish group had made two errors in their assumptions, one of which resulted in the inclusion of structures that were impossible, the other of which resulted in the exclusion of structures that were perfectly feasible. First, they had limited themselves to structures with integral numbers of residues per turn, although Huggins had clearly pointed out in 1943 that this was not necessary. Second, they had failed to realize that the peptide bond between amino acids was planar, and therefore the polypeptide chain could not rotate around this bond – although Pauling had first proposed this in the early 1930s and it had been confirmed by Corey's X-ray diffraction analysis of diketopiperazine.

Pauling later wrote: 'Bragg, Kendrew and Perutz were physicists working in a physics department[d] . . . None of them, I judge, knew very much about structural chemistry . . . I thought it very likely, however, that in the course of time they would learn enough chemistry to see what [sic] peptide group had a planar structure, and would discover the alpha [3.7] helix.'[79] Despite his misgivings about the 0.3 Å discrepancy between the 3.7 residue helix and α-keratin, he could wait no longer. In November 1950, a short letter from Pauling and Corey was published in the *Journal of the American Chemical Society*. However, this announced only the parameters necessary to unambiguously characterize the 3.7 and 5.1 helices: the number of residues per turn, the axial translation per residue and the pattern of hydrogen bonding. It was almost as if Pauling wanted to stake his claim without drawing too much attention to structures that might be incorrect.

Shortly thereafter, Pauling heard that the Courtauld chemical company had synthesized an artificial polypeptide consisting of a single type of amino acid – X-ray diffraction of this did not give the 5.1 Å reflection characteristic of α-keratin. Pauling now felt justified in publishing the full details of his protein helices. A paper entitled 'The structure of proteins: two hydrogen-bonded helical configurations of the polypeptide chain' was submitted to *Proceedings of the National Academy of Sciences* on 28 February 1951 – Pauling's fiftieth birthday.

Pauling and co-workers used assumptions about bond lengths and angles similar to those of Huggins and the Cavendish group. Unlike Huggins,

d   Perutz and Kendrew were actually trained as chemists.

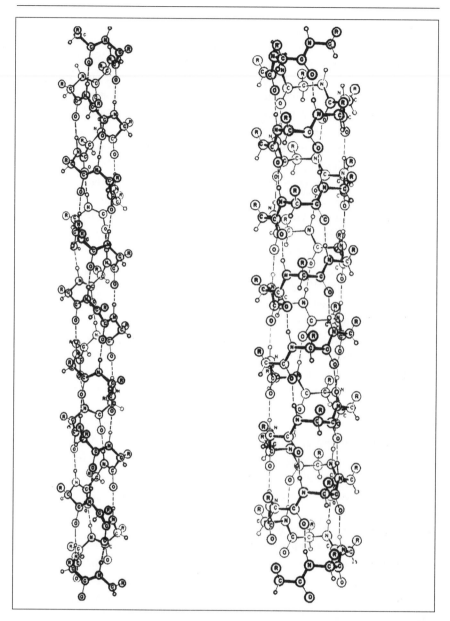

Figure 11.3: Linus Pauling's two helical protein structures, viewed perpendicular to the rotation axis. In the α-helix (left), there are 3.7 amino acids per 360° turn of the helix, with the carbonyl (C=O) group of each amino acid hydrogen-bonded (dotted lines) to the amino (N–H) group of the amino acid four positions away in the polypeptide chain. In the γ-helix (right), there are 5.1 amino acids per turn, with the N–H group of each amino acid hydrogen-bonded to the C=O group of the amino acid four positions away in the polypeptide chain. Both are right-handed helices in which the peptide bonds are planar and the hydrogen bonds are near-colinear. Reproduced from Pauling, L., Corey, R. B. and Branson, H. R. (1951) *Proceedings of the National Academy of Sciences* **37**, 207

however, they stipulated that the hydrogen bond must be within 30° of the N–H bond; unlike Bragg and his co-workers, they assumed that the peptide bond was planar. Using these conditions, only two satisfactory structures were found, the 3.7 and 5.1 helices (Figure 11.3). Both of these appeared highly plausible. Unlike the $2_{13}\frac{1}{3}$ structures favored by Astbury and Bragg, in which only one-third of the possible hydrogen bonds were formed, the 3.7 and 5.1 helices had every N–H hydrogen-bonded to a C=O group. Unlike the 'spiral' structures proposed by Huggins, all three bonds in the hydrogen bond system (N–H . . . O=C) were nearly colinear. This was because, in the Pauling helices, N–H groups sat directly over or under C=O groups, whereas in the rational helices N–H groups must sit directly over or under other N–H groups.

Of the two new structures, only the 3.7 helix had the correct degree of coiling to give a 100% increase in length on stretching. Pauling *et al.* wrote:

> It is our opinion that the structure of α-keratin, α-myosin, and similar fibrous proteins is closely represented by our 3.7-residue helix, and that this helix also constitutes an important structural feature in hemoglobin, myoglobin and other globular proteins, as well as of synthetic polypeptides. We think that the 5.1-residue helix may be represented in nature by supercontracted keratin and supercontracted myosin.[81]

They therefore proposed the names 'alpha-helix' for the 3.7-residue type and 'gamma-helix' for the 5.1-residue type.

Pauling could not resist the temptation of pointing out to Bragg the error of his ways:

> The reason for the difference in results obtained by other investigators and by us through essentially similar arguments is that both Bragg and his collaborators and Huggins discussed in detail only helical structures with an integral number of residues per turn, and moreover assumed only a rough approximation to the requirements about interatomic distances, bond angles, and planarity of the conjugated amide group. We contend that these stereochemical features must be very closely retained in stable configurations of polypeptide chains in proteins, and that there is no special stability associated with an integral number of residues per turn in the helical molecule.[81]

The publication of the two helical structures opened the floodgates; the results of a decade of model building could now be interpreted. The May issue of the *Proceedings* contained no less than seven papers from Pauling's group. These described a 'new' hydrogen-bonded sheet structure, a triple-stranded helical structure for collagen, and interpretations of the X-ray diffraction patterns of hair, feather, muscle and silk in terms of the various helical and sheet structures. As the Pauling biographer Thomas Hager put it, 'It was as though a single composer had debuted seven symphonies on the same day.'

Perutz read the Pauling series of papers one Saturday morning. He immediately realized that, if the α-helical structure were present in α-keratin, there should be an additional X-ray reflection at 1.5 Å representing the spacing between neighboring residues, which would not be seen on the flat plates normally used at the Cavendish. Perutz went straight to his laboratory and mounted a single horse hair in the X-ray diffractometer. When he developed the film, the 1.5 Å spot was there, exactly as predicted.[e]

Bragg reacted to Pauling's paper by taking it and his own 1950 paper to Alexander Todd, since 1944 the professor of organic chemistry at Cambridge. To his surprise, Todd, who had remained friendly with Pauling ever since the latter had tried to recruit him to Cal Tech in 1938 (see Chapter 8), already had a copy of the Pauling paper. When Bragg asked him his opinion about the proposed structures, Todd took one look and ruled in favor of the α-helix, on the grounds that Pauling, but not Bragg, had assumed a planar peptide bond. Todd later recalled bluntly telling Bragg: 'I think that, given the evidence, any organic chemist would accept Pauling's view. Indeed, if at any time since I have been in Cambridge you had come over to the Chemical Laboratory, I – or for that matter any of my organic colleagues – would have told you that. Don't you take any chemical advice when you do this kind of work?'[82] Bragg did – notably from the theoretical chemist Charles Coulson – but the chemical advice obtained had not directed him onto the right track.

The Wizard of Pasadena had done it again. This time Bragg had not merely been scooped by Pauling, as he had in 1929 (see Chapter 8), he had been publicly embarrassed. It was the low point of Bragg's professional life. He later wrote: 'I have always regarded this [1950] paper as the most ill-planned and abortive in which I have ever been involved.'

Bragg's ignorance of the planarity of the peptide bond was a source of

---

e   Perutz recorded that when he informed Pauling of this finding, 'he attacked me furiously, because he could not bear the idea that someone elso had thought of a test for the α-helix of which he had not thought himself.'

considerable personal discomfiture, but it was not the reason why he had missed the α-helix. Rather, it was by limiting themselves to helical structures with integral numbers of residues per turn that Bragg, Kendrew and Perutz excluded the α and γ helices from consideration. Here Bragg's tremendous knowledge of crystallography had limited his thinking. To his credit, he did not assume that the rules of classical crystallography automatically applied to proteins – he recognized, for example, that protein helices could have a five-fold screw axis of symmetry, impossible in a true crystal - but he could not envisage such a radical concept as an irrational helix. Pauling, who thought in terms of molecules rather than crystals, suffered from no such limitations.

Once it was pointed out, the α-helix was such an elegant solution to the problem of protein folding that it was accepted without significant resistance. The discrepancy between the pitch of the α-helix and the axial repeat distance of α-keratin was resolved in 1952 when Francis Crick and Pauling independently discovered that a further coiling of the helix would produce a reflection at 5.1 Å instead of 5.4 Å. The following year, Jerry Donohue, who worked with Pauling at Cal Tech, performed an extensive analysis of helical polypeptide conformations. This produced four additional helices, one of which was similar to a structure described by Huggins in 1943. Of the six conformations, the α-helix represented the most energetically favorable configuration of the polypeptide backbone.

'A Complex Interlocking System of Coiled Polypeptide Chains'

For Perutz, still toiling away at the analysis of the hemoglobin molecule, 1951 must have been even more dispiriting than it was for Bragg. The crystallographic analysis of hemoglobin had been motivated, from the first, by the expectation that the globular proteins would exhibit a high degree of internal symmetry. Progress had been very slow, however, and the anticipated molecular symmetry was proving elusive. By 1942, Perutz's studies had shown that horse methemoglobin (the form of the protein with no oxygen bound) formed monoclinic crystals of space group C2, with two molecules per unit cell. The individual molecules were 'platelets' 36 Å thick, 64 Å long and 'probably somewhat shorter' in width. The first two-dimensional Fourier maps of hemoglobin had also been obtained.[f]

---

f   Dorothy Crowfoot wrote in 1979: 'I remember how Max rang me up the day he got the first projection of the hemoglobin structure drawn out and I just got in the car and drove straight over to Cambridge to see it, and I knew I would see essentially nothing because the projection was not interpretable down [the] 63 angstroms [thickness of the b axis]. But I could not help going to look at it.'[83]

Five years later, the packing of methemoglobin molecules within the crystal could be analyzed. The crystals appeared to consist of alternating layers of protein and solvent parallel to the *ab* plane, the base of the unit cell. The hemoglobin molecules now appeared to be cylinders of 57 Å diameter and 34 Å height – about the same shape as a hat-box (Figure 11.4). One-dimensional Fourier projections showed four lines of high electron density around the circumference of the cylinder. Perutz wrote: 'it is tempting to propose a four-layered structure with the backbones of the polypeptide chains in the plane of the layers and the side chains protruding above and below . . .'.

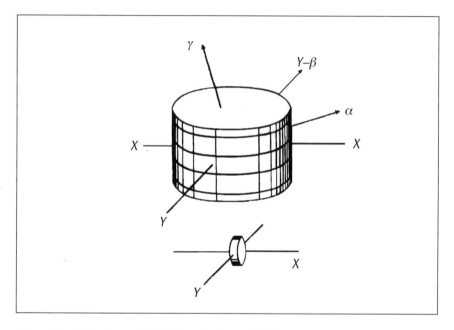

Figure 11.4: Max Perutz's 1947 'hat-box' model of hemoglobin structure. The lines encircling the cylinder represent concentrations of X-ray scattering matter that Perutz thought might be 'four layers of polypeptide chains'. The small disk below represents the relative size and orientation of the oxygen-binding heme group. Reproduced from Boyes-Watson, J., Davidson, E. and Perutz, M. F. (1947) *Proceedings of the Royal Society of London A* **191**, 123, with permission

It was clear that the internal symmetry of the hemoglobin molecule would not be discernible from two-dimensional Patterson maps. The thickness of the molecule along the projection axis was some forty atomic diameters, and therefore the contours on the Patterson map represented the superimposition of many internal features. Following the course of the polypeptide chain in a two-dimensional map was like trying to read the text printed on a number of transparent sheets that were sandwiched together. Although it would increase the labor immensely, the only solution seemed to be to construct three-dimensional maps.

The first three-dimensional Patterson synthesis of methemoglobin, involving thirty-one sections through the molecule, was completed in 1949. This required the analysis of all 7840 non-symmetry-related reflections that occurred at 2.8 Å or greater: 'The photographing, indexing, measuring, correcting and correlating of some 7000 reflexions was a task whose length and tediousness it will be better not to describe,' Perutz wrote in the resulting paper. To his frustration, little new information was obtained. The layers within the hemoglobin molecule now appeared as a series of rods, 10.5 Å apart and with a prominent 5 Å vector along the rod axis – possibly the same axial reflection that Astbury had observed in the fibrous proteins?

Perutz was acutely aware that determining the structure of hemoglobin by X-ray methods might well be hopeless: 'The actual chances of interpretation depend largely on the kind of molecular structure which the protein may be supposed to possess. For instance, if the globin molecule consisted of a complex interlocking system of coiled polypeptide chains where interatomic vectors occur with equal frequency in all possible directions, the Patterson synthesis would be unlikely to provide a clue to the structure.'[84] To continue or not to continue? 'It remains to be seen whether the X-ray analysis of haemoglobin can be carried further or whether future progress lies in the analysis of simple proteins of smaller molecular weight which are now being studied by several workers.' Perutz was probably thinking in particular of insulin, which was still under investigation by Dorothy Crowfoot and, at 6000 daltons, was less than one-tenth the size of hemoglobin.

In the acknowledgments section of this paper, Perutz wrote: 'I wish once again to record my indebtedness to Sir Lawrence Bragg for lending me his support and encouragement in a venture in which the chances of success seemed forlorn to most others.' Little did he suspect that even Bragg regarded the chances of success at this time as being 'indistinguishable from zero'.

Things were only to get worse. An analysis in 1950 by Crick showed that no arrangement of parallel rods could explain the diffraction pattern of hemoglobin. According to Robert Olby, this 'marked the end of the belief in regular geometric structures for the globular proteins'. As Crowfoot later wrote: 'The course of the main peptide chain through lysozyme or even through myoglobin is so complicated and irregular, it is not surprising for a brief time we thought that it was not there at all.' On top of this prospect of failure of the hemoglobin project came Pauling's achievement of one of the main prizes of protein structure, the α-helix. The Cavendish crystallographers were left with the crumbs from Pauling's table – Perutz's confirmation of the α-helical structure in keratin and Crick's explanation of the 5.1 Å axial reflection.

If hemoglobin was, as Perutz feared, 'a complex interlocking system of coiled polypeptide chains', its structure could only be solved by determining the position of every atom. This would require an electron-density map at a resolution of approximately 1.2 Å – an enormous task, as every two-fold improvement in resolution involved the measurement of eight times as many reflections. Such an effort could not even begin unless some way were found to determine the phases of the reflections.

For inorganic crystals, the phases were determined by the technique of isomorphous replacement, substituting one ion in the crystal with another of higher atomic weight and determining the effect on the diffraction pattern. The fact that certain proteins contained metal ions – insulin contained zinc, hemoglobin iron – suggested that isomorphous replacement could also be used to solve the phases of organic crystals. However, it was assumed that the X-ray scattering by a couple of metal atoms would be insignificant compared to the ten thousand atoms of hemoglobin. In 1951, Perutz decided to measure the absolute intensity of the X-rays diffracted by hemoglobin crystals and found that it was a very low proportion of the incident radiation. This suggested that a heavy metal atom might have a relatively large effect on the diffraction pattern of a protein crystal. However, the addition of such a metal atom would have to be isomorphous – that is to say, not change the structure of the crystal.

A means of performing isomorphous replacement of hemoglobin was presented to Perutz in 1953:

> One day I received a set of reprints . . . from an unknown man at Harvard called Austin Riggs. He . . . examined the effect of PMB [para-mercuribenzoate] on the oxygen equilibrium curve [of hemoglobin] and found that heme–heme interaction was largely preserved. I got very excited by this observation, because it suggested that you can attach molecules of PMB to hemoglobin without changing its structure significantly . . . When I developed the first precession picture of PMB-hemoglobin and compared it with that of native hemoglobin, I saw that the two crystals were isomorphous and that the intensity changes were just of the magnitude that my measurements of the absolute intensities had led me to expect. Madly excited, I rushed up to Bragg's room and fetched him down to

the basement dark room. Looking at the two pictures in the viewing screen, we were confident that the phase problem was solved.[83]

As Bragg put it, adding a few mercury atoms to hemoglobin had no more effect on the molecule than painting a gold star on its forehead would have on an elephant.

In 1954, Perutz and his colleagues published a series of papers comparing the X-ray diffraction pattern of horse methemoglobin with the mercury and silver derivatives of oxyhemoglobin. The pessimistic tone of the previous papers was now replaced by one of unmistakable excitement: 'It was a triumph to find that there were no inconsistencies. The signs determined by the isomorphous replacement method exactly fitted the loops and nodes which had been so laboriously worked out in the course of the previous two years, and confirmed the great majority of the sign relations established by the transform method.'[85]

It was an odd kind of triumph. Fifteen years of work had largely been wasted, as the hat-box model was 'clearly wrong'. The 1954 structure revealed neither the heme group nor the polypeptide chains, as the poor resolution meant that single atoms were smeared out to a diameter of 9 Å. The prospect of finding internal molecular symmetry seemed more remote than ever: 'There are no regularly spaced layers; the deceptive regularity of the seven peaks in the one-dimensional Fourier is purely accidental in origin.' Nonetheless, the phase problem had been solved and Perutz knew that, even if there were no symmetry at all, the elucidation of the three-dimensional structure of the hemoglobin molecule was now only a matter of time.

Chapter 12
Our Thread of Ariadne

'The Janus Molecule That is the Gene'

As described in Chapter 6, the idea that nucleic acids could be the material of heredity, although popular in the late nineteenth century, had been largely superseded in the early decades of the twentieth century by the idea that genes were autocatalytic proteins. By the 1940s, however, the ideological winds had shifted again – the predominant concept of the gene was now that both nucleic acids and proteins contributed in some way to the hereditary function.

A number of factors appear to have contributed to the rise of what Robert Olby called 'the nucleoprotein theory of the gene'. It will be recalled that Edmund Wilson and his contemporaries cited two reasons why nucleic acids could *not* contribute to genetic specificity. First, chromatin was not visible in the nucleus during the interphase of the cell division cycle and therefore could not be responsible for genetic continuity between generations. Second, as the nucleic acids of all cells were identical, these molecules could not determine the differentiation of tissues. Both these arguments could be countered by observations made by Torbjörn Caspersson in the 1930s. Beginning in 1936, Caspersson used a microscope fitted with an ultraviolet light source and quartz optics to quantify the nucleic acid contents of different areas of the cell at different stages of cell division. From these studies, and also from the use of the DNA-specific Feulgen stain, it was clear that nucleic acids were present in the nucleus at all stages of the cell cycle. In 1938, Caspersson also showed that DNA is a macromolecule (see Chapter 8). This subtly changed the genetic potential of nucleic acids. If DNA was not a tetranucleotide, then the nucleic acids of different tissues could contain the same *amounts* of the four bases but quite different *sequences*. Just as the same amino acids could be combined in different ways to give different proteins, the same bases, sugars and phosphates could be combined in different ways to give different polynucleotides.

By 1940, most workers appeared to think that nucleic acids (literally and metaphorically) played only a supporting role – as Erwin Chargaff put it,

'The nucleic acids were only thought of as the coat-hangers for the all-important proteins.' At the Cold Spring Harbor symposium of 1938, William Astbury noted that chromosomes are 'compounds of protamines and nucleic acids', but 'Knowing what we know now from X-ray and related studies of the fibrous proteins . . . it is but natural to assume, as a first working hypothesis at least, that they form the long scroll on which is written the pattern of life.'[a] The following year, Astbury wrote: 'When proteins are born, other molecules assist at their birth; and perhaps chief among them are the nucleic acids.' Cyril Darlington wrote in 1942 that: 'thymonucleic acid [is] indispensable in the reproduction of the chromosome.' Jack Schultz, who had moved from Thomas Morgan's laboratory to Caspersson's in 1937, believed that 'the Janus molecule that is the gene' replicated its nucleic acid component during the prophase of cell cycle and replicated its protein during interphase: 'Thus we maintain as our unit of synthesis a nucleoprotein.'

Throughout the 1940s, it became increasingly common for biological chemists to believe that nucleic acids contributed to the genetic function, not only the structure, of the chromosome. In a 1943 review article on 'Chromosomes and nucleoproteins', Alfred Mirsky wrote, in the context of a discussion on viruses: 'The specificity of a nucleoprotein may reside in its nucleic acid as well as its protein moiety.' A similar position was taken by Jesse Greenstein of the National Cancer Institute in a 1944 review on nucleoproteins: 'Perhaps the specific properties of a nucleoprotein are conditioned by the molecule as a whole.'

The nucleoprotein theory of the gene gained enormous experimental support from Oswald Avery's 1944 demonstration that the pneumococcal transforming principle apparently consisted solely of DNA (see Chapter 10). It could not be denied that the transforming extracts might contain very small amounts of protein. However, the fact that the activity was destroyed by deoxyribonuclease suggested that, even if protein were involved in transformation, DNA was also necessary. The surprising finding of the 1944 paper was not that nucleic acid was required for transformation, but rather that protein was *not* required.

---

a   Because the axial repeat distance of DNA appeared to be approximately seventeen times the length
    of a single nucleotide in the chain, Astbury commented: 'The four nucleotides can therefore hardly
    follow one another always in the same order.' In the discussion following his presentation, Stuart
    Mudd of the University of Pennsylvania noted that 'it seems apparent that these piles of nucleotide
    units, by slight changes in the order in which nucleotides occur, or possibly by other changes in
    configuration, give us an adequate base for specificity.' Mudd was talking about *serological* specificity; it does not seem to have occurred to him or to Astbury that changes in the order of nucleotides
    could be the basis of *genetic* specificity.

An important figure in the re-evaluation of the genetic role of nucleic acids was John Masson Gulland, Sir Jesse Boot professor of chemistry at Nottingham University. In a 1945 review article, Gulland noted that a nucleic acid molecule containing equimolar amounts of the four bases could be either a 'structural tetranucleotide', in which the same sequence of bases was repeated redundantly along the chain (e.g., ACGTACGTACGTACGT . . .), or a 'statistical tetranucleotide', in which the sequence was random but all four bases were equally represented (e.g., CAGGATCGTCAACTGT . . .). Gulland strongly favored the latter, stating that there was: 'no evidence to justify the recognition of the existence of the "structural tetranucleotide"'. For Gulland, the recognition of the macromolecular nature of DNA had knocked out a key pillar supporting the tetranucleotide theory: 'Had the sizes of nucleic acid molecules been realized at an earlier date, it is doubtful whether the hypothesis of the structural tetranucleotide would have gained such a firm hold as is apparently the case.' This was correct – when Phoebus Levene obtained clear evidence that nucleic acids were macromolecules, he suggested that these substances were 'polymers of the tetranucleotide' (Gulland's 'structural tetranucleotides') (see Chapter 7). The concept of the structural tetranucleotide therefore arose directly from that of the 'literal' tetranucleotide.

If nucleic acids were statistical rather than structural tetranucleotides, as Gulland believed, the sequence of nucleotides could easily generate enough variety for DNA to be Erwin Schrödinger's 'aperiodic crystal' of heredity. Kurt Stern, who worked at the Polytechnic Institute of Brooklyn, and appears to have been influenced by the views of the polymer chemist Herman Mark, wrote in 1947:

> It is generally assumed, without sufficient experimental proof, that the four bases present in thymonucleic acid . . . alternate in a regular and constant periodicity along the entire length of the polynucleotide chain. If one admits the possibility that the bases may vary in their sequence or orientation with reference to the backbone, this would constitute a principle of *modulation* which does not affect the over-all stoichiometry of composition of the nucleoprotein chain, and which is fully capable of accounting for all possible genotypes [emphasis in original].[86]

By this time, however, Gulland did not believe that DNA was a tetranucleotide of either the structural or statistical type. The 1947 Cold Spring Harbor symposium on nucleic acids, held midway between the publication of Avery's paper and the Hershey and Chase experiment (see below), represented a watershed in thinking about the biological functions of these substances. At this conference, Gulland noted that the chemical evidence for equimolar proportions of the four bases 'does not bear the test of close scrutiny'. His own titration data and chemical analyses of hydrolyzed DNA suggested that 'desoxypentose nucleic acid does not conform to the statistical tetranucleotide ratio'; therefore, 'there does not seem at present to be any indisputable chemical evidence that individual nucleotides are arranged in the polynucleotide in a fixed or regular sequence.'

Sadly, this was John Gulland's last contribution to science; in October 1947, the Scottish chemist was killed in a train crash in the north of England. The credit for demonstrating that there was no equimolar ratio or repeating sequence of bases in DNA – in other words, that it was an aperiodic polymer like the proteins – would fall instead to Erwin Chargaff.

'Chargaff's Rules'

Erwin Chargaff was born in 1905 in Czernowitz, Austria, moving to Vienna with his family at the end of World War I. He appears to have enjoyed a classical education, and at an early age acquired his love of poetry, drama and opera. It was for practical reasons that he studied chemistry at the University of Vienna, graduating in 1928, but he also took courses in the history of literature and English philology.

Conditions were difficult in Austria in the late 1920s, and Chargaff decided to take a position at Yale University, although not without some misgivings about going to a country that was 'younger than most of Vienna's toilets'. Initial impressions were not favorable, as he spent his first few days on American soil in detention on Ellis Island. After his release, Chargaff spent two years in the chemistry department at Yale working on the tubercle bacillus. In 1930, in a 'rare example of a rat returning to a sinking ship', he took up a position as assistant to Martin Hahn in the bacteriology department at the University of Berlin. As an assistant to Hans Buchner, Hahn had developed the grinding conditions that allowed Eduard Buchner to demonstrate cell-free fermentation (see Chapter 4), and had subsequently risen to the exalted status of *Geheimrat* (privy councillor).

In Berlin, Chargaff worked on bacterial lipids, completing his *Habilitationsschrift* (PhD thesis) at the beginning of 1933. It was, as he later recalled, 'the happiest time of my life'. However, the appointment of Adolf Hitler as chancellor of Germany disrupted the career of Chargaff, who was Jewish on his mother's side. He took a position at the Pasteur Institute in Paris, but this did not prove as happy an experience as Berlin, and in 1934 Chargaff sailed for the USA once more. After a brief sojourn at the Mount Sinai

Hospital in New York, he joined the Department of Biochemistry at Columbia University. Chargaff had found a home, although not always a happy one; he would remain at Columbia until his retirement in 1974.

During his first decade at Columbia, Chargaff worked mainly on lipids and blood clotting as part of a program of research he described as 'cell chemistry'. Although the two works that influenced him to switch to research on nucleic acids, Avery's paper and Schrödinger's book, both appeared in 1944, it was only two years later that Chargaff started work on these molecules. At the Cold Spring Harbor symposium of 1947, Chargaff adopted the now mainstream position that the unit of heredity was a nucleoprotein rather than protein or nucleic acid alone: 'The specific nature of a nucleoprotein would then be vouchsafed not only by the structure of the protein but also by that of the nucleic acid.'

As described above, Gulland presented data at this conference suggesting non-equimolar base ratios in DNA. If this were the case, the nucleic acids of different species might differ not just in base sequence but also in base composition. Perhaps influenced by this, Chargaff decided, like Levene almost fifty years earlier, to measure the base contents of nucleic acids from different sources. The fact that such species-specific differences had not been detected in the past could be explained by the crude analytical methods used. However, Chargaff would require techniques that could separate and accurately quantify each of the four bases.

In this, he had two strokes of good fortune. The recently developed technique of paper chromatography proved as useful for separating the nucleic acid bases as it was for separating amino acids (see Chapter 11). If the bases could be separated, it was only necessary to be able to measure the amount present. The development of ultraviolet light sources meant that the intense absorption of ultraviolet radiation at wavelengths around 260 nm by nucleic acid bases could be used as a means of quantitation.

Chargaff was also lucky with the DNA samples he initially chose to study. He started with DNA from yeast, which has an abnormally high content of adenine and thymine, and DNA from the tubercle bacillus, which has an abnormally high content of guanine and cytosine. Had he used the bacterium beloved of phage geneticists, *Escherichia coli*, he would have found base contents consistent with a statistical tetranucleotide structure.

By 1950, Chargaff had studied the nucleic acid base compositions of a sufficient number of different organisms to be able to discern some consistent

Table 12.1: Chargaff's base compositions of DNA, 1950–1

Species	% total base			
	Adenine	Cytosine	Guanine	Thymine
Cow	29.5	19.6	22.9	28
Human	30.5	18.2	19.5	31.8
Yeast	31.6	17.2	18.4	32.8
Mycobacterium tuberculosis	15	34	37.4	13.6
Serratia marcescens	19.5	31	27.9	21.6
Bacillus sp.	20.6	31.3	29.4	18.7
Salmon	29.7	20.4	20.8	29.1

features. One was that DNA did not contain equal amounts of the four bases; however, the amounts of guanine and cytosine, and of adenine and thymine, were approximately equal (Table 12.1). In the case of salmon, for example, the contents of the bases were 30% adenine, 20% cytosine, 21% guanine and 29% thymine. Further, RNA had a composition different from that of the DNA of the same organism or tissue.

Chargaff summarized these data in a 1950 review article. In a note apparently added in proof, he concluded: 'These results serve to disprove the tetranucleotide hypothesis. It is, however, noteworthy – whether this is more than accidental, cannot yet be said – that in all desoxypentose nucleic acids examined thus far the molar ratios of total purines to total pyrimidines, and also of adenine to thymine and of guanine to cytosine, were not far from 1.'[87]

Like his hero Avery, Chargaff was very careful not to let his conclusions get out ahead of his data. However, he clearly believed that the base sequence of DNA was responsible for genetic specificity: 'We must realize that minute changes in the nucleic acid, e.g. the disappearance of one guanine molecule out of a hundred, could produce far-reaching changes in the geometry of the conjugated nucleoprotein; and it is not impossible that rearrangements of this type are among the causes of the occurrence of mutations.'[87]

The fact that the bases were not present in DNA in equimolar amounts may have been inconsistent with a statistical tetranucleotide, but did not preclude a periodic distribution of bases. As Chargaff pointed out in 1951:

If one postulated an ideal case in which a desoxypentose nucleic acid exhibited ratios of adenine to guanine and of thymine to cytosine that were both 1.4 and ratios of adenine to thymine, of guanine to cytosine, and of purines to pyrimidines, all equalling one, a simple construction could, for instance, assume that a subunit consisting of 24 nucleotides contained 7 dinucleotides, in which adenylic acid [the nucleotide of adenine] was linked to thymidylic acid, and 5 dinucleotides, in which guanylic acid and cytidylic acid were united, all these distributed in a certain pattern.[88]

In other words, if the bases were not present in equimolar amounts, DNA could not consist of a repeating tetranucleotide unit. However, it could consist of a repeating unit greater than a tetranucleotide. The base composition Chargaff mentioned, which was similar to that of the salmon DNA described above, was consistent with a repeating structure such as (ATG-CATGCATGCATGCATGCATAT)$_n$.

To prove that such patterns did not exist, Chargaff performed the ingenious experiment of measuring the ratio of thymine to cytosine in a sample of DNA during a timed digestion with the enzyme deoxyribonuclease. If there were a regular distribution of bases along the polynucleotide chain, the four bases would be released at constant rates throughout the incubation. If, on the other hand, the bases were randomly distributed, the rates of release of the four bases would vary – much more adenine, for example, would be liberated when the enzyme reached an adenine-rich stretch of the sequence. In fact, the base ratios varied widely during the incubation period, 'suggesting a very complex pattern of the sequence of individual nucleotides in the original nucleic acids'.

In another paper that appeared in 1951, Chargaff wrote: 'Not only the ratio of purines to pyrimidines but also that of adenine to thymine and of guanine to cytosine equals 1. As the number of examples of such regularity increases, the question will become pertinent whether it is merely accidental or whether it is an expression of certain structural principles that are shared by many desoxypentose nucleic acids, despite far reaching differences in their individual composition and the absence of a recognizable periodicity in their nucleotide sequence.'[89]

Chargaff had invalidated the tetranucleotide hypothesis in both its 'statistical' and 'structural' forms. Yet there was a regularity in the base composition, one that determined that certain pairs of bases were always present in equal amounts.

Just as the recognition of the macromolecular nature of DNA made possible the nucleoprotein theory of the gene, Chargaff's disproof of the tetranucleotide structure made possible the nucleic acid theory of the gene. Although anticipated, both theoretically and experimentally, by Gulland, Chargaff's work conclusively demonstrated that DNA was an aperiodic polymer. The sequence of bases along the polynucleotide chain of DNA could therefore as easily encode genetic information as the sequence of amino acid side-chains along the polypeptide chains of proteins. In disproving both the 'structural' and 'statistical' tetranucleotide hypotheses, Chargaff converted DNA from a 'stupid' molecule to a 'smart' one, and thereby removed the last theoretical obstacle to its acceptance as the genetic material.

'A Little Hypodermic Needle Full of Transforming Principles'

Anyone asking Max Delbrück in 1951 whether he had yet solved the riddle of life would have found that the answer was still negative, as it had been five years earlier. However, the combined efforts of the members of the Phage Group were at least starting to unravel the secrets of the bacteriophage life cycle. Working at Cold Spring Harbor, Augustus Doermann had shown that there were no infectious particles present in bacteria for a ten-minute period following bacteriophage infection, suggesting that phage replication involved a temporary inactivation of the virus. Electron microscopic studies on phage conducted at the University of Pennsylvania by Thomas Anderson had revealed tadpole-shaped objects consisted of an icosahedral 'head' and a cylindrical 'tail'. The tail appeared to be the site of attachment to the bacterial cell. Osmotic shock treatment removed the contents of the heads and rendered the bacteriophage non-infectious. Roger Herriott of Johns Hopkins University showed that such bacteriophage 'ghosts' lacked DNA.

Anderson discussed with Alfred Hershey in 1950 or 1951 'the wildly comical possibility that only the viral DNA finds its way into the host cell, acting there like a transforming principle in altering the synthetic processes of the cell'. In November 1951, Herriot wrote to Hershey: 'I've been thinking – and perhaps you have, too – that the virus may act like a little hypodermic needle full of transforming principles.' Clearly the phage community had been struck by the similarity between the infection of *E. coli* by phage and the transformation of pneumococcus by DNA.

Hershey *had* been thinking along similar lines, and had conceived of a way in which the role of DNA in bacteriophage replication could be studied. Radioisotopes of several elements were now available for research purposes, and Hershey realized that these could be used to determine whether the viral genetic material was protein or nucleic acid. In association with Martha Chase, he radiolabeled bacteriophage T2 with sulfur 35, which is incorporated into protein but not DNA, and phosphorus 32, which is incorporated into DNA but not protein. The radioactive phage were allowed to attach to bacterial cells, and then a high-speed blender was used to shear apart the phage particles and the bacteria. By determining whether the bacterial cells had picked up ^{35}S or ^{32}P during co-culture, Hershey and Chase should be able to tell whether protein or DNA was transferred during infection.

The results, published in 1952, were not exactly clear-cut: 75–80% of the ^{35}S was removed from the bacterial cells by shearing off the attached phage, but only 21–35% of the ^{32}P. This suggested that DNA was more likely to be involved in infection than protein, but of course did not exclude the possibility that some bacteriophage protein in the 20–25% remaining with the cells was the phage genetic material. A much more convincing demonstration that the DNA was the infectious agent came from following the fate of the radioisotopes in the phage that emerged from the infected cells: 30% of the incorporated ^{32}P, but only 1% of the ^{35}S, was transmitted to progeny phage.

Historians and scientists agree that the evidence presented in Hershey's and Chase's paper in favor of a genetic role for DNA was far inferior to that offered by Avery and co-workers. Robert Olby wrote of Hershey and Chase: 'Their evidence was not all that convincing, certainly their chemical data was inferior to that of Avery, MacLeod and McCarty.' While Joseph Fruton hailed the 'miraculous detail' of Avery's paper, he wrote that 'the results of Hershey and Chase did not constitute proof of the identity of the infective material with DNA.' Francis Crick wrote that the Hershey and Chase study was 'a somewhat dirty experiment', and Joshua Lederberg that it was 'quantitatively less rigorous that McCarty and Hotchkiss' prior work on the pneumococcus'. Even Gunther Stent, a member of the Phage Group, wrote that Avery's study was 'experimentally much more clear cut than Hershey's'.

Perhaps aware of the ambiguity of their data, Hershey and Chase were as guarded in their conclusions as Avery et al. had been: 'Protein probably has no function in the growth of intracellular phage. The DNA has some function. Further chemical inferences should not be drawn from the experiments presented.'

Hershey and Chase did not cite Avery's 1944 study. As the historian Nicholas Mullins has noted, 'Phage workers tended to read and use ideas only from other phage workers.' However, the above quotations from Anderson and Herriot show that phage workers were influenced by Avery's work on the transforming principle. Also, the Hershey and Chase experiment was pure biochemistry – a far cry from what Max Delbrück had in mind when he conceived of bacteriophage as 'the hydrogen atom of biology'. The confirmation of DNA as the genetic material involved metabolic labeling and separation techniques, not the elegant and mathematical plaque-counting that was supposed to lead to a new scientific paradox. Much to Delbrück's disgust, the most significant achievement of the Phage Group had come from its despised 'subcommittee on biochemistry'.

The Structure of DNA

The nucleoprotein theory of the gene gave an impetus to studies on the structure of nucleic acid. By 1945, the chemical analysis of the nucleic acids was in some ways at a similar stage to that of proteins. Proteins were known to be polypeptides, and nucleic acids were polynucleotides. Proteins contained twenty or so different amino acids, nucleic acids four different bases. The amino acids of proteins were connected by amide (peptide) bonds, the nucleotides of the nucleic acids by phosphodiester bonds between the phosphates and sugars of the 'backbone'. Analysis of the three-dimensional structure of the DNA molecule, however, was still in its infancy.

At the 1947 Cold Spring Harbor symposium on nucleic acids, Gulland described his studies on potentiometric titration of DNA. Earlier nucleic acid chemists such as Walter Jones, Robert Feulgen and Phoebus Levene had measured the change in pH as acid or alkali was added to a solution of DNA in order to identify the chemical groups that 'buffer' (counteract) the change in pH. The buffering groups were found to be the phosphate groups of the backbone and the amino ($-NH_2$) and hydroxyl ($-OH$) groups of the bases. Gulland's titration analysis showed that certain pH curves were obtained when DNA was titrated from pH 7 (neutrality) with acid or alkali, but different curves were obtained when these solutions were back-titrated to 7. He interpreted this as evidence for hydrogen bonding between the hydroxyl and amino groups of the bases, which could in theory be on the same or different polynucleotide molecules. However, the addition of acid or alkali to DNA also caused a decrease in viscosity and flow birefringence, suggesting that the 'native' structure of the molecule was only stable at neutral pH. Therefore, 'at least a large proportion of the hydrogen bonds unite nucleotides in neighbouring polynucleotide chains'.

This work on the titration of DNA added an important new clue to the conformation of the molecule. If Gulland's interpretation were correct, the DNA molecule consisted of two or more polynucleotide chains that were held together by hydrogen bonds between bases.

At the same conference, Astbury presented his most recent data on the X-ray diffraction of DNA. He was now convinced that the polynucleotide chain must be folded up into a regular pattern analogous to the folding of α-keratin. The diffraction patterns were not of sufficiently high quality to determine how many nucleotides were contained within each structural repeating unit, but he suspected eight or sixteen.

Like that of α-keratin, the periodicity of DNA represented a repeating *arrangement* of subunits, not necessarily a repeating *sequence* of subunits. Nonetheless, the observation that DNA appeared to contain equal amounts of four different bases taken together with the observation that the periodicity of the molecule was a multiple of four bases suggested to Astbury that the bases 'must follow one another in some definite order'.

It is typical of Astbury's style of casual hypothesizing that his 1938 estimate of seventeen nucleotides in the repeat unit of DNA was evidence *against* a structural tetranucleotide (see above), while his slightly revised 1947 estimate of sixteen nucleotides per repeat unit was evidence *in favor* of a structural tetranucleotide. However, the latter conclusion was also supported by an analysis of thymic acid, DNA from which the purine bases had been removed. The X-ray diffraction pattern of thymic acid showed a prominent reflection at 6.8 Å instead of the 3.4 Å reflection in the parent molecule, suggesting that purine and pyrimidine bases alternated along the polynucleotide chain.

Desmond Bernal's 'gentlemen's agreement' with Astbury precluded him from studying the nucleic acids, which were fibrous, but not the crystalline nucleosides. When the Norwegian Sven Furberg arrived at Birkbeck to study for a PhD with Bernal, he was assigned the project of determining the crystal structure of cytidine. Furberg later provided this impression of Bernal's department: 'It was an extraordinary place: the laboratories were an old four-floor apartment house with just a couple of small rooms on each floor; Bernal's flat on the top floor; all in bad shape because of a German bomb which had removed the house next door.' Bernal's apartment was an extraordinary place in its own right – Paul Robeson sang there and a drawing by Pablo Picasso adorned one wall.

The structure of cytidine was solved in 1949 – the most complex molecule yet analyzed by isomorphous replacement. The major finding was that the sugar and the base were mutually perpendicular – a significant change from Astbury's and Bell's 1938 model, in which these were parallel (see Figure 8.4).

In his PhD thesis, Furberg proposed two possible structures for DNA that were based on the periodicities reported by Astbury – a 'pitch' (repeat distance along the axis) of 27 Å and a 3.4 Å spacing between adjacent bases – as well as his own findings on the relative orientation of bases and sugars. In model I, the deoxyribose–phosphate backbone formed a spiral with eight nucleotides per turn, and the pyrimidine and purine rings lying parallel to one another and 3.4 Å apart. This structure had favorable van der Waals interactions between the bases, but some of the hydrogen atoms in adjacent sugars were uncomfortably close to one another. In model II, the ribose rings and the phosphate groups zigzagged along the axis of the molecule, with adjacent bases protruding from opposite sides of the plane of the backbone. Because bases on each side of the molecule were 6.8 Å apart, no van der Waal's interactions could occur, but there were no steric problems. The two DNA structures were only published in 1952, but, as noted by Horace Judson, John Randall's group at King's College had 'early access to Furberg's model'.

In May 1950, Rudolf Signer visited London to attend a meeting of the Faraday Society. He brought with him DNA extracted from cells by a new method, of far higher quality than anything previously purified. Einar Hammarsten and Torbjörn Caspersson's 1938 material had an estimated molecular weight of 500000–1000000; Signer's 1950 preparation was approximately 7000000. Signer generously distributed his DNA to anyone who wished to study it. Maurice Wilkins took a sample for optical analysis. Raymond Gosling, a student working for Alexander Stokes, was the only person at King's using X-ray diffraction. He performed the first X-ray analysis of the Signer DNA. Stokes realized that the diffraction pattern obtained was consistent with DNA being a helix with a turn of about 40° per nucleotide.

This set Wilkins and Stokes thinking about the relationship between helices and X-ray diffraction. If it were possible to predict the diffraction pattern that a particular helical structure would give, a proposed structure could be tested against the actual diffraction pattern. According to Wilkins, Stokes 'worked it all out on the train between London and Welwyn Garden City'.

Stokes' helical diffraction theory provided an important tool for the X-ray analysis of biological fibers. It could be used to predict the diffraction pattern generated by specific helical structures, although not to derive a unique structure from a specific diffraction pattern. Thus, its utility was in testing helical models that previously could only be validated in a general way.

Meanwhile, the King's biophysics unit had acquired a real crystallographer. Rosalind Elsie Franklin was born in London in 1920 and graduated in physical chemistry from Newnham College, Cambridge, in 1941. From 1942 to 1947, she studied the physical structure of coal at the British Coal Utilisation Research Association. From 1947 to 1950, Franklin performed X-ray analyses of graphite at the *Laboratoire central des services chimique de l'état* in Paris. When she arrived at King's in January 1951, therefore, Franklin was a trained crystallographer but had no background in biology. Gosling was put under her supervision, and together they continued the analysis of the Signer DNA.

In the summer of 1951, Franklin and Gosling made a crucial discovery – the same DNA specimen could give two different X-ray diffraction patterns, depending upon the water content. At 75% relative humidity or lower, a 'crystalline' diffraction pattern predominated, but at 95% or higher a 'paracrystalline' pattern was observed. These crystalline and paracrystalline forms of DNA became known as A and B, respectively.

In the fall of 1951, a disagreement occurred between Wilkins and Franklin concerning the 'ownership' of the DNA project. Randall ruled that the Signer DNA was for Franklin's exclusive use. Wilkins was free to study sperm heads or a sample of inferior DNA that he had obtained from Chargaff.

At this point the last of the major players entered the scene. James Dewey Watson was born in 1928 and attended the University of Chicago at the age of fifteen on a special program that included literature, science, mathematics and philosophy. On graduating in 1947, he moved to the University of Indiana to work for his PhD with Salvador Luria on the X-ray inactivation of bacteriophage. During his time at Indiana, Watson attended a course on proteins and nucleic acids taught by Felix Haurowitz, the man who had persuaded Max Perutz to study hemoglobin (see Chapter 8). He also took Hermann Muller's course on 'Mutation and the gene'; in 1954, Muller described Watson as 'the brightest student I've ever had'.

On graduating in 1950, Watson went to Copenhagen to do postdoctoral work on nucleic acid chemistry, but found it profoundly uninteresting. In the spring of 1951, having met Wilkins in Naples and read Pauling's series of papers on protein structure, Watson decided that X-ray crystallography would provide the way to the secret of the gene. With the help of Luria, he managed to arrange that his postdoctoral fellowship be transferred to Cambridge. In the fall of that year, Watson arrived at the MRC group and started work with John Kendrew on the structure of myoglobin.

Watson found that Francis Crick shared his interest in nucleic acids. No-one in Cambridge was doing X-ray diffraction of DNA, but Wilkins and Crick were good friends, and the latter was therefore well informed about the progress made at King's. François Jacob provided descriptions of the two men:

> Tall, florid, with long sideburns, Crick looked like the Englishman seen in illustrations to nineteenth-century books about Phileas [sic] Fogg or the English opium eater. He talked incessantly . . . Breaking up his sentences with loud laughter . . . He had no taste for experimentation, for manipulation.

> Jim Watson was an amazing character. Tall, gawky, scraggly . . . shirttails flying, knees in the air, socks down around his ankles . . . his eyes always bulging, his mouth always open, he uttered short, choppy sentences punctuated by 'Ah! Ah!'.[74]

There was, however, another side to Watson and Crick, as recorded by Perutz: 'To say that they did not suffer fools gladly would be an understatement. Crick's comments could hit out like daggers at *non sequiturs*, and Watson demonstratively unfolded his newspaper at seminars that bored him.'

In November 1951, a colloquium on nucleic acid structure was held at King's. Watson attended, and heard Franklin present her data on the X-ray diffraction of Signer's DNA. The notes she made for her presentation show that she favored a helical structure with more than one chain. Watson took no notes, but, on the basis of his recollections of Franklin's data, he and Crick came up with a structure for the DNA molecule and built a physical model. This had three chains, held together by electrostatic interactions between phosphate groups and sodium ions. Despite the importance of hydrogen bonding in Pauling's α-helix and Gulland's evidence for inter-chain hydrogen bonds in DNA, this form of bonding was explicitly rejected.

When the King's group viewed the three-chain model, the fallibility of Watson's memory, if not his grasp of chemistry, immediately became apparent. Franklin ridiculed the proposed structure, pointing out that it was incompatible with her density measurements and did not take into account the hydration of sodium ions. To make things worse, Crick also found out that the method of calculating X-ray diffraction patterns of helices that he had developed with William Cochran and Vladimir Vand had already been discovered by Stokes.

Lawrence Bragg was furious. He had no desire for the Cavendish Laboratory to be associated with any more incorrect structures, and he did not approve of Watson and Crick poaching on what he perceived as King's territory. As a result of a conversation between Bragg and Randall, it was agreed that the structure of DNA would henceforth be left to the King's group. The chastened Watson and Crick went back to their work on proteins.

Watson's funding was due to end in the fall of 1952, and Delbrück had arranged that he would come to Cal Tech on a fellowship from the National Foundation for Infantile Paralysis. In the early part of the year, however, Watson told Delbrück that he needed more time at Cambridge. Delbrück asked Pauling to inform the Foundation that the facilities for Watson's research were better at the Cavendish; the first year of the fellowship was then transferred to Cambridge. It was therefore thanks to Pauling that Watson was still in Cambridge when the structure of DNA was solved.

The final breakthrough in the X-ray analysis of DNA came in May, 1952, when Franklin obtained high-quality images of the B form. These showed clearly that the 3.4 Å reflection was on the tenth layer line. If the molecule were a helix, therefore, it probably had a pitch of 34 Å and ten nucleotides per turn. By now, Franklin realized that the diffraction patterns obtained by Astbury were from specimens containing a mixture of the A and B forms. The structures proposed by Astbury and by Furberg, which both had eight nucleotides per turn, were based on the 27 Å pitch of the crystalline A form. However, it was the B form that should be present in the aqueous environment of the living cell. Indeed, Wilkins' studies on the sperm head showed that its DNA was of the B form.

In May 1952, Erwin Chargaff visited the Cavendish Laboratory. He was not impressed with Watson and Crick, who struck him as 'two pitchmen in search of a helix'. According to Chargaff, he described the curious ratios of the bases in DNA, which he had felt represented the presence of adenine next to thymine and guanine next to cytosine in the polynucleotide chain. However, the nuclease digestion of DNA did not support this interpretation. Crick had not read Chargaff's papers, but on learning of the 1:1 base ratios immediately associated this with Pauling and Delbrück's idea of complementarity.

In a 1948 lecture entitled 'Molecular architecture and the processes of life', Pauling had further developed this idea:

> I believe that the genes serve as a template on which are moulded the enzymes that are responsible for the chemical characters of the organisms, and that they

also serve as templates for the production of replicas of themselves . . . If the structure that serves as a template (the gene or virus molecule) consists of, say, two parts, which are themselves complementary in structure, then each of these parts can serve as the mould for the reproduction of a replica of the other part, and the complex of two complementary parts thus can serve as the mould for the production of duplicates of themselves.[90]

Pauling thought of these self-replicating, enzyme-producing molecules as proteins. He was finally disabused of this notion in the summer of 1952, when he attended a conference in Royaumont, France, and heard Hershey present his studies with Chase. Pauling may have been slow to realize the importance of nucleic acids in the transfer of genetic information, but he was quick to realize the significance of the Hershey and Chase experiment. By the time he returned to Cal Tech, he had decided to try to solve the structure of DNA.

The fall of 1952 was very late for anyone to join the quest for the DNA structure. However, if anyone could make up the lost ground it was Linus Pauling. As the above quotations show, he was already thinking in terms of a gene consisting of paired complementary molecules. He was the leading structural chemist in the world, with the α-helix already under his belt. And the opposition was in disarray. The King's group was crippled by the animosity between Wilkins and Franklin; Watson and Crick had been, as it were, sent to their bedrooms.

But Pauling's eleventh-hour effort was handicapped by bad luck and missed opportunities. In 1938, he had tried and failed to attract Alexander Todd to Cal Tech (see Chapter 8); by 1952, Todd was the leading authority on nucleic acid chemistry. Another opportunity to recruit a top nucleic acid chemist had come in 1945, when Chargaff inquired about the possibility of a position in the 'chemical biology' program Pauling was assembling with George Beadle, then in the process of moving to Cal Tech from Stanford. Pauling's 'curt' reply was that there was no suitable position available in either chemistry or biology. In 1947, when Pauling was sailing to England on the *Queen Mary* to start his sabbatical at Oxford, he discovered that Chargaff was a fellow passenger. According to Pauling's son Peter, Pauling disliked Chargaff and did not pay much attention to what he said.[b] A third opportunity was lost in May, 1952, when Pauling was

---

b   Chargaff could not have told Pauling about the A–T and G–C equivalencies in DNA, as the evidence for this was not published until 1950. However, he could perhaps have persuaded Pauling of the importance of nucleic acids in genetics.

denied a passport to attend a Royal Society symposium in London. Robert Corey went in his place, and visited King's College. Franklin showed Corey diffraction images of the B form of DNA. Unluckily for Pauling, this was only a few days before she obtained the famous image 51 that allowed her to determine that the axial repeat distance was 34 Å.

Pauling had been refused a passport because his public support for left-wing causes had irked the various US authorities investigating communist and 'un-American' activities. In December 1952, the political pressure on Pauling increased when he was named as a communist before a committee of the US Congress. It was against this background of intense personal stress that he attempted to solve the structure of DNA.

Pauling did not know about the existence of the A and B forms of DNA. He did not have access to DNA of the quality that Franklin was using, and the diffraction images he obtained were a mixture of the A and B forms. Consequently, he tried, as Astbury and Furberg had done, to accommodate both the 3.4 Å spacing between nucleotides of the B form and the 27 Å pitch of the A form. It took him less than a week to come up with what he felt was a plausible structure; not long, but then the α-helix had only taken him two hours. However, whereas he had waited three years before he felt sure enough to publish his protein helix, Pauling submitted his DNA manuscript straight away.

From Peter Pauling, a physicist who had come to work at the Cavendish in the fall of 1952, Watson and Crick heard the tantalizing news that Pauling and Corey had submitted a paper on the DNA structure. A copy of the manuscript arrived in Cambridge in mid-January of 1953. To the immense relief of Watson and Crick, the proposed structure was similar to their much-ridiculed three-chain model of 1951. In the Pauling and Corey structure, each of the three chains had 3.43 nucleotides per turn (24 in 7 turns) and these were intertwined in such a way that the entire triple-helical structure repeated every 27.2 Å (Figure 12.1). The sugar–phosphate backbones of the polynucleotide chains were in the middle, with the purine and pyrimidine bases radiating outwards, and no mechanism was suggested to neutralize the electrostatic repulsion between the phosphate negative charges.

The paper was, for Pauling, strangely tentative, with none of the triumphal certainty of the α-helix paper. The correspondence between theory and experiment was nowhere more than 'satisfactory', and in some instances even less: 'It is found that it is very difficult to assign atomic positions in such a way that the residues can form a bridge between an outer oxygen

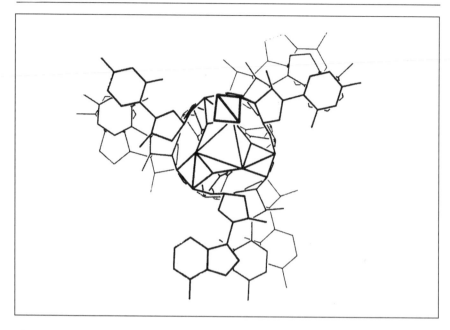

Figure 12.1: Linus Pauling's 1953 structure of DNA, viewed down the rotation axis. The molecule is a three-stranded helix with the deoxyribose–phosphate 'backbones' forming a core from which the bases radiate out. Reproduced from Pauling, L. and Corey, R. B. (1953) *Proceedings of the National Academy of Sciences USA* **39**, 92

atom of one phosphate group and an outer oxygen atom of a phosphate group in the layer above, without bringing some atoms into closer contact than is normal.'[91] Was it not for exactly this that Pauling had criticized Dorothy Wrinch's cyclol structure of proteins? (See Chapter 8.) 'There is reasonably satisfactory agreement with the experimental values; on the other hand, similar agreement might be given by any cylindrical molecule with approximately the same diameter.' Was not Pauling's fabled 'stochastic method' supposed to give unambiguous verification of proposed structures? According to Bernal, the triple-helical structure of DNA 'was an exceptional case in which Pauling's chemical intuition got ahead of his crystallographic dimensional approach.'

Watson and Crick knew that the Pauling structure was incorrect because it was based on many of the same mistaken assumptions that they had made in 1951. Like them, and like Astbury and Furberg before, Pauling had assumed a pitch of 27 Å. Like Watson and Crick in 1951, he had used incorrect values for the density of the molecule. Even better than the fact that Pauling had failed to devise the correct structure for DNA was the fact that his attempt brought Watson and Crick back into the game. Now that another group was actively working on DNA structure, Bragg no longer felt obliged to leave the King's group a clear field. No doubt the fact that the third party was Pauling made this decision easier for Bragg. He would not keep Watson and Crick on the side-lines while Pauling grabbed the glory once again; the deal with Randall was off – Watson and Crick were free to pursue the DNA structure.

Watson only found out about the B form of DNA when he visited King's in January 1953, and Wilkins showed him some of Franklin's diffraction patterns. At the beginning of February, more of Franklin's data fell into the hands of the Cavendish group when Perutz gave Crick a copy of an MRC report on the King's biophysics unit. This included no diffraction patterns, but did contain values for the water content and the crucial axial repeat distance of the B form. Even more importantly, the MRC report described the characteristics of the unit cell of the B form. From the relative lengths and angles, Crick realized that it was a monoclinic unit cell of space group C2. By good fortune this was the same space group as the horse methemoglobin crystals Crick was working on, so he knew that the C2 unit cell had a dyad axis of symmetry. In the case of hemoglobin, this symmetry arose from the distribution of the two α and two β subunits around the centre of the molecule. What, however, was the structural significance of the C2 space group in a one-dimensional fiber like DNA? The answer had been supplied by Astbury in 1931, when he proposed that a protein composed of two antiparallel chains would have C2 symmetry (see Chapter 7). Crick concluded that DNA contained two (or a multiple of two) chains running in opposite directions.

Another attempt at model-building was made. Unlike the abortive 1951 model, the new structure had two polynucleotide chains, and these were held together by hydrogen bonds rather than electrostatic interactions. While he waited for proper metal pieces to be constructed, Watson made cardboard cut-outs of the four bases and tried to find ways in which they could form bonds. He could only produce 'like with like' bonds, which would not explain the Chargaff regularities. He showed these models to Jerry Donohue, a former student of Pauling's now sharing an office at the Cavendish with Watson and Crick. Donohue pointed out that the nucleic acid bases occur in different isomeric forms. Adenine, cytosine and guanine could exist in either amino ($-NH_2$) or imino ($=NH$) form; cytosine, guanine and thymine could exist in either keto ($C=O$) or hydroxy ($C-OH$) form. Watson was using the wrong isomeric forms of all the bases except adenine. When he remade his cut-outs with the bases in the amino and keto forms, Watson found that he could make hydrogen bonds between adenine and thymine and between guanine and cytosine – but only if these pairs were inverted relative to one another. At a stroke, this explained Chargaff's findings and Franklin's evidence for antiparallel chains. It also vindicated Pauling's idea of molecular complementarity, as the specific base-pairing meant that each chain was complementary to its partner.

A new model could now be built. This was a double-stranded helix with the two polynucleotide chains running in opposite directions. The angle of turn per nucleotide was 36°, resulting in one complete turn per ten nucleotides, corresponding to a distance of 34 Å along the fiber axis (Figure 12.2). The bases were in the middle of the helix, such that every adenine in one chain was hydrogen-bonded to a thymine in the other chain, and every guanine in one chain was hydrogen-bonded to a cytosine in the other chain (Figure 12.3)

Franklin was only two years into her three-year fellowship, but she had decided to move to Bernal's department at Birkbeck. While Watson and Crick measured the atomic coordinates of their DNA model, a letter from

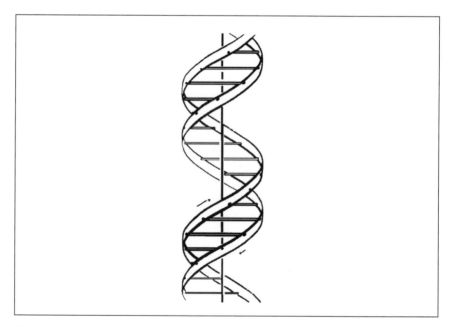

Figure 12.2: James Watson's and Francis Crick's 1953 helical structure of DNA, viewed perpendicular to the rotation axis (vertical line). The molecule consists of two antiparallel polynucleotide chains with hydrogen bonds between the bases holding the two chains together, and ten base-pairs per 360° turn of the helix. Reproduced from Watson, J. D. and Crick, F. H. C. (1953) *Nature* **171**, 737

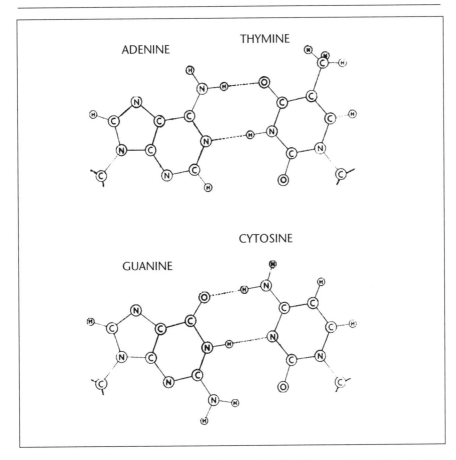

Figure 12.3: Hydrogen bonding between bases in the 1953 Watson–Crick DNA helix. The hydrogen bonds between adenine and thymine and between guanine and cytosine require that the polynucleotide chains be in opposite orientations – for example, the adenine-containing chain runs in the 5´-3´ direction and the thymine-containing chain in the 3´-5´ direction, or vice versa. Note that a third hydrogen bond can be formed between guanine and cytosine, involving the amino (–NH₂) group of the former and the carbonyl (C=O) group of the latter. Reproduced from Watson, J. D. and Crick, F. H. C. (1953) *Nature* **171**, 966

Wilkins arrived. It contained the news that the 'dark lady' was leaving: 'At last the decks are cleared and we can put all hands to the pumps! It won't be long now.' It was clearly time to inform the King's group that a possible solution had already been found. It was agreed that Watson's and Crick's paper would be published jointly with papers by Franklin and Wilkins.

Franklin had been very close to the double helix structure. In a paper submitted for publication before she learned of the Watson–Crick structure, she had described the A form of DNA as follows: 'in the crystalline structure, the purine and pyrimidine groups will be linked to similar groups of neighbouring chains by hydrogen bonds . . . in solution there is evidence for the existence of small stable aggregates of molecules linked by hydrogen bonds between their base groups, and having their phosphate groups exposed to the aqueous environment.' In this paper, she also noted that the X-ray diffraction pattern of the B form was similar to that proposed by Cochran *et al.* for a helix. By this time, she had realized that the B form consisted of two or more coaxial chains each containing ten nucleotides per turn.

No-one was in a better position to solve the DNA structure than Rosalind Franklin. She had discovered the B form and had measured its axial repeat. Only she had a correct crystallographic characterization of the physiological form of DNA. What Franklin lacked was the intuitive genius of a Linus Pauling or a Francis Crick. Her approach was to solve the DNA structure by the deductive approach, using Fourier methods. She did not appreciate, or was not temperamentally inclined towards, the short-cut approach that could be used for molecules with periodic structures.

The three papers on DNA structure appeared in the 25 April 1953 issue of *Nature*. For the purposes of publication, Bragg and Randall had arranged a shotgun wedding of the three groups involved in the elucidation of the DNA structure. Franklin's paper with Gosling, entitled 'Molecular configuration in sodium thymonucleate' was the only one of the three to contain any data relevant to the double-helical structure – the beautiful image of the B form of DNA obtained in May 1952. They interpreted this in terms of the Crick helical diffraction theory, and showed that it was consistent with a helix having a pitch of 34 Å, ten nucleotides per turn and a diameter of 20 Å. On crystallographic grounds, Franklin and Gosling concluded that the phosphates must be on the outside of the molecule, and from the density measurements that the helix must consist of two or three chains: 'Thus our general ideas are not inconsistent with the model proposed by Watson and Crick in the preceding communication.' In their paper, entitled 'Molecular structures of nucleic acids', Watson and Crick showed a diagram of the double-helical structure and discussed two aspects of it not mentioned by Franklin and Gosling as these were not directly derived from the X-ray data: the antiparallel orientation of the two polynucleotide chains and the presence of hydrogen bonds between bases in the middle of the molecule. Some careful phraseology was used to describe the source of the data upon which their model was based:

> The previously published X-ray data on deoxyribose nucleic acid are insufficient for a rigorous test of our structure. So far as we can tell, it is roughly compati-

ble with the experimental data, but it must be regarded as unproved until it has been checked against more exact results. Some of these are given in the following communications. We were not aware of the *details* of the results presented there when we devised our structure, which rests *mainly though not entirely* on published experimental data and stereochemical arguments [author's emphases].[92]

Such coyness was necessary to obscure the awkward fact that the model of the double helix was largely based upon data that were not only unpublished, but officially unavailable to Watson and Crick: the Franklin diffraction images that Wilkins had shown Watson, and the MRC report that Perutz had given Crick.

Wilkins' paper, co-authored with Stokes and Wilson and entitled 'Molecular structure of deoxypentose nucleic acids', mainly consisted of a description of the Stokes helical diffraction theory – still unpublished two years after its elucidation. This was used to interpret a poor image of the B form of DNA from *E. coli*. From this analysis Wilkins and co-workers concluded that 'there appears to be reasonable agreement between the experimental data and the kind of model described by Watson and Crick'. They also noted that diffraction patterns similar to those of the bacterial DNA had been obtained from sperm heads, bacteriophage and the pneumococcal transforming principle.

It was a motley collection of papers with which to announce a revolution in biology. Franklin and Gosling had the only significant crystallographic data, from which some, but not all, of the features of the double helix could be deduced. Watson and Crick had the model, but were in the awkward position of not having been officially aware of the data upon which it was based. Wilkins and his co-workers had neither model nor data; a hard-eyed reviewer might have rejected their paper on the grounds that it contained little, in terms of experimentation or theory, that was not in the paper by Franklin and Gosling.

Much better for all concerned to look forwards rather than backwards. Watson and Crick wrote: 'It has not escaped our notice that the specific [base] pairing we have postulated immediately suggests a possible copying mechanism for the genetic material.' This suggestion was expanded upon in a subsequent *Nature* paper:

> The phosphate–sugar backbone of our model is completely regular, but any sequence of the pairs of bases can fit into the structure. It follows that in a long molecule many different permutations are possible, and it therefore seems likely

that the precise sequence of the bases is the code which carries the genetical information. If the actual order of the bases on one of the pair of chains were given, one could write down the exact order of the bases on the other one, because of the specific pairing. Thus one chain is, as it were, the complement of the other, and it is this feature which suggests how the deoxyribonucleic acid might duplicate itself.[93]

Many streams of thought were thus united in the double helix: Muller's idea of autosynthesis, Schrödinger's hereditary codescript, Pauling's concept of complementarity. The nucleoprotein theory of the gene was swiftly abandoned as it became clear that DNA contained within its structure a means of self-replication and a means of encoding the sequence of proteins.

While the first *Nature* paper was in press, Pauling arrived in Cambridge to visit his son and was shown the DNA model. He realized immediately that the Watson–Crick structure must be correct. It was the greatest set-back of his brilliant career, but Pauling reacted with great dignity. A few days later, he attended a Solvay conference at which Bragg announced the double helix. Pauling stood up and told the group of eminent physicists: 'Although it is only two months since Professor Corey and I published our proposed structure for nucleic acid, I think that we must admit that it is probably wrong. Although some refinement may be made, I feel that it is very likely that the Watson–Crick structure is essentially correct.'[94]

Not everyone reacted so favorably to the double helix. The 88-year-old physicist George Searle asked to see the Watson and Crick model in March 1953, and commented that if this was the basis of heredity, 'no wonder we're such a queer lot'! Crick wrote in 1993: 'How was the double helix received? . . . Jean Brachet, the Belgian biochemist and embryologist . . . thought ours was another silly theoretical idea, best ignored. Arthur Kornberg also thought nothing of it . . . The reactions of many biochemists, such as Joseph Fruton, ranged from coolness to muted hostility.' François Jacob, then working on bacteriophage at the Pasteur Institute in Paris, has recorded that: 'the first article by Watson and Crick on the structure of DNA . . . had not electrified me or anyone else in the laboratory. I had only skimmed through this article . . . It was only some weeks later, at a Cold Spring Harbor Symposium, organized that year by Max Delbrück, that I appreciated the virtues of the double helix.'[74] Chargaff's verdict was characteristically Delphic: 'If DNA is really our thread of Ariadne, the labyrinths out of which it is expected to lead us are truly inscrutable.'

The Cold Spring Harbor symposium mentioned by Jacob was actually on viruses, but Delbrück had made a last-minute change to the program to allow his disciple, Watson, to present his and Crick's structure of DNA. Watson described the evidence in favor of the double-helical structure of DNA and discussed some of the difficulties involved in the proposed replication mechanism. No mention was made of how DNA might code for protein, but Watson did briefly discuss the physical basis of mutation, which he felt might be due to a shift between the isomeric forms of the bases during the process of DNA replication. For example, the imino form of adenine could base-pair with cytosine instead of with thymine.

The discovery of the structure of DNA represented a watershed not only in the history of biology, but also in the lives of many of the scientists involved. Crick went on sabbatical leave to the Polytechnic Institute of Brooklyn. When he returned to Cambridge in the summer of 1954, the MRC offered him a seven-year contract rather than a tenured appointment. Watson became a research fellow in the division of biology at Cal Tech. After spending the year 1955–6 back in Cambridge, he became professor of biology at Harvard University. Rosalind Franklin was already gone from King's by the time the double helix structure was published. In the more hospitable surroundings of Birkbeck College, she performed elegant studies on the structure of the tobacco mosaic virus. She developed cancer, perhaps as a result of her exposure to X-rays, and died in 1958. Of the major figures involved in the discovery of the double helix, only Francis Crick would make any further significant contribution to biological science. By 1955, he was bemoaning the 'comparative isolation of Cambridge'.[c]

In 1954, Lawrence Bragg left Cambridge to become, like his father before him, resident professor at the Royal Institution. It was the final stop on a remarkable career. According to Perutz, 'if we think of Bragg as an artist and compare him to, say, Giotto, it is as though he had himself invented three dimensional representation, and then lived through all the styles of European painting from the Renaissance to the present day, to be finally confronted by computer art.'[95] Despite the tension with his father in the

---

c   In 1960, Lawrence Bragg nominated Max Perutz, John Kendrew and Dorothy Crowfoot for the Nobel prize for physics, and James Watson, Francis Crick and Maurice Wilkins for the chemistry prize. Bragg wrote to a number of eminent scientists asking for their support of his nominations. Linus Pauling replied that he thought the nomination of Watson and Crick was premature and that Wilkins did not deserve a Nobel prize. However, Bragg had made a deal with Randall that credit for the DNA work would be shared between the Cavendish Laboratory and King's College groups. Rosalind Franklin, who had made crucial contributions to the double helix, could not be honored posthumously. Although Wilkins' contribution had been minor, he was next in line for King's share of the credit. Bragg's suggestions were largely adopted by the Nobel Foundation. The 1962 Nobel prize for physiology or medicine was awarded to Watson, Crick and Wilkins, and the chemistry prize to Perutz and Kendrew. Crowfoot was awarded the Nobel prize for chemistry in 1964.

early days of the development of X-ray diffraction, his problems in Manchester and the nervous collapse of 1930, the mantle of scientific prodigy rested lightly on the shoulders of Lawrence Bragg. He 'always had time for his family . . . Bragg was a genial person whose creativity was sustained by a happy home life.' A remarkably unpretentious man, Bragg 'was intrigued when the greengrocer woman in Soho told him that he was "the spitting image of a man she saw on the telly last night" and modestly signed the bill for her to keep as a souvenir'. In London, Bragg was as self-effacing as ever, as recorded by Crick:

> When he moved in 1954 from his large house and garden in West Road, Cambridge, to London, to head the Royal Institution in Albemarle Street, he lived in the official apartment at the top of the building. Missing his garden, he arranged that for one afternoon each week he would hire himself out as a gardener to an unknown lady living in The Boltons, a select inner-London suburb. He respectfully tipped his hat to her and told her his name was Willie. For several months all went well till one day a visitor, glancing out of the window, said to her hostess, 'My dear, what *is* Sir Lawrence Bragg doing in your garden?'[96]

The discovery of the DNA structure also represented a watershed in the career of Erwin Chargaff. Because of his abhorrence for the research style of its proponents, Chargaff found himself in the difficult position of simultaneously belittling the double helix and claiming credit for the importance of 'Chargaff's rules' in its discovery. The discovery of the structure of DNA made Chargaff realize that the style of science he loved would be replaced by one that was larger-scale, less humanistic and further separated from its philosophical roots. Like everything else he loved in human civilization, science was in an irrevocable state of moral and intellectual decline: 'in our epoch . . . what goes as art and literature and science is only an artificial, youthful-looking blooming skin stretched tight over a crumbling skeleton.' Pnina Abir-Am wrote: 'Chargaff conceived of science as a branch of philosophy, an art, a creative expression of the human intellect, and admits to having belatedly realized that science nowadays lost these attributes and became an enormous "problem-solving machine".' One of the founders of molecular biology thus became one of its most persistent critics. Chargaff's writings, which range from book reviews to a Socratic dialog, are unique in the scientific literature for their combination of wit and erudition, and are sprinkled with memorable epigrams: 'In science, there is always one more Gordian knot than there are Alexanders'; 'Life is what's lost in the test-tube'; 'to make a scientific revolution, one must break many eggheads'; and, most memorable of all, 'molecular biology is essentially the practice of biochemistry without a license'.

For Max Delbrück, the discovery of the double helix marked the end of the quest for the paradox of life. As Donald Fleming put it: 'He [Watson] and Crick had indeed found complementary relations at the heart of biology, but it was a totally mundane complementarity, exemplified in locks and keys, and sharing nothing but a verbal echo with the complementarity principle of Bohr and Delbrück.' On 1 February 1953, a few weeks before the structure of DNA was solved, Delbrück wrote to the phage geneticist Seymour Benzer to tell him that he was giving up work on bacteriophage in favor of studying the light response of the fungus *Phycomyces blakesleeanus*. The following year, he gave a seminar on the *Phycomyces* work in Copenhagen. Following the seminar, Delbrück wrote to his mentor Bohr, the man who had inspired him to begin the search for complementarity twenty years earlier:

> I talked about this system as something analogous to a gadget of physics, and explained at some length why it seemed more hopeful to me to analyze this gadget in great detail, rather than the many other biological gadgets which have been the subject of conventional research for many years. What I failed to stress was my suspicion, you might almost say hope, that when this analysis is carried sufficiently far, it will run into a paradoxical situation analogous to that into which classical physics ran in its attempt to analyze atomic phenomena. This, of course, has been my ulterior motive in biology from the beginning.[97]

Delbrück shared the 1969 Nobel prize for physiology or medicine with Salvador Luria and Alfred Hershey, 'for their discoveries concerning the replication mechanism and the genetic structure of viruses'.[d] It was a generous award, considering the unevenness of Delbrück's research career. Both the 'Delbrück model' of mutation and the 1941 mechanism of protein autocatalysis had had to be discarded. His 1954 mechanism of DNA replication was also incorrect. Delbrück's main research program, the Phage Group, was, as Elof Carlson described the target theory, a 'successful failure'. Studies on bacteriophage had helped identify DNA as the material basis of heredity and would play an important role in the discovery of messenger RNA (see Chapter 13). Phage would also prove useful in the development of techniques for fine-structure genetic mapping and in the discovery of the chain-termination codons of the genetic code. However, Delbrück's goal of finding the 'paradox' behind the workings of the gene had proved fruitless.

---

d   From Stockholm, Delbrück traveled to the USSR to visit his old friend Nikolai Timoféeff-Ressovsky, who had recently retired from the Institute of Medical Radiology in Obninsk. In 1981, suffering from multiple myeloma, Delbrück started work on his autobiography. He died three days after starting work on the chapter entitled 'Light and life'. Timoféeff died later the same year.

Arguably, the major achievement of the Phage Group was to introduce the bacterium *Escherichia coli* into biochemistry.

As Nicholas Mullins observed of Delbrück, 'his importance to phage work and the eventual development of molecular biology lies elsewhere than in his intellectual accuracy.' Where else? Horace Judson described Delbrück as 'one of the most seductive intellects of our time'. His importance to the development of molecular biology was that his 'seductive intellect' attracted scientists to the discipline and inspired his colleagues to seek the secret of the gene. Like Ezra Pound in poetry, Delbrück's most important role was a mentoring one. His former student Gunther Stent wrote in his obituary of Delbrück: 'he did not make any spectacular breakthrough discoveries with which the names of very great scientists are normally associated. Rather . . . Delbrück provided the ideological and spiritual fountainhead for the discipline that would eventually style itself molecular biology.'

The effects were felt not only in the 'informational' tradition of bacterio-phage genetics but also in the 'conformational' tradition of macromolecular crystallography. The 'Delbrück model' inspired Erwin Schrödinger, and Schrödinger in turn inspired a generation of biophysicists. It is the Green Paper, not the Phage Group, that represents Max Delbrück's greatest con-tribution to molecular biology.

Chapter 13
Nature is Blind and Reads Braille

'The Outstanding Unsolved Problem'

The 'one gene–one enzyme' hypothesis of George Beadle and Edward Tatum was widely and approvingly cited throughout the 1940s and 1950s, albeit sometimes in such modified forms as 'one gene–one protein' or 'one cistron–one polypeptide'. Although Alfred Hershey reported witnessing an acrimonious debate on the one gene–one enzyme hypothesis at a conference as late as 1951, more typical was the 1947 view of Salvador Luria: 'The concept has become widely accepted that a gene affects a character by determining the presence and specificity of one of the enzymes whose action is necessary for the appearance of the character.' It therefore appears that the central message of biochemical genetics, that the function of genes was to make proteins, quickly became adopted.

However, the means by which DNA directed the synthesis of proteins was completely unknown. The Cambridge biochemist Ernest Gale pointed out in 1957 that biosynthetic pathways had been elucidated for virtually every other class of biological molecules, including fats, carbohydrates, nucleic acid bases, amino acids and vitamins, but these seemed to give no clue as to how amino acids could be linked together in the specific arrangements of proteins. The mechanism of protein synthesis was 'the outstanding unsolved problem before biochemists today'.

The idea that protein synthesis could occur by proteolytic enzymes working in reverse had been championed in the late 1930s by Max Bergmann (see Chapter 8). Some experimental evidence in favor of this mechanism was obtained. As the formation of peptide bonds was thermodynamically unfavorable, however, proteinases could only polymerize amino acids under very constrained circumstances, such as when amino acids were present in massive excess or when the peptide product was highly insoluble. Accordingly, the eminent bioenergeticist Fritz Lipmann of Harvard University proposed in 1949 that the carboxylate groups of amino acids could be 'activated' by transfer of a phosphate group from the high-energy molecule adenosine triphosphate (ATP). Similar activation reactions were involved in the formation of amide bonds in hippuric acid, a compound of benzoic acid and glycine, and in the tripeptide glutathione.

Amino acid activation solved the problem of polymerization, but not the problem of what Lipmann called 'patternization'. Even if peptide bond formation were thermodynamically favorable, what determined the linking of amino acids into specific sequences? Unlike other biosynthetic processes, the synthesis of proteins could not be assumed simply to be the result of the action of specific enzymes. Because proteinases tended to be non-specific in terms of amino acid sequence, as Cyril Hinshelwood noted in 1950, polypeptides formed by reverse proteolysis would have a 'completely indiscriminate' amino acid sequence. In any case, as Hubert Chantrenne of the Free University of Brussels pointed out in 1953, an enzymatic mechanism of protein synthesis would mean that 'we explain the formation of a protein by more proteins'.

However, most workers appeared to accept that the synthesis of proteins somehow involved RNA. The evidence in favor of this came principally from two groups: that of Jean Brachet in the Laboratory of Animal Morphology at the Free University of Brussels, and that of Torbjörn Caspersson in the Institute for Cell Research at the Karolinska Institute in Stockholm. In 1940, Brachet developed a cytochemical technique for staining RNA with methyl green and pyronine – a much-needed counterpart to the DNA-specific Feulgen reaction. Using ribonuclease to control for the specificity of staining, Brachet was able to show that RNA levels were high in actively dividing and secretory tissues, but low in muscle and kidney. Yeast cells starved of phosphate were poorly basophilic, indicating a low nucleic acid content, but feeding them phosphate increased their basophilia by up to fifteen times. Subcellular localization studies showed that most RNA was present in the cytoplasm, but some occurred in a specific area of the nucleus, the nucleolus. By 1942, Brachet had concluded that DNA was the genetic material, while RNA was involved in protein synthesis.

That same year, closure of Brussels University by the occupying Germans interrupted Brachet's studies, although he worked clandestinely until he was discovered and arrested. After the liberation of Belgium, Brachet was able to obtain strong support for a role of RNA in protein synthesis by showing that enucleated amoebae and algae can synthesize protein and even regenerate some differentiated structures. By 1956, he had concluded that RNA was 'probably the best candidate as a cytoplasmic carrier of genetic information'. However, protein synthesis was not restricted to any particular subcellular location: 'Considering all the data available at present, it would seem that RNA as well as protein can be formed concurrently and independently in the nucleus and in the cytoplasm; either part of the cell contains the complete system necessary for their synthesis . . .'

Caspersson studied the distribution of nucleic acid within the cell by ultraviolet microscopy. This not only allowed nucleic acids to be specifically visualized by the absorbance of the bases at 260 nm, but also gave better spatial resolution than visible-light microscopy. In a 1947 article, Caspersson concluded: 'The nucleus itself is a cell organelle organized especially for being the main centre of the cell for the formation of proteins.' Brachet and Caspersson agreed that protein synthesis occurred both in the nucleus and in the cytoplasm; in the former compartment it was mediated by DNA, in the latter by RNA.

Subsequent evidence for a role of RNA in protein synthesis came from studies on tobacco mosaic virus (TMV). In 1956, Heinz Fraenkel-Conrat of the University of California at Berkeley showed that reconstitution of RNA from one strain of TMV with coat protein from another strain resulted in hybrid viruses that were immunologically of the protein-donating strain but produced progeny containing the protein of the RNA-donating strain. Therefore, RNA carried the strain-specific characters of TMV, just as DNA did for the pneumococcus.

The relationships between DNA, RNA and protein were by no means clear. The high RNA contents of rapidly dividing cells led some workers to believe that RNA was a storage form of nucleic acid that was drawn upon when rapid DNA synthesis occurred. Another possibility, raised in 1946 by Solomon Spiegelmann of Washington University in St Louis, was that nucleoproteins were 'specific energy donators which made possible reactions leading to protein and enzyme synthesis'.

Yet another possible role for RNA in the synthesis of proteins was that it acted as a template upon which specific arrangements of amino acids were polymerized. The idea that nucleic acids could direct the assembly of proteins appears to have arisen in the early 1940s from theoretical models of gene replication. Both Hermann Muller and J. B. S. Haldane had spec- ulated on how an identical copy of the gene could be produced from its free subunits (see Chapter 9). Max Delbrück proposed in 1941 that the replica- tion of a protein gene could occur by the mutual resonance-stabilization of imide-bonded intermediates. That same year, the University of Chicago geneticist Sewall Wright wrote: 'The only mechanism by which a given organism can produce particular ones in the array of possible proteins would seem to be autosynthesis by a preexistent molecule acting as a model. This is the same mechanism that seems required for the duplication of genes and suggests that each protein molecule is formed on a model carried in a chromosome.'[98] As described in Chapter 12, most biologists believed that genes were proteins. George Beadle was therefore thinking more about autocatalysis than about heterocatalysis when he wrote in 1945: 'Perhaps the most widely held view is that the gene somehow acts as a master molecule or templet [*sic*] in directing the final configuration of the protein molecule as it is put together from its component parts.' However, Jesse Greenstein of the National Cancer Institute was possibly the earliest scientist to come up with something like the modern scheme for protein synthesis. Discussing the work of Caspersson in a 1941 review, Greenstein wrote: 'It is probable therefore that the pentosenucleic acid [RNA] synthesized in the heterochromatin combines with histones in the various cells and by diffusing into the cytoplasm stimulates or perhaps acts as template by which the production of proteins is effected.'

Template mechanisms were clearly in the air, but there was no unanimity as to whether these applied to the synthesis of protein on protein, DNA or RNA; nor as to whether these mechanisms permitted the copying of the template (autocatalysis) or the generation of a different polymer (heterocatalysis). The demonstration by Erwin Chargaff around 1950 that DNA had a non-repeating distribution of bases along the polynucleotide chain (see Chapter 12) showed that nucleic acids had the sequence specificity necessary to direct the synthesis of the myriad different proteins. It now became possible to speculate on how sequences of about twenty different amino acids could be encoded by sequences of four different bases.

The earliest statement of what became known as the 'coding problem' appears to have been a 1950 publication from the laboratory of Cyril Hinshelwood, professor of physical chemistry at Oxford University. Hinshelwood noted that RNA consisted of five different components: four bases and 'ribose-phosphate', while proteins contained 'about 23' different amino acids. Therefore, all the amino acids could be specified by 'twenty-five different internucleotide arrangements'. It is hard to see how twenty-five different internucleotide arrangements could be achieved, unless it were assumed, in defiance of all known RNA chemistry, that equivalent positions along the polynucleotide could be occupied by either a base or by ribose-phosphate. However, Hinshelwood's main point was that there was a reciprocal relationship between the sequence of bases in RNA and the sequence of amino acids in proteins:

Autosynthesis depends essentially upon a coordination of the following kind: in the synthesis of protein, the nucleic acid, by a process analogous to crystallization, guides the order in which the various amino-acids are laid down; in the formation of nucleic acid the converse holds, the protein molecule governing the order in which the different nucleotide units are arranged.[99]

The different streams of thought that contributed to template theories of protein synthesis were shown by the mechanism proposed by Alexander Dounce of the University of Rochester in 1952. Dounce had become interested in the problem of protein synthesis as a result of a question asked by his PhD oral examiner, James Sumner, the discoverer of crystalline enzymes (see Chapter 7). In his 1952 paper, he assumed that there was one nucleic acid molecule for each protein molecule in the cell, and that 'the specific arrangement of amino acid residues in a given peptide chain is derived from the specific arrangement of nucleotide residues in a corresponding specific nucleic acid molecule . . .' Dounce must have been aware of the work of Brachet and Caspersson, as his template, like that of Hinshelwood, was RNA rather than DNA. Another important element of the Dounce mechanism was the recognition that peptide bond formation required an input of energy, which he felt must come from nucleic acid. Here he may have been influenced by Lipmann or by studies by Carl and Gerti Cori showing that the biosynthesis of glycogen used as a substrate the 'high-energy' compound glucose 1-phosphate rather than glucose itself.

To take into account all these aspects, the Dounce mechanism of RNA-directed protein synthesis was rather complicated. In essence, it proposed that amino acids reacted with a phosphorylated form of RNA, which brought them into the correct juxtaposition, at a high enough energy level, to polymerize. Replication of the RNA molecule occurred from the same phosphorylated intermediate, which in this case reacted with nucleotide monophosphates rather than amino acids (Figure 13.1).[a]

The most influential aspect of the Dounce mechanism of protein synthesis was the idea of a triplet code. In order that twenty different amino acids could be specified, a sequence of at least three bases was required ($4^2 = 16$; $4^3 = 64$). Assuming that the polarity of the RNA triplet was not important – in other words, that the base sequence ACG specified the same amino acid

---

a   Had Dounce used nucleotide diphosphates (or triphosphates) as the raw material for RNA synthesis instead of monophosphates, it would have obviated the necessity for prior phosphorylation of the template and produced a scheme very similar to the contemporary mechanism of *DNA* replication. However, this would have weakened his argument for a common mechanism of protein and RNA synthesis.

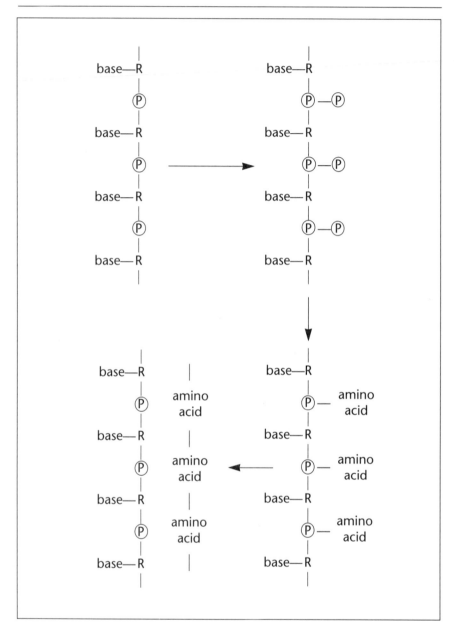

Figure 13.1: Simplified representation of Alexander Dounce's 1952 model of protein synthesis. An RNA molecule consisting of ribose (R), phosphate (P) and bases (top left) becomes 'activated' by addition of extra phosphate groups (top right). Free amino acids then replace the extra phosphate groups (bottom right). The specific amino acid attached depends upon a sequence of three bases in the vicinity of the attachment site. Formation of peptide bonds breaks the bonds attaching them to the RNA, regenerating the RNA molecule and producing a polypeptide chain (bottom left)

as the sequence GCA – forty different triplet combinations of four different bases were possible.[b] Dounce therefore proposed that the enzymes that
attached amino acids to phosphorylated RNA recognized not only the
adjacent base but also those on either side of it. After the first fumbling
attempt of Hinshelwood, the hypothesis that base sequences in RNA
corresponded to amino acid sequences in proteins had now been formally
stated. As Chargaff later put it, 'nature is blind and reads Braille.' If this
were correct, the nature of genetic information and the relationship between
genotype and phenotype could be finally understood.

The discovery of the double-helical structure of DNA did not impinge
directly upon the problem of template-directed protein synthesis. It was the
subunit sequence of the purported template that was important for producing a specific protein, not its conformation. Also, it appeared to be RNA,
not DNA, that was responsible for protein synthesis, and the former lacked
a fiber diffraction pattern. In their 1953 paper entitled 'Genetical implications of the structure of deoxyribonucleic acid', James Watson and Francis
Crick discussed DNA replication in some detail, but only mentioned in
passing that 'it therefore seems likely that the precise sequence of the bases
is the code which carries the genetical information'.

The Watson–Crick structure did, however, attract George Gamow to the
problem of encoding genetic information. A theoretical physicist who had
been Max Delbrück's room-mate in Copenhagen (see Chapter 9), Gamow
had gone on to revolutionize cosmology by proposing in 1948 the 'big bang'
theory of the origin of the universe. As Horace Judson put it, he 'came
through biology in the 1950s – on a highly eccentric orbit'. In a short paper
published in 1954, Gamow noted that the DNA helix contained diamond-
shaped cavities enclosed by four bases. Considering all the possible combinations of different bases, he calculated that there were exactly twenty
different diamonds. Gamow speculated that protein synthesis occurred by
binding of different amino acids to specific holes in the DNA helix, followed
by polymerization (Table 13.1).

An important feature of Gamow's 'diamond' code was that each amino acid
binding site shared two nucleotides with each of its neighbors, ensuring that
only certain dipeptide combinations could be formed. The Gamow code was

---

b   As he was writing more than half a century after Emil Fischer's 'lock and key' mechanism of enzyme-
     substrate recognition (see Chapter 4), it is hard to see how Dounce could think that the interaction
     between a sequence of nucleic acid bases and an optically active amino acid could *not* be sensitive
     to polarity.

Table 13.1: Proposed systems of genetic coding

System	Rule	Amino acid coding units
Diamond code (Gamow, 1953)	Amino acids recognize groups of four DNA bases arranged at the vertices of a diamond	(see diagrams below)

```
      A        C        G        T        A        C        G
   A     C, A     C, A     C, A     C, G     T, G     T, G     T,
      A        C        G        T        A        C        G
   ─────────────────────────────────────────────────────────────
      T        A        A        A        C        C        G
   G     T, A     C, A     C, A     C, A     C, A     C, A     C,
      T        C        G        T        G        T        T
   ─────────────────────────────────────────────────────────────
      A        A        A        C        C        G
   G     T, G     T, G     T, G     T, G     T, G     T
      C        G        T        T        G        T
```

System	Rule	Amino acid coding units			
Combination code (Gamow and Yčas, 1955)	Amino acids recognize specific groups of three RNA bases; the order of bases within the group is irrelevant	(AAA), (CCA), (GGA), (UUA), (ACG),	(AAC), (CCC), (GGC), (UUC), (ACU),	(AAG), (CCG), (GGG), (UUG), (AGU),	(AAU), (CCU), (GGU), (UUU), (CGU)
Comma-free code (Crick et al., 1957)	Amino acids recognize specific sequences of three RNA bases, but do not recognize any base sequences in other reading frames	ACA, AGG, AUU, CUA, GUA,	ACC, AUA, CGA, CUC, GUC,	AGA, AUC, CGC, CUG, GUG,	AGC, AUG, CGG, CUU, GUU

therefore, like the Dounce triplet code, an 'overlapping' one. An important aspect of overlapping codes was the restrictions placed upon amino acid sequence, which meant that any such code could be tested by comparison with known protein sequences.

Gamow's code was based on a DNA template, envisaged a three-dimensional interaction and ignored the unfavorable energetics of peptide bond formation. It was therefore conceptually inferior to that of Dounce. However, it did have an important effect on the development of the field; just as Watson's and Crick's 1953 paper had interested Gamow in coding, Gamow's mechanism stimulated Crick to turn his attention to this problem. On sabbatical at the Polytechnic Institute of Brooklyn in 1953–4, Crick was able to show that the diamond code was inconsistent with existing amino acid sequence data.

Over the next decade, Crick was to serve as the major ideologist of the genetic code, and was responsible for many of the key concepts (and terminology) of biological information flow. Once it was accepted that the sequence of bases in RNA specifies the sequence of amino acids in protein, the coding problem became a formal one – merely a matter of determining which sequences of bases correspond to which amino acids. Crick's remarkable intuitive sense of how nature works came into full flower in his thinking about the coding mechanism; more than the double helix, the genetic code is the measure of Crick's scientific stature. If, as claimed by Judson, the structure of DNA was 'Francis Crick's homage to Linus Pauling', the genetic code was his tribute to Erwin Schrödinger.

Thus began what Gale described as 'a new parlor game for biophysicists and chemists'. The rules of the game were very simple; a sequence of four different bases must specify a sequence of twenty different amino acids. The number of amino acids used in protein synthesis was not known exactly, but twenty seemed about right. The identities of the amino acids involved was not a crucial issue at this stage – Gamow's twenty included hydroxyproline and 'canine', and excluded glutamine and asparagine – but the list generated by Crick in 1953 and included in his unpublished manuscript of 1955 turned out to be exactly correct. The necessity of having a unique sequence of bases for each of twenty amino acids seemed to indicate that at least three bases were required. The 'coding ratio' was therefore assumed to be three. However, a three-base code would have sixty-four different triplets, or forty if polarity were ignored. This could mean that a given amino acid was specified by more than one triplet. Crick borrowed the term 'degenerate' from quantum physics to describe such a code. Alternatively, twenty of the sixty-four triplets could code for amino acids, the rest being 'nonsense'. Neither of these alternatives appearing particularly elegant solutions to the coding problem, many attempts, of which Gamow's diamond code was the first, were made to reduce an apparent sixty-four triplets down to an actual twenty.

At its heart, then, the genetic code was a mathematical puzzle of permutations and combinations. Possible solutions to the 'game' of genetic coding were circulated among the members of the 'RNA Tie Club', founded by Gamow and consisting of twenty members, one for each amino acid. Later, theoretical and experimental studies in this area tended to be published in the *Proceedings of the National Academy of Sciences*, which for some reason became the house journal of the genetic code.

In single-stranded RNA, a diamond code is not an option. Therefore, as the name of his 'club' suggests, Gamow quickly became reconciled to the idea that the template for protein synthesis was RNA rather than DNA. However, he was still interested in trying to find ways of mathematically reducing sixty-four triplet combinations of bases to twenty amino acids. Accordingly, Gamow and Martynas Yčas proposed in 1955 a code in which any combination of the same three bases coded for the same amino acid (Table 13.1). This 'combination' code had a certain mathematical elegance, as it turns out that there are exactly twenty possible combinations. It was also non-overlapping, so there were no restrictions upon amino acid sequence. Even more intriguing, Gamow and Yčas were able to show that the frequencies of such combination triplets in viral RNA closely matched the relative abundancies of amino acids in proteins. The major disadvantage of the combination code seemed to be the implausible assumption that an amino acid that can recognize the sequence ACG would also recognize one so stereochemically dissimilar as CAG, but not one as similar as ACA. However, Gamow was able to get around this objection, albeit at the cost of making another gratuitous assumption, by suggesting that the bases of the triplets were arranged in a triangle.

Another ingenious way of reducing the number of triplets to twenty was suggested in 1957 by Crick. This was motivated by the fact that a non-overlapping code could be read three different ways, depending upon which base was chosen as the start point. For example, the sequence ACGACGAC-GACG could be read as ACG-ACG-ACG . . . , CGA-CGA-CGA . . . or GAC-GAC-GAC . . . These three 'reading frames' would presumably correspond to different amino acid sequences. Crick's solution to the problem of alternative reading frames was to construct the code in such a way that all overlapping triplets were 'nonsense'; if so, there was only one way in which the message could be read. It turned out there were many such solutions, but the maximum number of amino acids that could be encoded in this way was exactly twenty (Table 13.1).

Overlapping codes were formally disproved a few months later in a paper written by Sydney Brenner, Crick's colleague at the MRC group in Cambridge. Brenner noted that a fully overlapping triplet code, in which each coding unit shares two bases with each of its neighbors, reduces the possible number of dipeptide combinations that can occur in proteins from 400 (20^2) to 256. Sequencing of proteins had revealed fewer than 256 different dipeptides, but Brenner found another way to disprove the existence of overlapping codes:

Proof: Since successive triplets share two nucleotides in common, any given triplet can be preceded by only four different triplets and succeeded by only four different triplets. In an amino acid sequence *j.k.l.*, we call *j* an N-neighbour, and *l* a C-neighbour, of *k*. For every four different N-neighbours (or C-neighbours) or part thereof, *k* must have one triplet assigned to it. Thus the *minimum* number of triplet representations for each amino acid can be counted from a table of neighbours.[100]

Performing this calculation for known amino acid sequences gave a total of 70, more than the 64 possible triplet combinations of bases: 'We conclude, then, that all overlapping triplet codes are impossible.'

Several more exotic methods of encoding protein sequence information in nucleic acids were also proposed, generally to address specific aspects of the coding problem. These included a comma-free code in which the complement of all coding and nonsense sequences was nonsense, ensuring that protein synthesis could only occur from one polynucleotide chain of DNA; and the 'biorthogonal code', with 24 coding units of six bases each, which allowed sufficient differences in base sequence that even changes in two bases permitted the coding unit to be recognized. Although ingenious, these turned out to be wrong-headed. Ultimately closer to the mark was Robert Sinsheimer's 'two-symbol code', which postulated that adenine and cytosine were equivalent for coding purposes, as both have amino groups, as were guanine and uracil, which both have keto groups. This mechanism necessitated that at least some coding units consist of five bases, and was therefore wrong in detail; however, the idea that chemically similar bases could not be distinguished by the code-reading machinery did point the way to a possible chemical basis of degeneracy.

'On Protein Synthesis'

In a symposium on the biological replication of macromolecules held in London in 1957, Crick gave a talk entitled 'On protein synthesis'. This summarized his thinking on the genetic code and provides a telling insight into his unique approach to biological problems:

I shall also argue [in addition to stating that the main function of proteins is to act as enzymes] that the main function of the genetic material is to control (not necessarily directly) the synthesis of proteins. There is little direct evidence to support this, but to my mind the psychological drive behind this hypothesis is at the moment independent of such evidence. Once the central and unique role of proteins is admitted there seems little point in genes doing anything else.[101]

Ex cathedra statements of this type occur frequently in Crick's writings; it is as if he knew certain things were true without being aware of how he knew it. Almost invariably, Crick's instincts were correct; for example, his 1957 belief that 'folding of polypeptides is simply a function of the order of the amino acids' would not be proved until Christian Anfinsen's 1962 study on the refolding of denatured ribonuclease (see below).

The major theoretical concepts of Crick's 1957 article were what he called the 'sequence hypothesis' and the 'central dogma'. The former: 'assumes that the specificity of a piece of nucleic acid is expressed solely by the sequence of its bases, and that this sequence is a (simple) code for the amino acid sequence of a particular protein'. The sequence hypothesis had been implicitly accepted by the proponents of all previous template theories of protein synthesis, although, as Crick noted, the evidence for it was 'negligible'. However, support for the idea of a direct relationship between nucleic acid and protein sequences came that year from another of Crick's Cambridge colleagues, Vernon Ingram. In 1949, Linus Pauling had shown that hemoglobin from individuals with the hereditary blood disorder sickle-cell anemia migrated differently from normal hemoglobin in the Tiselius electrophoresis apparatus. Working in Frederick Sanger's laboratory, Ingram now discovered that normal and sickle cell hemoglobin differed only by a single amino acid.[c]

Crick's second general principle of genetic coding was the 'central dogma':

> This states that once information has passed into protein *it cannot get out again*. In more detail, the transfer of information from nucleic acid to nucleic acid, or from nucleic acid to protein may be possible, but transfer from protein to protein, or from protein to nucleic acid, is impossible. Information here means the *precise* determination of sequence, either of bases in the nucleic acid or of amino acid residues in the protein [emphases in original].[101]

The transfer of information from nucleic acid to nucleic acid was freely reversible, as the copying mechanism presumably involved base complementarity. Why was the transfer of information from nucleic acid to protein not reversible? Crick may already have been convinced, intuitively, that the genetic code was degenerate; if so, a particular amino acid would correspond to more than one nucleotide triplet, and thus protein-to-nucleic acid transfer

---

c   Were it still necessary, Ingram's finding also represented support for the non-overlapping nature of the genetic code; in an overlapping code, a single base change would affect either two or three amino acids in the encoded protein.

would not fulfill Crick's criterion of '*precise* determination of sequence'. The central dogma may also have been influenced by general biological considerations, such as the continuity of the germ-plasm (see Chapter 3) and the lack of inheritance of acquired characteristics.

Taken together, Crick's sequence hypothesis and central dogma represent a fusion of two overarching themes of biological chemistry: the concept of the aperiodic polymer, derived from Albrecht Kossel's *Baustein* hypothesis; and the concept of genotype and phenotype, derived from Claude Bernard's legislative and executive forces. The last major restatement of these ideas had been in Schrödinger's 1944 book *What is Life?*, the reading of which had inspired Crick to enter the field of biophysics. Schrödinger had conceived of the gene as an aperiodic crystal or solid, probably protein in nature, the subunit sequence of which represented the 'hereditary codescript'. By 1957, it was clear that genetic information was carried by nucleic acid, not protein, and it had been shown that both these classes of macromolecules were one-dimensional aperiodic structures. It was also clear, from the 'one gene–one enzyme' hypothesis and similar concepts, that genes acted by directing the synthesis of specific polypeptides. Crick was therefore able to come up with a grand scheme of biological information flow. The DNA was the legislative force, or genotype; protein the executive force, or phenotype. The genotype produced the phenotype by virtue of a unidirectional conversion of DNA base sequences into protein amino acid sequences.

Crick's 1957 article was a remarkable theoretical synthesis comparable to Hermann Muller's 1936 paper on gene replication (see Chapter 9). However, it marked the high-water mark of code theory; it was becoming clear that, while the mathematical approach could generate plausible solutions and exclude others, it could not provide a unique solution. From now on, these ingenious speculations would have to be matched by equal ingenuity in experimentation.

'A Biochemical Bog'

By the time 'On protein synthesis' appeared, a great deal of information about the mechanism of protein synthesis had been derived from *in vitro* systems. However, these systems had been developed by biochemists interested in specific cellular processes and largely inspired by bioenergetic concepts. The biochemical subdiscipline of *in vitro* protein synthesis therefore had little in common with the molecular biological subdiscipline of genetic coding. As Mahlon Hoagland put it: 'We biochemists tended to see the molecular biologists as imperious invaders from an alien culture and sensed that they viewed us as drab, if industrious, blue-collar workers.' A similar tension occurred between the chemists and geneticists involved in the development of 'chemical genetics' in the 1930s (see Chapter 9). Short-term antagonism between disciplines is the price science pays for long-term conceptual cross-fertilizations.

A leading center for experimental studies on protein synthesis was the laboratory of Paul Zamecnik at the Huntington Memorial Hospital in Boston. Zamecnik had not planned to study the mechanism of protein synthesis, but rather was initially trying to identify metabolic differences between normal and cancerous cells. Using ^{14}C-labeled glycine and alanine, he showed in 1948 that slices of rat liver incorporate amino acids *in vitro*. This incorporation was abolished by addition of the mitochondrial uncoupling agent[d] dinitrophenol, suggesting that it was an energy-requiring process.[e]

Homogenized liver preparations were being explored as an alternative to tissue slices, but in 1950 Zamecnik described this system as 'a biochemical bog in which much effort is being expended to reach firm ground'. The following year, however, it was shown that the liquid produced by low-speed centrifugation of liver extracts contained no intact cells or nuclei but incorporated amino acids into protein when supplied with ATP. By 1954, Zamecnik had concluded that *in vitro* protein synthesis required 'a microsome-rich fraction into the proteins of which the amino acids are bound by a linkage as stable as the peptide linkages of the protein' and also 'a soluble, heat-labile, non-dialyzable fraction which facilitates the incorporation of amino acids into the microsome protein'.

These microsomes were shown by electron microscopy to consist of membranes and small, electron-dense particles. By treatment with detergents, the membranes could be removed leaving particles 20 nm in diameter consisting of about 50% RNA. In 1958, the name 'ribosome' was suggested for these particles. It seemed likely that the ribosomes represented the cytoplasmic site of protein synthesis previously inferred by Brachet and Caspersson. The ribosomal nucleic acid could therefore be the template RNA predicted by Greenstein, Hinshelwood and Dounce.

---

d   These agents do not inhibit mitochondrial respiration, but uncouple it from the production of the high-energy compound adenosine triphosphate (ATP).

e   By 1949, Zamecnik was sufficiently interested in protein synthesis *per se* to study the synthesis of silk fibroin by silkworms. This appeared to be a much 'cleaner' system than rat liver, as fibroin is virtually the only protein of silkworm cocoons and is highly enriched in glycine and alanine, two amino acids then available in radioactive form. However, the small size of the silk glands and their resistance to homogenization made this ingenious system experimentally unfeasible.

The soluble factor required for *in vitro* protein synthesis came under investigation when Hoagland joined Zamecnik's group in 1953. Hoagland had previously trained in Lipmann's bioenergetics laboratory, and his studies on the high-speed supernatant showed that it contained an activity that attached amino acids to ATP. Subsequent work showed that the supernatant fraction contained a series of 'activating' enzymes, each of which catalyzed the addition of a specific amino acid to the adenosine monophosphate (AMP) moiety of ATP. The resulting activated amino acids appeared likely to be high-energy precursors of protein biosynthesis that could overcome the free energy barrier to polypeptide formation.

The microsomal supernatant was known to contain a small amount of RNA, but this was thought to represent ribosomal RNA released during homogenization or centrifugation. However, Zamecnik showed in 1957 that amino acids could be transferred from AMP to this 'soluble' RNA.

In the course of these experiments, Zamecnik learned from Watson, by then professor of biology at Harvard, that Crick had several years earlier proposed that an 'adaptor' molecule was necessary to match a specific amino acid to a specific nucleotide sequence. In a manuscript entitled 'On degenerate templates and the adaptor hypothesis' circulated to members of the RNA Tie Club in 1955, Crick had noted that it was extremely unlikely on chemical grounds that amino acids could undergo specific interactions with sequences of bases in RNA. How, for example, could a hydrophilic (water-soluble) molecule like RNA form hydrophobic (water-insoluble) 'pockets' that were capable of discriminating between closely related amino acids like leucine and isoleucine? Instead, he proposed: 'Each amino acid would combine chemically, at a specific enzyme, with a small molecule which, having a specific hydrogen-bonding surface, would combine specifically with the nucleic acid template.'

The disproof of overlapping triplet codes provided another argument in favor of Crick's adaptor hypothesis. The distance between amino acids in a protein was similar to that between nucleotide units in DNA (and presumably RNA), suggesting a one-to-one correspondence; however, a minimum of three nucleotide units were required to encode twenty different amino acids, suggesting a three-to-one correspondence. In 1957, Crick proposed that the binding of amino acids to RNA templates was mediated by molecules that were capable of binding both to particular amino acids and to specific trinucleotide sequences of the template. As Crick noted, these adaptors could be proteins, like the enzymes that Dounce had proposed catalyzed bond formation between amino acids and RNA: 'But there is one possibility which seems inherently more likely than any other – that they might contain nucleotides. This would enable them to join on to the RNA template by the same "pairing" of bases as is found in DNA, or in polynucleotides.' Crick's adaptors were conceived of as trinucleotides complementary to template RNA triplets and to which amino acids could be attached by specific enzymes.

Zamecnik and Hoagland's soluble RNA – in 1958 renamed 'transfer RNA' – seemed to be Crick's adaptor: it reacted with activated amino acids, which were subsequently incorporated into protein. However, transfer RNA was not a trinucleotide – it was at least ten times larger than that. For Zamecnik, at least, this was not an insuperable obstacle to identifying the adaptor as transfer RNA; in a 1960 review article, he pointed out that 'a fairly large number of mononucleotides' of transfer RNA may be required for recognition by the corresponding amino acid-activating enzymes.

By 1960, then, a fairly detailed outline of the biochemistry of protein synthesis had emerged. Zamecnik's review contained a diagram of the process divided into six stages: amino acid activation by reaction with ATP; preparation of the transfer RNA by addition of two cytidine and one adenine monophosphate units; attachment of the amino acid to the terminal adenine of the transfer RNA; arrangement of transfer RNAs on the microsomal RNA template; peptide bond formation; and folding of the protein into its three-dimensional conformation.

Messenger RNA

In another 1960 review article, Hoagland wrote: 'The essential riddle which the newer knowledge poses for us is: why are the two distinct kinds of ribonucleic acids involved in protein synthesis, and how does deoxyribonucleic acid exert its ultimate control when it does not participate in the act of synthesis itself?' This pointed out the weak points in the proposed schemes for protein biosynthesis – the identity of the RNA template and the mechanism of nuclear control. Specifically, ribosomal RNA did not appear to have the properties required of a protein-specifying template molecule. First, it came in only two sizes, whereas the template molecule should be heterogeneous in length, reflecting the different sizes of proteins. Second, ribosomal RNA, unlike DNA, did not vary much in base composition between species. Third, it was metabolically very stable, which appeared to preclude changes in cellular protein synthesis in response to hormones and other growth factors.

The requirement that the template of protein synthesis be metabolically labile had been noted, on genetic grounds, by Sewall Wright as early as 1941:

An alternative view is that duplicates or partial duplicates of genes reach the cytoplasm when the nuclear membrane disappears in mitosis and that these can produce duplicates in turn, and so on, permitting exponential increase. But such particles, having the essential property of genes, would give practically pure cytoplasmic heredity unless it be supposed that their genic property is subject to decay.[98]

The problem identified here is similar to the one confronted by the nineteenth-century 'factor theorists' when they tried to explain how the adult organism could be composed of differentiated tissues but capable of producing undifferentiated germ cells (see Chapter 6). However, the problem was now in the cytoplasm, not the nucleus. If nuclear genes exported to the cytoplasm templates for the synthesis of a particular protein, how could the production of that protein subsequently be turned off? The only solution seemed to be that the templates were short-lived. In this case, unless a particular template were being continuously produced by the nuclear genes, synthesis of the protein specified by that template would not be continued.

It became possible to measure rates of RNA metabolism only in the postwar period, when the radioactive phosphorus isotope ^{32}P became commercially available. In 1954, Alfred Hershey published a long paper on nucleic acid metabolism in phage-infected bacteria, in which he noted in passing that 'an extremely rapid incorporation of P^{32} into RNA occurred, reaching a peak . . . at 5 minutes after infection'. Elliot Volkin and Lazarus Astrachan, working at the Oak Ridge National Laboratory, had early access to [^{32}P]-phosphate that contained little non-radioactive phosphorus (^{31}P). Following up on Hershey's study, Volkin and Astrachan showed in 1956 that RNA formed in E. coli soon after infection with bacteriophage T2 was rich in adenine and uracil, unlike the host cell DNA. The following year, they noted that the base composition of the rapidly synthesized RNA was similar to that of the infecting phage DNA. However, Volkin and Astrachan thought that this RNA was a bacterial product that was related to 'a general reorganisation of cellular material' rather than to 'the synthesis of new protein'.

This line of investigation exemplifies the tunnel vision of phage researchers and, perhaps, the emerging disciplinary divergence between biochemists and molecular biologists. In 1950, a paper by Cyrus Barnum and Robert Huseby had cited seven studies conducted since 1944 demonstrating a higher turnover of RNA than DNA. Their data on ^{32}P incorporation by mouse liver *in vivo* led Barnum and Huseby to conclude that: 'Incorporation of P^{32} into nuclear PNA [pentose nucleic acid] is extremely rapid and at 45 min. the relative specific activity of this fraction is greater than that of the phospholipides [*sic*], and more than 200 times that of the bulk of the cellular PNA which occurs in the microsome fraction.' A rapidly turning-over species of RNA in the nucleus seems like an obvious candidate for the proposed template RNA, but Barnum and Huseby made no such suggestion and their findings seem not to have been noticed by those interested in the coding problem. In their 1956 paper, Volkin and Astrachan referred neither to Barnum and Huseby's paper nor to any of the previous studies cited therein.

However, the idea that labile RNA may act as a genetic messenger had been mooted as early as 1955. At a symposium on 'Structure of enzymes and proteins' held at the Oak Ridge Laboratory, Walter Vincent of the State University of New York at Syracuse reported that a small fraction of the nucleolar RNA in starfish oocytes was rapidly labeled with ^{32}P and rapidly degraded. Vincent, who had previously worked with Brachet, stated that: 'One exciting implication of the active, or labile, form would be that it is involved in the transfer of nuclear "information" to the synthetic centers of the cytoplasm.'

A completely independent line of evidence for a labile RNA species came from studies on sugar utilization in bacteria. At the Pasteur Institute, Jacques Monod had been studying the preferential use of sugars by bacteria since the late 1930s. By the mid-1950s, it was clear that growth of *E. coli* in the presence of the disaccharide lactose resulted in the induction of β-galactosidase and two other enzymes that were required to metabolize lactose. Similar studies on *Staphylococcus aureus* led Gale to conclude in 1957 that 'β-galactosidase is an example of an enzyme-forming mechanism involving a highly unstable RNA component which must be continually synthesized during formation of the specific protein.'

Monod's studies on the 'lactose operon' of *E. coli* were joined in 1957 by François Jacob, who had previously worked on the transfer of genetic material between bacteria by the process of conjugation. Conjugation allowed bacteria to be studied by the techniques of classical genetics; by crossing strains of *E. coli* containing various mutations affecting lactose metabolism, Jacob and Monod were able to show that the induction of the lactose operon involved the inactivation of a repressor molecule that normally prevented expression of the 'structural' genes encoding the lactose-metabolizing enzymes. In their classic paper of 1961, they wrote: 'the structural message must be carried by a very short-lived intermediate both rapidly formed and rapidly destroyed during the process of information transfer.' Jacob and Monod believed that this hypothetical 'messenger RNA' would be a highly labile polynucleotide fraction of heterogeneous molecular weight with a

DNA-like base composition and should be associated, at least transiently, with ribosomes. Neither ribosomal nor transfer RNA had these properties, but the species studied by Volkin and Astrachan had at least some of them.

In the summer of 1960, Jacob went to Max Delbrück's laboratory at Cal Tech to look for messenger RNA in phage-infected *E. coli*. These studies were performed in collaboration with Matthew Meselson and Sydney Brenner; two years earlier, Meselson had developed a technique for centrifuging DNA on cesium chloride density gradients in order to demonstrate that DNA replication occurred by polymerization of complementary mononucleotides on each parental strand. Brenner, Jacob and Meselson now proposed to use this technique to determine whether the Volkin–Astrachan RNA was associated with ribosomes. Bacteria were grown in medium containing the heavy isotopes of carbon and nitrogen, ^{13}C and ^{15}N, then infected with bacteriophage and transferred to medium containing ^{12}C and ^{14}N. The RNA and ribosomes present at various times thereafter were separated on cesium chloride gradients. It was found that a rapidly turning-over RNA species with a DNA-like base composition was added to pre-existing ribosomes, indicating that the DNA-like RNA, not ribosomal RNA, must be the template for protein synthesis. Similar findings were obtained by François Gros, another member of the Pasteur team, working in the laboratory of James Watson.

The year 1961 also brought strong evidence that the genetic code was based on nucleotide triplets. This came from Crick, making a rare foray into the laboratory. Studies on the rII region of bacteriophage T4 had identified two classes of mutations, '+' and '−'. The '−' mutations were thought to delete a base from the phage DNA, and the '+' ones to add one. Recombinants between '+' and '−' strains ('+/−') usually exhibited the wild-type phenotype, while '+/+' and '−/−' recombinants never did. However, Crick and associates were able to show that '+/+/+' recombinants exhibited a wild-type or pseudo-wild-type phenotype. Their interpretation was that addition of one or two bases shifted the reading frame of the protein-specifying base sequence, but addition of three bases restored the proper reading frame, allowing the remainder of the protein to be synthesized normally. This study also suggested that the genetic code was degenerate, as otherwise the '+/−' recombinants would often result in nonsense triplets, stopping synthesis of the protein at that point.

'The Code for Phenylalanine'

As noted by Hans-Jörg Rheinberger, Paul Zamecnik hoped to solve the genetic code by sequencing the transfer RNAs for each amino acid; Seymour

Benzer and Francis Crick by fine-mapping mutations in T4; Heinz Fraenkel-Conrat by correlating base changes in TMV RNA with amino acid changes in its coat protein. However, the actual solution came from a completely different direction. Attending the International Congress of Biochemistry in Moscow in August 1961, Crick found that the first letter of the code had been solved by a young man of whom he had never heard.

Marshall Warren Nirenberg, working at the National Institute of Arthritis and Metabolic Diseases in Bethesda, was interested in how a nucleic acid sequence could direct the synthesis of a protein sequence. Nirenberg and his postdoctoral fellow Heinrich Matthaei used an *in vitro* system originally developed by Zamecnik in which *E. coli* cells were ground with alumina. Several technical improvements were introduced, the most important of which was the addition of the enzyme deoxyribonuclease to eliminate the endogenous incorporation of amino acids. This meant that the *E. coli* system could be used as an assay for template activity. Addition of ribosomal RNA to the deoxyribonuclease-treated ribosomes resulted in a low level of protein synthesis, but addition of yeast RNA resulted in higher levels and TMV RNA was more effective still.

In the process of testing the effects of different RNAs in this system, Nirenberg and Matthaei discovered that polyuridylic acid, a polymer of uridine monophosphate, caused massive incorporation of phenylalanine into protein, but had little or no effect on other amino acids. The properties of the product of the reaction were consistent with it being polyphenylalanine. Nirenberg and Matthaei had discovered the first 'word' of the genetic code: 'One or more uridylic acid residues therefore appears to be the code for phenylalanine.' In a note added in proof, they mentioned that polycytidylic acid stimulated the incorporation of proline. Assuming that it was a triplet code, incorporation of phenylalanine and proline into protein must be specified by the RNA sequences UUU and CCC, respectively.[f] From this point onward, it was accepted that *in vitro* translation of synthetic polynucleotides held the key to the genetic code, and other methods would only be used for confirmation. The biochemists studying protein synthesis and the molecular biologists studying genetic coding were finally on the same track.

In principle, the way ahead was open. One needed only to synthesize polynucleotides composed of repeats of the sixty-four different triplets and then determine which amino acid was incorporated when each polynucleotide

---

f These findings disproved Crick's elegant 'comma-free' code, in which all triplets composed of identical bases were disallowed.

was used to direct protein synthesis. It quickly became apparent, however, that it was not going to be that simple. Polyadenylic acid and polyguanylic acids did not produce the incorporation of any amino acid, and the polycytidylic acid-directed proline incorporation proved to be poorly reproducible. If the code were fully or partly degenerate, of course, AAA and GGG could be nonsense triplets. A far more serious problem, however, was that there was no way to make polynucleotides of defined sequence (except, of course, for the four homopolymers). A direct solution of the genetic code was therefore not feasible.

While polynucleotides of defined sequence could not be made, copolymers of random sequence could. In 1955, Marianne Grunberg-Manago and Severo Ochoa had discovered the first polynucleotide-synthesizing enzyme, polynucleotide phosphorylase, in the bacterium *Azotobacter vinelandii*. This enzyme had the activity of producing polynucleotides from nucleotide diphosphates. When fed a mixture of different nucleotides, therefore, polynucleotide phosphorylase synthesized a random copolymer. For this discovery, Ochoa shared the Nobel prize for physiology or medicine in 1959.

When added to Nirenberg's system, random nucleotide copolymers caused incorporation of more than one amino acid. This was to be expected, as such polymers will contain all possible permutations of their constituent subunits; copolymers composed of two, three and four different bases will contain 8, 27 and 64 different triplets, respectively. However, the composition of triplets coding for particular amino acids could be estimated from the predicted frequency with which certain combinations of bases occurred in copolymers made from known amounts of two different nucleotides. For example, the copolymer made from U and G at a ratio of 5:1 should contain 5 times as many UUU triplets as UUG, UGU or GUU triplets; 25 times as many as UGG, GUG or GGU triplets; and 125 times as many as GGG triplets. The incorporation of phenylalanine produced by this copolymer was 5 times greater than that of cysteine and valine, 8 times that of leucine, 20 times that of tryptophan and 24 times that of glycine. This suggested that the triplets coding for cysteine, valine and leucine were likely to have the composition 2U + 1G, while those for glycine and tryptophan were likely to be 1U + 2G.

Using this approach, Ochoa and his co-workers at New York University were able by 1962 to predict the compositions of triplets coding for all the amino acids except glutamine. Essentially identical results for fifteen of these amino acids were obtained by Nirenberg's group (Table 13.2). In confirmation of these findings, most of the assignments could explain observed

Table 13.2: Correspondences between trinucleotide sequences in RNA and amino acid incorporation into proteins (the genetic code)

Amino acid	Speyer et al., 1962	Matthaei et al., 1962	Nirenberg et al., 1965	Khorana et al., 1966	Modern
Alanine	(UCG)	(UCG)	GCU		GCU, GCC GCA, GCG
Arginine	(UCG)	(UCG)	CGC, CGA, AGA		CGU, CGA, CGC, CGG, AGA, AGG
Asparagine	(UAA) (UAC)?		AAU, AAC	AAC	AAU, AAC
Aspartic acid	(UAG)		GAU, GAC	GAU	GAU, GAC
Cysteine	(UUG)	(UUG) or (UGG)	UGU	UGU	UGU, UGC
Glutamic acid	(UAG)	(UAG)	GAA		GAA, GAG
Glutamine			CAA, CAG	CAA	CAA, CAG
Glycine	(UGG)	(UGG)	GGU, GGC, GGA, GGG		GGU, GGC, GGA, GGG
Histidine	(UAC)		CAU, CAC	CAU	CAU, CAC
Isoleucine	(UUA)	(UUA)	AUU, AUC	AUC	AUU, AUC, AUA
Leucine	(UUC) (UUG)? (UUA)?	(UUC) (UUG)	UUG, CUU CUC, CUG	UUA, UUG CUU, CUA	UUA, UUG CUU, CUC CUA, CUG
Lysine	(UAA)	(UAA)	AAA, AAG		AAA, AAG
Methionine	(UAG)	(UGA)	AUG	AUG	AUG
Phenylalanine	(UUU)	(UUU)	UUU, UUC	UUC	UUU, UUC
Proline	(UCC)	(UCC)	CCU, CCC, CCA		CCU, CCC, CCA, CCG
Serine	(UUC)	(UUC) (UCG)	AGU, AGC, UCU, UCC, UCG	AGU, UCU, UCA	AGU, AGC, UCU, UCC, UCA, UCG
Threonine	(UAC) (UCC)?		ACU, ACC, ACA, ACG	ACU, ACA	ACU, ACC, ACA, ACG
Tryptophan	(UGG)	(UGG)	UGA		UGG
Tyrosine	(UUA)	(UUA)	UAU, UAC	UAU, UAC	UAU, UAC
Valine	(UUG)	(UUG)	GUU	GUU, GUA	GUU, GUC, GUA, GUG

amino acid substitutions in natural mutations of human hemoglobin and nitrous acid-induced mutations of TMV coat protein by changes in a single base.

The random copolymer studies showed that more than one base composition was associated with the incorporation of certain amino acids, notably leucine, suggesting that the genetic code was at least partially degenerate. This conclusion was supported by Robert Holley's 1962 finding of two different leucine transfer RNAs – these were both capable of acting as acceptors for leucine, but bound to different ribonucleotide copolymers. Another indication of degeneracy came from the fact that all the copolymers used to direct protein synthesis contained some uracil. As messenger RNAs were not particularly rich in uracil, however, there were presumably additional coding triplets without U, which would further increase the degeneracy of the code.[g]

'Beautifully Precise Copying Machines'

In a 1963 review article, Crick concluded that 'the weight of evidence certainly suggests that it is a *non-overlapping triplet code, heavily degenerate in some semisystematic way, and universal, or nearly so*' [emphasis in original]. To define the extent of the degeneracy, Crick suggested two approaches: fractionation and characterization of all the transfer RNAs, and 'synthesis of polynucleotides with defined or partly defined sequence'. The latter approach would not only answer the question of degeneracy, but would solve the code itself. However, because of the number of reactive groups, the chemical synthesis of defined polynucleotide sequences was a nightmarish prospect. Enzymatic synthesis did not seem any more hopeful: polynucleotide phosphorylase could only produce random sequences; DNA polymerase, discovered by Arthur Kornberg in 1957, could only copy an existing sequence.

The leading expert on the chemical synthesis of polynucleotides was Har Gobind Khorana. As a postdoctoral fellow with Alexander Todd at Cambridge in 1950–2, Khorana had worked on the synthesis of nucleotides. Todd's confirmation of the chemical structures of DNA and RNA during that period made it possible, in principle, to synthesize polynucleotides, and Khorana started to work on this when he subsequently moved to the British

---

g   A variety of technical reasons underlay the apparent inability of U-free copolymers to direct amino acid incorporation. These included the low affinity of poly-C for ribosomes, the helix-forming tendency of poly-G and the solubility of some amino acid polymers under the conditions used for recovery of protein.

Columbia Research Council laboratories in Vancouver. Because of its extra ribose hydroxyl group, RNA was more difficult to synthesize than DNA. By 1963, when Khorana was at the Institute for Enzyme Research at the University of Wisconsin, Madison, it was possible to make oligoribonucleotides and short polydeoxyribonucleotides. At this time, the work on random copolymers had reached its limits of interpretation, and Khorana felt that 'the responsibility for complete elucidation of the genetic code now essentially rested with the chemist'.

To break the code, however, required polyribonucleotides, and these could not be produced by chemical means. To learn about DNA polymerase, Khorana made 'many pilgrimages' to Kornberg's laboratory at Washington University in St Louis. He also studied RNA polymerase, an enzyme that synthesizes RNA complementary to a DNA template. Using these 'beautifully precise copying machines', Khorana came up with a method of producing large polyribo- and deoxyribonucleotides of defined sequence. Chemical methods were used to synthesize oligodeoxyribonucleotides (maximum length eighteen nucleotides) with repeating di-, tri- and tetra-nucleotide sequences, and their complementary antiparallel partners. These short synthetic DNAs were 'amplified' by the action of DNA polymerase to produce high-molecular-weight (300000–1000000) polynucleotides of identical sequence; RNA polymerase was then used to synthesize a polyribonucleotide complementary to one of the 'DNA' strands.

In cell-free translation systems, Khorana's dinucleotide polymers directed the synthesis of polypeptides composed of alternating amino acids: $(UC)_n$ produced (serine-leucine)$_n$, $(AG)_n$ produced (arginine-glutamic acid)$_n$, $(UG)_n$ produced (valine-cysteine)$_n$, and $(AC)_n$ produced (threonine-histidine)$_n$. Trinucleotide polymers normally directed the synthesis of three different homopolypeptides, corresponding to the three different reading frames. For example, $(UUC)_n$ produced phenylalanine, serine and leucine polymers, and $(AAG)_n$ produced lysine, glutamic acid and arginine polymers. However, $(GUA)_n$ produced only valine and serine polymers, suggesting that one reading frame (actually UAG) consisted of a 'nonsense' codon. From these data and the previous compositional studies, the amino acid assignments of each of the sixty-four codons could be determined (Table 13.2). Studies with tetranucleotide polymers were not necessary for code-breaking purposes, but did show that the reading of the message starts from the 5' end of RNA.[h]

---

h   It is ironic that repeating polymers of four nucleotides – Gulland's 'structural tetranucleotides' – should play a role in the elucidation of the genetic function of nucleic acids. For technical reasons, however, Khorana's tetranucleotides contained a maximum of three different bases.

Khorana's elegant system not only determined the nature of the code, but also recapitulated *in vitro* the whole process of biological information transfer from DNA to RNA to protein, vindicating Crick's central dogma.

Independent confirmation of these codon assignments came from Nirenberg, who had realized that polynucleotides of defined sequences were not required to break the genetic code. If the amino acid-specifying nucleotide sequences, for which Crick had recently introduced the term 'codon', were triplets, perhaps trinucleotides could be used to determine amino acid assignments. A trinucleotide would not, of course, direct the incorporation of amino acids into protein – but it should be capable of binding to a particular transfer RNA.

The synthesis of a trinucleotide was a much easier proposition than that of a polynucleotide, but at this time only twenty to thirty of the possible trinucleotides had been prepared. Nirenberg devised a technique for adding a particular nucleotide to a chemically synthesized dinucleotide using polynucleotide phosphorylase. With Philip Leder, he developed an assay in which the trinucleotides were used to bind transfer RNAs to ribosomes. Using radioactive amino acids, the transfer RNA binding to a particular trinucleotide could be determined, and thus another 'word' of the genetic code. The system worked as planned: for example, valine-transfer RNA was bound to ribosomes in the presence of the trinucleotide GUU, but not in the presence of UGU or UUG. Using this approach, the amino acids specified by all sixty-four codons were determined (Table 13.2).[i]

Of the sixty-four nucleotide triplets, only three, UAA, UAG and UGA, did not specify a particular amino acid; these codons were used to terminate the polypeptide chain. The code was, as Crick had suspected, degenerate; whereas methionine and tryptophan had only one codon, leucine and arginine had six. This degeneracy resulted in part from the existence of multiple transfer RNAs for the same amino acid, as shown by Holley in 1962, but mostly from the existence of transfer RNAs that bound to more than one codon, as shown by Nirenberg in 1965. Codons specifying the same amino acid generally differed only in the base at the third position, suggesting that the codon-anticodon interaction was weakest at that point. The rules governing this special form of base-pairing were described in Francis Crick's last contribution to the theory and terminology of the genetic code – his 1966 'wobble hypothesis'.

---

i   For their roles in breaking the genetic code, Nirenberg, Khorana and Holley shared the 1968 Nobel prize for physiology or medicine.

'The Two Great Polymer Languages'

As Crick put it in a 1966 review article, 'the genetic code . . . is, in a sense, the key to molecular biology because it shows how the two great polymer languages, the nucleic acid language and the protein language, are linked together.' The code provided a means of translating the sequence of bases in a polynucleotide into the sequence of amino acids in a polypeptide (although, because of degeneracy, not vice versa). In doing so, it also demonstrated the central importance in biology of aperiodic polymers. However, it had become clear that there was a significant difference between the two 'polymer languages'. Whereas the information-bearing properties of nucleic acids resided in their one-dimensional sequences of bases, the biological properties of proteins were dependent upon their three-dimensional structures.

John Kendrew and Max Perutz had finally succeeded in determining the three-dimensional structure of myoglobin – electron density maps at 1.4 Å resolution, sufficient to place every atom, were completed in 1963. There were eight segments of α-helix separated by two sharp corners and five non-helical regions. The overall shape of the molecule was similar to a triangular prism. In hemoglobin, four of these prisms, corresponding to the two α-globin and two β-globin subunits, were symmetrically disposed around the center of the molecule.[j]

In 1965, Kendrew delivered the Herbert Spencer lecture at Oxford University. In his lecture, entitled 'Information and conformation in biology', he pointed out that the base sequence of a nucleic acid represents genetic information stored in the form of a one-dimensional array. Proteins are also linear macromolecules, but spontaneously form folded structures, and thus the 'operational unit' is three-dimensional. Because of the large number of bonds around which the polypeptide chain can be rotated, there are huge numbers of three-dimensional structures possible – but only one of these is the actual functional protein.[k] Therefore, the genotype corresponds to nucleic acid *information* and the phenotype to protein *conformation*.

---

j   An atomic-level structure of hemoglobin was not published until 1968. After seeing it, Lawrence Bragg, who had recently retired from the Royal Institution, wrote to Desmond Bernal: 'I marvel that these solutions are possible although several have now come out. It seems so amazing that diffraction effects should reveal the positions of individual groups of atoms in such a maze.' Bernal and Bragg both died in 1971.

k   The importance of conformation had been dramatically demonstrated by Christian Anfinsen, who had shown in the early 1960s that, by denaturing and then renaturing the enzyme ribonuclease in the presence of reagents that break the disulfide bonds between cysteine residues, it was possible to produce a variety of 'scrambled' structures, none of which had catalytic activity.

Kendrew's information and conformation were, of course, Bernard's legislative and executive forces updated for the era of molecular biology.

Kendrew also provided a new twist on the concept of the aperiodic polymer: in that a huge number of different polymers could be assembled from a small number of subunits, nucleic acids and proteins were like digital computers. Biological macromolecules could serve a large number of information-coding and reaction-catalyzing functions in the same way that a digital computer could be used to solve any numerical problem. The same biological mechanisms thereby underlay all forms of life. Kossel's idea of proteins as 'mosaics' or 'railroad trains' of *Bausteine*, which in Schrödinger's hands became the 'aperiodic crystal', now found new metaphorical life as Kendrew's binary code.

'This Riddle of Life Has Been Solved'

For Crick, the 1966 Cold Spring Harbor symposium on the genetic code 'marked the end of classical molecular biology'. This feeling of *fin de siècle* was shared by other molecular biologists. Max Delbrück stated in his 1969 Nobel lecture: 'Molecular genetics, our latest wonder, has taught us to spell out the connectivity of the tree of life in such palpable detail that we may say in plain words, "This riddle of life has been solved".' The following year, Delbrück's former student Gunther Stent wrote: 'Molecular genetics now presents an integral canon of biological knowledge which must be preserved and passed on to succeeding generations in the academies . . . its appeal as an arena for strife against the Great Unknown is gone . . . the would-be explorer of uncharted territory must direct his attention elsewhere.'[102]

No wonder that Stent, like Alexander, cried because there were no more worlds to conquer! Recent findings had not only fulfilled the ambitious postwar agenda of molecular biology, but had also provided answers to questions that had occupied biological chemists and geneticists for over a century – questions about the material basis of heredity and how heritable properties were transmitted, about the structures of biological macromolecules, about the means by which a complex organism arose from a single cell. It was the end of what Judson called the 'golden age, the age of innocence'.

Delbrück had written in 1949: 'any living cell carries with it the experiences of a billion years of experimentation by its ancestors. You cannot expect to explain so wise an old bird in a few simple words.' By 1970, however, his friend Alfred Hershey felt that he could do just that:

If the overall plan is simple, it ought to be possible to put it into a few words, which I attempt as follows. First, the genotype resides in DNA – more importantly, in the linear sequence of the four nucleotides in single DNA strands. Second, nucleotide sequences in single DNA strands are transcribable into complementary sequences according to simple one-to-one rules: the four bases form only two interstrand pairs, guanine-cytosine and adenine-thymine. This code or its equivalent is used for DNA replication, gene transcription, and synthesis of ribosomal and transfer RNAs. It also regulates the structure of typical double stranded DNA molecules.

Third, sequences in one of the two complementary strands transcribed into messenger RNA, are translatable into amino-acid sequences in proteins according to a second code unrelated to the first. This is a non-overlapping triplet code (three bases per amino acid) usually called the genetic code.

Fourth, the phenotype, to the extent that it is understood at all, depends on the structures of species-specific proteins, structures that depend in turn on linear sequences of amino-acids.[103]

This simple scheme made many things clear – the nature and replication of the gene, the diversity of species, the development of specialized tissues, the physiological role of RNA. Vindicated were Beadle's one gene–one enzyme hypothesis, Muller's autosynthesis of the gene, Crick's central dogma; discarded, soon to be forgotten, were protein genes, the colloid theory, autocatalysis and the tetranucleotide structure of DNA.

At the chemical level, the phenomena of life were seen to be the result of specific interactions between macromolecules. So far as the transmission and expression of hereditary properties were concerned, the key macromolecules were the nucleic acids and the proteins: the nucleic acids by encoding precise instructions for the construction of proteins; and the proteins in turn by acting as specific catalysts for the synthesis of all other cellular components, including the nucleic acids. By this cycle of nucleic acid to protein, and protein to nucleic acid, were closed the individual links of the great chain of being.

The solving of the genetic code thereby revealed the relationship between genotype and phenotype, between DNA and protein, between structure and function. Many fundamental biological processes were still mysterious, including the origin of life on earth, the mechanism of embryonic development, and the functioning of the brain, but the great project of molecular biology, the creation of Crick's 'chemical physics of biology', had essentially been completed – and, with it, the biochemical revolution that had begun with Antoine Lavoisier's demonstration of the common nature of chemistry and metabolism. Even such unexpected findings as catalytic RNA and infectious proteins would only augment, not fundamentally change, Hershey's description of the molecular basis of life. In the future, less attention would be paid to understanding how nature works at the molecular level and more to how these molecular mechanisms could be manipulated. By around 1980, the age of molecular biology would have given way to the age of genetic engineering.

At the center of the new molecular biology were two concepts that had underlain much of biochemical research for the previous century. The first was the aperiodic nature of biological polymers, first articulated by Albrecht Kossel, popularized by Erwin Schrödinger, and proved for proteins and nucleic acids by Frederick Sanger and Erwin Chargaff, respectively. The second was the distinction between the legislative and the executive forces, first articulated by Claude Bernard, subsequently restated as phenotype and genotype by Wilhelm Johannsen, as genes and enzymes by George Beadle, and as information and conformation by John Kendrew. Linear sequences of *Bausteine* not only allowed nucleic acids to encode genetic information, but allowed proteins to express it. The realization that the legislative and executive forces are both mediated by aperiodic polymers therefore represents the apotheosis of biological chemistry.

In the sense that the biochemical revolution of 1770–1970 was a revolt of scientific rationalism against vitalism, the solving of the genetic code marked its end. As Crick wrote in 1966, 'It will be difficult, after this, for doubters not to accept the fundamental assumptions of molecular biology which we have been trying to prove for so many years.' The ranks of the 'doubters' had grown thin by 1966. The vast majority of biological scientists were philosophical materialists who accepted without reservation that living systems operated by the same physical and chemical laws that governed the behavior of inanimate matter. The discovery of the electrical activity of the brain, the isolation of neurotransmitters and the synthesis of mind-altering chemicals suggested that even biology's last 'black box' would prove amenable to a physico-chemical explanation. Vitalism had not so much been disproved as become unnecessary.

The living cell, which Max Delbrück had likened in 1949 to a 'magic puzzle box', had now yielded up most of its secrets. And yet, the elucidation of the molecular basis of life in no way diminished – perhaps even enhanced – the sense of wonder Delbrück had expressed so lyrically. If the gene was like clockwork, as Schrödinger had stated in *What Is Life?*, then the clock was now in a transparent case. Rather than wondering at what mechanism could produce such regularity, the observer now marvelled at the intricacy of the cogwheels and springs laid bare to view.

Delbrück may have been wrong in thinking that a paradox would be found in the workings of the hereditary clockwork, but his skepticism about the reductionist agenda of biochemistry had nonetheless been justified. Even the detailed knowledge of the molecular mechanisms of life available in 1970 did not allow an explanation of why these particular mechanisms had arisen or a prediction of the complex actions they produced. As Kendrew wrote in 1965: 'At every level of complexity there is something new, something which in one sense cannot be explained in terms of the level below, and yet which in no way destroys the validity of the conceptions appropriate to that lower level.' What was true then remains true today; we may understand life, but we cannot explain it.

References

1. Priestley, J. (1790) *Experiments and Observations on Different Kinds of Air.* Thomas Pearson, Birmingham.
2. Leicester, H. M. and Klickstein, H. S. (1952) *A Source Book in Chemistry, 1400–1900.* McGraw-Hill, New York.
3. Cavendish, H. (1784) Experiments on air. *Philosophical Transactions* **74**, 119–153.
4. Lavoisier, A. (1790) *Elements of Chemistry in a New Systematic Order, Containing All the Modern Discoveries.* William Creech, Edinburgh.
5. Holmes, F. L. (1985) *Lavoisier and the Chemistry of Life: An Exploration of Scientific Creativity.* University of Wisconsin Press, Madison.
6. Dalton, J. (1808) *A New System of Chemical Philosophy*, Vol. 1 Part I. Bickerstaff, London.
7. Dalton, J. (1827) *A New System of Chemical Philosophy*, Vol. 2. Bickerstaff, London.
8. Tilden, W. A. (1921) *Famous Chemists: The Men and their Work.* Books for Libraries Press, Freeport, New York.
9. Thorpe, T. E. (1894) *Essays in Historical Chemistry.* Books for Libraries Press, Freeport, New York.
10. Fisher, N. W. (1973) Organic classification before Kekulé. *Ambix* **20**, 106–131.
11. Liebig, J. (1847) *Animal Chemistry, or Organic Chemistry in its Application to Physiology and Pathology.* J. M. Campbell, Philadelphia.
12. Cannizzaro, S. (1947) Sketch of a course of chemical philosophy. *Alembic Club Reprints* **18**, 1–55.
13. Fruton, J. S. (1972) *Molecules and Life: Historical Essays on the Interplay between Chemistry and Biology.* Wiley-Interscience, New York.
14. Lipman, T. O. (1964) Wöhler's preparation of urea and the fate of vitalism. *Journal of Chemical Education* **41**, 452–458.
15. Lipman, T. O. (1967) Vitalism and reductionism in Liebig's physiological thought. *Isis* **58**, 167–185.
16. Liebig, J. (1859) *Familiar Letters on Chemistry, in its Relations to Physiology, Dietetics, Agriculture, Commerce, and Political Economy.* Walton and Maberly, London.
17. Jørgensen, B. S. (1965) More on Berzelius and the vital force. *Journal of Chemical Education* **42**, 394–396.
18. Florkin, M. (1972) From forces-of-life to bioenergetics, *Comprehensive*

Biochemistry **30**, 215–249.

19. Baker, J. R. (1949) The cell theory; a restatement, history, and critique. Part II. *Quarterly Journal of Microscopical Science* **90**, 87–108.

20. Schwann, T. (1847) *Microscopical Researches into the Accordance in the Structure and Growth of Animals and Plants.* The Sydenham Society, London.

21. Virchow, R. (1860) *Cellular Pathology, as Based Upon Physiological and Pathological Histology.* Robert M. de Witt, New York.

22. Cranefield, P. F. (1957) The organic physics of 1847 and the biophysics of today. *Journal of the History of Medicine* **12**, 407–423.

23. Elkana, Y. (1974) *The Discovery of the Conservation of Energy.* Harvard University Press, Cambridge, Massachusetts.

24. Huxley, T. H. (1869) On the physical basis of life. *Fortnightly Review* **26**, 129–154.

25. Bernard, C. (1974) *Lectures on the Phenomena of Life Common to Animals and Plants.* Charles C. Thomas, Springfield, Illinois.

26. Florkin, M. (1972) The nature of alcoholic fermentation, the 'theory of the cell' and the concept of the cells as units of metabolism. *Comprehensive Biochemistry* **30**, 129–143.

27. de Mayo, P., Stoessl, A. and Usselman, M. C. (1990) The Liebig/Wöhler satire on fermentation. *Journal of Chemical Education* **67**, 552–553.

28. Dixon, M. (1971) The history of enzymes and of biological oxidations. In: Needham, J. (ed.) *The Chemistry of Life: Lectures on the History of Biochemistry*, pp. 15–37. Cambridge University Press, Cambridge.

29. Kottler, D. B. (1978) Louis Pasteur and molecular dissymmetry. *Studies in the History of Biology* **2**, 57–98.

30. Pasteur, L. (1863) The function of atmospheric oxygen in the destruction of animal and vegetal substances after death. *Chemical News* **7**, 280–282.

31. Pasteur, L. (1879) *Studies on Fermentation: The Disease of Beer, Their Causes, and the Means of Preventing Them.* McMillan, London.

32. Wotiz, J. H. and Rudofsky, S. (1989) Louis Pasteur, August Kekulé, and the Franco-Prussian War. *Journal of Chemical Education* **66**, 34–36.

33. Finegold, H. (1954) The Liebig-Pasteur controversy. *Journal of Chemical Education* **31**, 403–406.

34. Richardson, G. M. (1901) *Foundations of Stereochemistry: Memoirs by Pasteur, van't Hoff, LeBel and Wislicenus.* American Book Company, New York.

35. Lichtenthaler, F. W. (1994) 100 years 'Schlüssel -Schloss-Prinzip': what made Emil Fischer use this analogy? *Angewandte Chemie International Edition in English* **33**, 2364–2374.

36. Gabriel, M. L. and Vogel, S. (1955) *Great Experiments in Biology.* Prentice-Hall, Englewood Cliffs, New Jersey.

37. Buchner, E. (1966) Cell-free fermentation. In: *Nobel Lectures in Chemistry, 1901–1921*, pp. 103–120. Elsevier, Amsterdam.

38. Teich, M. and Needham, D. M. (1992) *A Documentary History of Biochemistry.* Fairleigh Dickinson Universities Press, Rutherford, New Jersey.

39. Greenstein, J. P. (1943) Friedrich Miescher, 1844–1895. *Scientific Monthly* **57**, 523–532.

40. Graham, T. (1861) Liquid diffusion applied to analysis. *Philosophical Transactions of the Royal Society of London* **151**, 183–224.

41. Kekulé, A. (1878) The scientific aims and achievements of chemistry. *Nature* **18**, 210–213.

42. Fruton, J. S. (1990) *Contrasts in Scientific Style: Research Groups in the Chemical and Biochemical Sciences.* American Philosophical Society, Philadelphia.

43. Kossel, A. (1911) The chemical composition of the cell. *Harvey Lecture Series* **7**, 33–51.

44. Weismann, A. (1892) *The Germ-Plasm: A Theory of Heredity.* Scribners, New York.

45. Wilson, E. B. (1896) *The Cell in Development and Inheritance.* McMillan, New York.

46. Sutton, W. S. (1903) The chromosomes in heredity. *Biological Bulletin* **4**, 231–251.

47. Bateson, W. (1902) *Mendel's Principles of Heredity: A Defence.* Cambridge University Press, Cambridge.

48. Galton, F. (1872) On blood-relationship. *Proceedings of the Royal Society of London* **20**, 394–402.

49. Bateson, W. (1909) *Mendel's Principles of Heredity.* Cambridge University Press, Cambridge.

50. Troland, L. T. (1914) The chemical origin and regulation of life. *The Monist* **22**, 92–133.

51. Troland, L. T. (1917) Biological enigmas and the theory of enzyme action. *American Naturalist* **51**, 321–350.

52. Muller, H. J. (1922) Variation due to change in the individual gene. *American Naturalist* **56**, 32–50.

53. Bragg, W. L. (1966) Reminiscences of fifty years' research. *Proceedings of the Royal Institution of Great Britain* **41**, 92–100.

54. Sumner, J. B. (1964) The chemical nature of enzymes. In: *Nobel Lectures in Chemistry, 1942–1962*, pp. 114–121. Elsevier, Amsterdam.

55. Vickery, H. B. and Osborne, T. B. (1928) A review of hypotheses of the structure of proteins. *Physiological Reviews* **8**, 393–446.

56. Wrinch, D. M. (1938) Is there a protein fabric? *Cold Spring Harbor Symposia on Quantitative Biology* **6**, 122–139.

57. Chibnall, A. C. (1942) Amino-acid analysis and the structure of proteins. *Proceedings of the Royal Society of London Series B* **131**, 136–160.

58. Mirsky, A. E. and Pauling, L. (1936) On the structure of native, denatured, and coagulated proteins. *Proceedings of the National Academy of Sciences USA* **22**, 439–447.

59. Astbury, W. T. (1939) X-ray studies of the structures of compounds of biological interest. *Annual Review of Biochemistry* **8**, 113–132.
60. Snow, C. P. (1964) J. D. Bernal, a personal portrait. In: Goldsmith, M. and Mackay, A. (eds) *The Science of Science*, pp. 19–31. Penguin, London.
61. Phillips, D. (1979) William Lawrence Bragg. *Biographical Memoirs of the Royal Society of London* **25**, 75–143.
62. Bohr, N. (1933) Light and life. *Nature* **131**, 421–423; 457–459.
63. Delbrück, M. (1949) A physicist looks at biology. *Transactions of the Connecticut Academy of Arts and Science* **38**, 173–190.
64. Muller, H. J. (1936) Physics in the attack on the fundamental problems of genetics. *Scientific Monthly* **44**, 210–214.
65. Haldane, J. B. S. (1938) The biochemistry of the individual. In: Needham, J. and Green, D. E. (eds) *Perspectives in Biochemistry*, pp. 1–10. Cambridge University Press, Cambridge.
66. Pauling, L. and Delbrück, M. (1940) The nature of the intermolecular forces operative in biological processes. *Science* **92**, 77–79.
67. Scott-Moncrieff, R. (1981) The classical period in genetics: recollections of Muriel Wheldale Onslow, Robert and Gertrude Robinson and J. B. S. Haldane. *Notes and Records of the Royal Society of London* **36**, 125–154.
68. Kohler, R. E. (1991) Systems of production: Drosophila, Neurospora, and biochemical genetics. *Historical Studies in the Physical Sciences* **21**, 87–130.
69. Beadle, G. W. and Tatum, E. L. (1941) Genetic control of biochemical reactions in Neurospora. *Proceedings of the National Academy of Science USA* **27**, 499–506.
70. Luria, S. E. (1966) Mutations of bacteria and of bacteriophage. In: Cairns, J., Stent, G. S. and Watson, J. D. (eds) *Phage and the Origins of Molecular Biology*, pp. 173–179. Cold Spring Harbor Laboratory Press, New York.
71. Schrödinger, E. (1944) *What is Life?* Cambridge University Press, Cambridge.
72. Hotchkiss, R. D. (1966) Gene, transforming principle, and DNA. In: Cairns, J., Stent, G. S. and Watson, J. D. (eds) *Phage and the Origins of Molecular Biology*, pp. 180–200. Cold Spring Harbor Laboratory Press, New York.
73. Olby, R. (1970) Francis Crick, DNA, and the central dogma. *Daedalus* **99**, 938–988.
74. Jacob, F. (1989) *The Statue Within: An Autobiography.* Basic Books, New York.
75. Delbrück, M. (1946) Experiments with bacterial viruses. *Harvey Lecture Series* **41**, 161–187.
76. Olby, R. C. (1979) The significance of the macromolecules in the historiography of molecular biology. *History and Philosophy of the Life Sciences* **1**, 185–198.

77. Sanger, F. and Tuppy, H. (1951) The amino acid sequence in the phenyl-alanine chain of insulin. 2. The investigation of peptides from enzymic hydrolyzates. *Biochemical Journal* **49**, 481–490.

78. Huggins, M. L. (1943) The structure of fibrous proteins. *Chemical Reviews* **32**, 195–218.

79. Pauling, L. (1993) How my interest in proteins developed. *Protein Science* **2**, 1060–1063.

80. Bragg, W. L., Kendrew, J. C. and Perutz, M. F. (1950) Polypeptide chain configurations in crystalline proteins. *Proceedings of the Royal Society of London Series A* **203**, 321–357.

81. Pauling, L., Corey, R. B. and Branson, H. R. (1951) The structure of proteins: two hydrogen-bonded helical configurations of the polypeptide chain. *Proceedings of the National Academy of Sciences USA* **37**, 205–211.

82. Todd, A. (1990) A recollection of Sir Lawrence Bragg. In: Thomas, J. M. and Phillips, D. (eds) *Selections and Reflections: The Legacy of Sir Lawrence Bragg*, pp. 95–96. Science Reviews Ltd., Northwood.

83. Hodgkin, D. C. (1979) Crystallographic measurement and the structure of protein molecules as they are. *Annals of the New York Academy of Sciences* **325**, 121–148.

84. Perutz, M. F. (1949) An X-ray study of horse methaemoglobin. II. *Proceedings of the Royal Society of London Series A* **195**, 474–499.

85. Green, D. W., Ingram, V. M. and Perutz, M. F. (1954) The structure of haemoglobin. IV. Sign determination by the isomorphous replacement method. *Proceedings of the Royal Society of London Series A* **225**, 287–307.

86. Stern, K. (1947) Nucleoproteins and gene structure. *Yale Journal of Biology and Medicine* **19**, 937–949.

87. Chargaff, E. (1950) Chemical specificity of nucleic acids and mechanism of their enzymatic degradation. *Experientia* **6**, 201–240.

88. Chargaff, E. (1951) Some recent studies on the composition and structure of nucleic acids. *Experimental Cell Research* **2**, 41–59.

89. Chargaff, E. Z., Lipshitz, R., Green, C. and Hodes, M. E. (1951) The composition of the deoxyribonucleic acid of salmon sperm. *Journal of Biological Chemistry* **192**, 223–230.

90. Pauling, L. (1970) Fifty years of progress in structural chemistry and molecular biology. *Daedalus* **99**, 989–1014.

91. Pauling, L. and Corey, R. B. (1953) A proposed structure for the nucleic acids. *Proceedings of the National Academy of Sciences USA* **39**, 84–97.

92. Watson, J. D. and Crick, F. H. C. (1953) Molecular structure of nucleic acids. *Nature* **171**, 737–738.

93. Watson, J. D. and Crick, F. H. C. (1953) Genetical implications of the structure of deoxyribonucleic acid. *Nature* **171**, 964–967.

94. Hager, T. (1995) *Force of Nature: The Life of Linus Pauling*. Simon & Schuster, New York.

95. Perutz, M. F. (1971) Sir Lawrence Bragg. *Nature* **233**, 74–76.

96. Crick, F. (1988) *What Mad Pursuit: A Personal View of Scientific Discovery.* Basic Books, New York.
97. Fischer, E. P. and Lipson, C. (1988) *Thinking About Science: Max Delbrück and the Origins of Molecular Biology.* W. W. Norton and Co., New York.
98. Wright, S. (1941) The physiology of the gene. *Physiological Reviews* **21**, 487–527.
99. Caldwell, P. C. and Hinshelwood, C. (1950) Some considerations on autosynthesis in bacteria. *Journal of the Chemical Society* 3156–3159.
100. Brenner, S. (1957) On the impossibility of all overlapping triplet codes in information transfer from nucleic acids to proteins. *Proceedings of the National Academy of Science USA* **43**, 687–694.
101. Crick, F. H. C. (1958) On protein synthesis. *Symposia of the Society for Experimental Biology* **12**, 138–163.
102. Stent, G. S. (1970) DNA. *Daedalus* **99**, 909–937.
103. Hershey, A. D. (1970) Genes and hereditary characteristics. *Nature* **226**, 697–700.

Selected Readings

General

Fruton, J. S. (1972) *Molecules and Life: Historical Essays on the Interplay between Chemistry and Biology*. Wiley-Interscience, New York.

Judson, H. F. (1979) *The Eighth Day of Creation: Makers of the Revolution in Biology*. Simon and Schuster, New York.

Teich, M. and Needham, D. M. (1992) *A Documentary History of Biochemistry*. Fairleigh Dickinson Universities Press, Rutherford, New Jersey.

Olby, R. C. (1994) *The Path to the Double Helix: The Discovery of DNA*. Dover, New York.

Morange, M. (1998) *A History of Molecular Biology* (Cobb, M., trans.). Harvard University Press, Cambridge, Massachusetts.

Chapter 1

Donovan, A. (1993) *Antoine Lavoisier: Science, Administration, and Revolution*. Blackwell, Oxford.

Holmes, F. L. (1985) *Lavoisier and the Chemistry of Life: An Exploration of Scientific Creativity*. University of Wisconsin Press, Madison.

Chapter 2

Holmes, F. L. (1967) Elementary analysis and the origins of physiological chemistry. *Isis* **54**, 50–81.

Brock, W. H. (1997) *Justus von Liebig: The Chemical Gatekeeper*. Cambridge University Press, Cambridge.

Friedman, H. B. (1930) The theory of types – a satirical sketch. *Journal of Chemical Education* **7**, 633–636.

Glas, E. (1983) Bio-science between experiment and ideology, 1835–50. *Studies in the History and Philosophy of Science* **14**, 39–57.

Vickery, H. B. and Schmidt, C. L. A. (1931) The history of the discovery of the amino acids. *Chemical Reviews* **9**, 169–318.

Chapter 3

Geison, G. L. (1969) The protoplasmic theory of life and the vitalist-mechanist debate. *Isis* **60**, 273–292

Teich, M. (1973) From 'enchyme' to 'cytoskeleton': the development of ideas on the chemical organisation of living matter. In: Teich, M. and Young, R. (eds) *Changing Perspectives in the History of Science*, pp. 439–471. Heinemann, London.

Welch, G. R. (1995) T. H. Huxley and the 'protoplasmic theory of life': 100 years later. *Trends in Biochemical Sciences* **20**, 481–485.

Holmes, F. L. (1974) *Claude Bernard and Animal Chemistry: the Emergence of a Scientist*. Harvard University Press, Cambridge, Massachusetts.

Chapter 4

Teich, M. (1981) Ferment or enzyme: what's in a name? *History and Philosophy of the Life Sciences* **3**, 193–215.

Harden, A. (1923) *Alcoholic Fermentation*. Longmans, Green and Co., London.

Geison, G. L. (1995) *The Private Science of Louis Pasteur*. Princeton University Press, Princeton, New Jersey.

Palladino, P. (1990) Stereochemistry and the nature of life: mechanist, vitalist, and evolutionary perspectives. *Isis* **81**, 44–67.

Conant, J. B. (1957) Pasteur's study on fermentation. In: Conant, J. B. and Nash, L. K. (eds) *Harvard Case Studies in Experimental Science*, pp. 439–485. Harvard University Press, Cambridge, Massachusetts.

Kohler, R. (1971) The background to Eduard Buchner's discovery of cell-free fermentation. *Journal of the History of Biology* **4**, 35–61.

Chapter 5

Portugal, F. H. and Cohen, J. S. (1977) *A Century of DNA*. MIT Press, Cambridge, Massachusetts.

Mirsky, A. E. (1968) The discovery of DNA. *Scientific American* **218**, 78–88.

Jones, M. E. (1953) Albrecht Kossel, a biographical sketch. *Yale Journal of Biology and Medicine* **26**, 80–97.

van Slyke, D. D. and Jacobs, W. A. (1943) Biographical memoir of Phoebus Aaron Theodor Levene. *Biographical Memoirs of the National Academy of Sciences USA* **23**, 75–126.

Hunter, G. K. (1999) Phoebus Levene and the tetranucleotide structure of nucleic acids. *Ambix* **46**, 73–103.

Fruton, J. S. (1990) *Contrasts in Scientific Style: Research Groups in the Chemical and Biochemical Sciences*. American Philosophical Society, Philadelphia.

Chapter 6

Olby, R. (1985) *Origins of Mendelism*. University of Chicago Press, Chicago, Illinois.

Sandler, I. and Sandler, L. (1986) On the origin of Mendelian genetics. *American Zoologist* **26**, 753–768.

Stern, C. and Sherwood, E. R. (1966) *The Origins of Mendelism: A Mendel Source Book*. W.H. Freeman and Co., San Francisco.

Ravin, A. W. (1977) The gene as catalyst, the gene as organism. *Studies in the History of Biology* **1**, 1–45.

Kohler, R. E. (1994) *Lords of the Fly: Drosophila Genetics and the Experimental Way of Life*. University of Chicago Press, Chicago.

Allen, G. E. (1978) *Thomas Hunt Morgan: the Man and His Science*. Princeton University Press, Princeton, New Jersey.

Chapter 7

Fruton, J. S. (1979) Early theories of protein structure. *Annals of the New York Academy of Sciences* **325**, 1–18.

Phillips, D. (1979) William Lawrence Bragg. *Biographical Memoirs of the Royal Society of London* **25**, 75–143.

Bernal, J. D. (1963) William Thomas Astbury. *Biographical Memoirs of the Royal Society of London* **9**, 1–35.

Hodgkin, D. M. C. (1980) John Desmond Bernal. *Biographical Memoirs of the Royal Society of London* **26**, 17–84.

Mark, H. F. (1993) From small organic molecules to large: a century of progress. In: Seeman, J. I. (ed.) *Profiles, Pathways, Dreams: Autobiographies of Eminent Chemists*. American Chemical Society, Washington, DC.

Claesson, S. and Pedersen, K. O. (1972) The Svedberg. *Biographical Memoirs of the Royal Society of London* **18**, 595–627.

Furukawa, Y. (1982) Hermann Staudinger and the emergence of the macro-molecular concept. *Historia Scientiarum* **22**, 1–18.

Olby, R. C. (1986) Structural and dynamical explanations in the world of neglected dimensions. In: Horder, T. J., Witkowski, J. A. and Wylie, C. C. (eds) *A History of Embryology*, pp. 275–308. Cambridge University Press, Cambridge.

Chapter 8

Hager, T. (1995) *Force of Nature: The Life of Linus Pauling*. Simon & Schuster, New York.

Kohler, R. E. (1976) The management of science: the experience of Warren Weaver and the Rockefeller Foundation programme in molecular biology. *Minerva* **14**, 279–306.

Abir-Am, P. G. (1987) The biotheoretical gathering, trans-disciplinary authority and the incipient legitimization of molecular biology in the 1930s: new perspective on the historical sociology of science. *History of Science* **25**, 1–70.

Chapter 9

Fischer, E. P. and Lipson, C. (1988) *Thinking About Science: Max Delbrück and the Origins of Molecular Biology*. W. W. Norton and Co., New York.

Carlson, E. A. (1966) *The Gene: A Critical History.* W. B. Saunders, Philadelphia.

Kay, L. E. (1986) W. M. Stanley's crystallization of the tobacco mosaic virus, 1930–1940. *Isis* 77, 450–472.

Carlson, E. A. (1981) *Genes, Radiation and Society: the Life and Work of H. J. Muller.* Cornell University Press, Ithaca, New York.

Kohler, R. E. (1991) Systems of production: Drosophila, Neurospora, and biochemical genetics. *Historical Studies in the Physical Sciences* 21, 87–130.

Chapter 10

Perutz, M. F. (1989) *Is Science Necessary?* E. P. Dutton, New York.

Cairns, J., Stent, G. S. and Watson, J. D. (1966) *Phage and the Origins of Molecular Biology.* Cold Spring Harbor Laboratory Press, New York.

Moore, W. (1989) *Schrödinger: Life and Thought.* Cambridge University Press, Cambridge.

Yoxen, E. J. (1979) Where does Schrödinger's 'What is life?' belong in the history of molecular biology? *History of Science* 17, 17–52.

Hotchkiss, R. D. (1979) The identification of nucleic acids as genetic determinants. *Annals of the New York Academy of Sciences* 325, 320–342.

McCarty, M. (1985) *The Transforming Principle: Discovering that Genes are Made of DNA.* W. W. Norton, New York.

Chapter 11

Sanger, F. (1988) Sequences, sequences, sequences. *Annual Review of Biochemistry* 57, 1–28.

Bragg, W. L. (1965) First stages in the X-ray analysis of proteins. *Reports on Progress in Physics* 28, 1–14.

Law, J. (1973) The development of specialities in science: the case of X-ray protein crystallography. *Science Studies* 3, 275–303.

Olby, R. C. (1985) The 'mad pursuit': X-ray crystallographers' search for the structure of haemoglobin. *History and Philosophy of Life Science* 7, 171–193.

Chapter 12

Chargaff, E. (1978) *Heraclitean Fire: Sketches From a Life Before Nature.* Rockefeller University Press, New York.

Abir-Am, P. (1980) From biochemistry to molecular biology: DNA and the acculturated journey of the critic of science Erwin Chargaff. *History and Philosophy of the Life Sciences* 2, 3–60.

Brock, T. D. (1990) *The Emergence of Bacterial Genetics.* Cold Spring Harbor Laboratory Press, New York.

Mullins, N. C. (1972) The development of a scientific specialty: the phage group and the origins of molecular biology. *Minerva* 10, 51–82.

Jacob, F. (1989) *The Statue Within: An Autobiography* (Philip, F., trans.). Basic

Books, New York.

Olby, R. (1970) Francis Crick, DNA, and the central dogma. *Daedalus* **99**, 938–988.

Watson, J. D (1968) *The Double Helix: A Personal Account of the Discovery of the Structure of DNA.* Penguin, London.

Crick, F. (1988) *What Mad Pursuit: a Personal View of Scientific Discovery.* Basic Books, New York.

Fleming, D. (1968) Émigré physicists and the biological revolution. *Perspectives in American History* **2**, 152–189.

Chapter 13

Rheinberger, H.-J. (1992) Experiment, difference and writing. I. Tracing protein synthesis. *Studies in the History and Philosophy of Science* **23**, 305–331.

Rheinberger, H.-J. (1992) Experiment, difference and writing. II. The laboratory production of transfer RNA. *Studies in the History and Philosophy of Science* **23**, 389–422.

Woese, C. (1967) *The Genetic Code.* Harper and Row, New York.

Ycas, M. (1969) *The Biological Code.* North Holland Publishing Company, Amsterdam.

Name Index

Subject Index

DATE DUE

MAY 0 1 2003			

GAYLORD

PRINTED IN U.S.A.